AF693010

Plant Responses to a Changing Climate: Ecological and Evolutionary Perspectives

Guest Editors

Sanja Manitašević Jovanović
Ana Vuleta
Yuval Sapir

MDPI

Basel • Beijing • Wuhan • Barcelona • Belgrade • Novi Sad • Cluj • Manchester

Guest Editors

Sanja Manitašević Jovanović
Department of Evolutionary Biology
University of Belgrade
Belgrade
Serbia

Ana Vuleta
Department of Evolutionary Biology
University of Belgrade
Belgrade
Serbia

Yuval Sapir
School of Plant Sciences and Food Security
Tel-Aviv University
Tel Aviv
Israel

Editorial Office
MDPI AG
Grosspeteranlage 5
4052 Basel, Switzerland

This is a reprint of the Special Issue, published open access by the journal *Plants* (ISSN 2223-7747), freely accessible at: https://www.mdpi.com/journal/plants/special_issues/8R12G95CI1.

For citation purposes, cite each article independently as indicated on the article page online and as indicated below:

Lastname, A.A.; Lastname, B.B. Article Title. *Journal Name* **Year**, *Volume Number*, Page Range.

ISBN 978-3-7258-4987-1 (Hbk)
ISBN 978-3-7258-4988-8 (PDF)
https://doi.org/10.3390/books978-3-7258-4988-8

Contents

Preface . . . . **vii**

Katarina Hočevar, Ana Vuleta and Sanja Manitašević Jovanović
Plastic Responses of *Iris pumila* Functional and Mechanistic Leaf Traits to Experimental Warming
Reprinted from: *Plants* **2025**, *14*, 960, https://doi.org/10.3390/plants14060960 . . . . **1**

Li Zhang, Qi Yu, Xin Yin, Laipan Liu, Zhentao Ren, Zhixiang Fang, et al.
Changes in the Stress Response and Fitness of Hybrids Between Transgenic Soybean and Wild-Type Plants Under Heat Stress
Reprinted from: *Plants* **2025**, *14*, 622, https://doi.org/10.3390/plants14040622 . . . . **21**

Zichen Zhang, Xin Yan, Chang Guo, Wenpan Dong, Liangcheng Zhao and Dan Liu
Changes in the Suitable Habitat of the Smoke Tree (*Cotinus coggygria* Scop.), a Species with an East Asian–Tethyan Disjunction
Reprinted from: *Plants* **2025**, *14*, 547, https://doi.org/10.3390/plants14040547 . . . . **37**

Shimeng Zhao, Zongxian Zhang, Changyu Gao, Yiding Dong, Zeyao Jing, Lixia Du and Xiangyang Hou
MaxEnt-Based Predictions of Suitable Potential Distribution of *Leymus secalinus* Under Current and Future Climate Change
Reprinted from: *Plants* **2025**, *14*, 293, https://doi.org/10.3390/plants14020293 . . . . **51**

David M. Meko, Dina F. Zhirnova, Liliana V. Belokopytova, Yulia A. Kholdaenko, Elena A. Babushkina, Nariman B. Mapitov and Eugene A. Vaganov
Central Asia Cold Case: Siberian Pine Fingers New Suspects in Growth Decline CA 1700 CE
Reprinted from: *Plants* **2025**, *14*, 287, https://doi.org/10.3390/plants14020287 . . . . **68**

Bruna Evelyn Paschoal Silva, Stefânia Nunes Pires, Sheila Bigolin Teixeira, Simone Ribeiro Lucho, Natan da Silva Fagundes, Larissa Herter Centeno, et al.
Physiological and Biochemical Responses of Pseudocereals with C3 and C4 Photosynthetic Metabolism in an Environment with Elevated CO_2
Reprinted from: *Plants* **2024**, *13*, 3453, https://doi.org/10.3390/plants13233453 . . . . **84**

Chunlei Yue, Hepeng Li and Xiaodeng Shi
Geographical Distribution Dynamics of *Acorus calamus* in China Under Climate Change
Reprinted from: *Plants* **2024**, *13*, 3352, https://doi.org/10.3390/plants13233352 . . . . **106**

Xu Xiao, Zhi Li, Zhaohui Ran, Chao Yan and Juyan Chen
Impact of Climate Change on Distribution of Endemic Plant Section *Tuberculata* (*Camellia* L.) in China: MaxEnt Model-Based Projection
Reprinted from: *Plants* **2024**, *13*, 3175, https://doi.org/10.3390/plants13223175 . . . . **122**

Maciej Bartold, Konrad Wróblewski, Marcin Kluczek, Katarzyna Dąbrowska-Zielińska and Piotr Goliński
Examining the Sensitivity of Satellite-Derived Vegetation Indices to Plant Drought Stress in Grasslands in Poland
Reprinted from: *Plants* **2024**, *13*, 2319, https://doi.org/10.3390/plants13162319 . . . . **140**

Lu Zhang, Beibei Jiang, Yu Meng, Yin Jia, Qian Xu and Yuanzhi Pan
The Influence of Climate Change on the Distribution of *Hibiscus mutabilis* in China: MaxEnt Model-Based Prediction
Reprinted from: *Plants* **2024**, *13*, 1744, https://doi.org/10.3390/plants13131744 . . . . **159**

Xian-Ge Hu, Jiahui Chen, Qiaoyun Chen, Ying Yang, Yiheng Lin, Zilun Jin, et al.
The Spatial Shifts and Vulnerability Assessment of Ecological Niches under Climate Change Scenarios for *Betula luminifera*, a Fast-Growing Precious Tree in China
Reprinted from: *Plants* **2024**, *13*, 1542, https://doi.org/10.3390/plants13111542 . . . . **174**

Changrong Deng, Qiwen Zhong, Dengkui Shao, Yanjing Ren, Quanhui Li, Junqin Wen and Jianling Li
Potential Suitable Habitats of Chili Pepper in China under Climate Change
Reprinted from: *Plants* **2024**, *13*, 1027, https://doi.org/10.3390/plants13071027 . . . . **192**

Qijing Zhang, Zhaoping Lu, Mingchen Guo, Jia Kang, Jia Li, Xiaojing He, et al.
Responses of Three *Pedicularis* Species to Geological and Climatic Changes in the Qinling Mountains and Adjacent Areas in East Asia
Reprinted from: *Plants* **2024**, *13*, 765, https://doi.org/10.3390/plants13060765 . . . . **207**

Preface

This multidisciplinary collection aims to promote a deeper understanding of the ecological and evolutionary responses of plants to climate change. It brings together a diverse body of research that collectively explores how plants respond, adjust, and potentially adapt to the unprecedented challenges posed by global environmental change. In doing so, it contributes to a growing knowledge base that is essential for anticipating ecological and evolutionary outcomes and for biodiversity conservation strategies worldwide.

We hope that this reprint will serve as a valuable resource for researchers, educators, and students of plant science, ecology, and evolutionary biology, and that it will inspire further study of the crucial question of how plant life is responding to the challenges of a changing climate.

Sanja Manitašević Jovanović, Ana Vuleta, and Yuval Sapir
Guest Editors

plants

MDPI

Article

Plastic Responses of *Iris pumila* Functional and Mechanistic Leaf Traits to Experimental Warming

Katarina Hočevar, Ana Vuleta and Sanja Manitašević Jovanović *

Department of Evolutionary Biology, Institute for Biological Research "Siniša Stanković"—National Institute of the Republic of Serbia, University of Belgrade, 11108 Belgrade, Serbia; katarina.hocevar@ibiss.bg.ac.rs (K.H.); ana.vuleta@ibiss.bg.ac.rs (A.V.)

* Correspondence: manitas@ibiss.bg.ac.rs

Abstract: Phenotypic plasticity is an important adaptive strategy that enables plants to respond to environmental changes, particularly temperature fluctuations associated with global warming. In this study, the phenotypic plasticity of *Iris pumila* leaf traits in response to an elevated temperature (by 1 °C) was investigated under controlled experimental conditions. In particular, we investigated important functional and mechanistic leaf traits: specific leaf area (SLA), leaf dry matter content (LDMC), specific leaf water content (SLWC), stomatal density (SD), leaf thickness (LT), and chlorophyll content. The results revealed that an elevated temperature induced trait-specific plastic responses, with mechanistic traits exhibiting greater plasticity than functional traits, reflecting their role in short-term acclimation. SLA and SD increased at higher temperatures, promoting photosynthesis and gas exchange, while reductions in SLWC, LDMC, LT, and chlorophyll content suggest a trade-off in favor of growth and metabolic activity over structural investment. Notably, chlorophyll content exhibited the highest plasticity, emphasizing its crucial role in modulating photosynthetic efficiency under thermal stress. Correlation analyses revealed strong phenotypic integration between leaf traits, with distinct trait relationships emerging under different temperature conditions. These findings suggest that *I. pumila* employs both rapid physiological adjustments and longer-term structural strategies to cope with thermal stress, with mechanistic traits facilitating rapid adjustments and functional traits maintaining ecological stability.

Keywords: experimental warming; thermal acclimation; phenotypic plasticity; functional leaf traits; mechanistic leaf traits; *Iris pumila* L.

1. Introduction

Human activity, primarily the production of greenhouse gas emissions, has undeniably driven global warming, with global surface temperatures rising 1.1 °C above pre-industrial levels between 2011 and 2020 [1]. Rising temperatures create new climatic conditions that affect the functioning and distribution of species. Human-induced climate change has already led to numerous negative impacts on natural ecosystems, including disruption of key ecological processes, changes in the functioning of ecosystems, and a decline in biodiversity [2–4]. As the climate changes, plant populations may no longer be optimally adjusted to new conditions and their survival depends on their capacity to adequately respond to altered environments [4]. If global warming continues, numerous species could face extinction, as the environments to which they have adapted over the course of time will change within a few decades [5]. Therefore, understanding plant responses to rapid climate change is essential for comprehending both the changes that have already occurred and those likely to take place in the near future.

To persist in a rapidly changing climate, plant organisms may exploit different coping mechanisms. They can move to more favorable habitats through seed dispersal, adapt through natural selection, or adjust to new conditions through phenotypic plasticity [6]. If the rate of environmental change is greater than the rate of evolutionary response, phenotypic plasticity is a likely mechanism for such a circumstance [7–9]. If traits are plastic, and if this plastic change shifts trait values in a direction that maintains fitness, this may be sufficient to anticipate or prevent extinction [9]. To date, most empirical evidence suggests that phenotypic plasticity is the predominant mechanism by which natural plant populations cope with rapid climate warming within their distribution range [10–12]. Phenotypic plasticity is considered to facilitate the survival of plant populations in novel environments, enhancing their ecological success and adaptability by enabling the development of phenotypes better suited to changing conditions [13–16]. Plastic responses in plants can have significant implications for various aspects of their life history, including reproductive success, dispersal patterns, and interactions between species [17].

Among plant traits, leaf traits are particularly important in associating plant resource use and biomass production with ecosystem functioning [18,19] and are, therefore, expected to be especially affected by rising temperatures due to global warming [20–22]. Functional leaf traits encompass morpho-physio-phenological characteristics that directly influence an individual's growth, reproduction, and survival by shaping its interactions with the environment. In contrast, mechanistic leaf traits are those with clearly defined physiological roles, providing a more specific understanding of processes like resource uptake and utilization [23]. Examples of functional traits include leaf area and specific leaf weight, while mechanistic traits encompass photosynthetic capacity and transpiration rates. This classification offers a comprehensive framework for investigating plant adaptive strategies under varying environmental conditions, which is essential for understanding ecological responses to climate warming.

Specific leaf area (SLA) is the major determinant of a plant's ecological function, since it reflects the expected return of previously captured environmental resources [18,24,25]. It is highly responsive to environmental factors, such as irradiance [26,27], water availability [28], atmospheric carbon dioxide [29], ozone concentration [30], soil fertility, and nutrient supply [31–33], while the relationship between SLA and temperature has not yet been sufficiently explored [34]. Leaf dry matter content (LDMC), another important functional trait, is related to leaf protein content and cytoplasmic volume, which reflect the structure of leaf cells [35]. LDMC is positively related to leaf longevity and negatively related to potential relative growth rate [19]. The relationship between LDMC and temperature is complex and varies considerably across different species and ecosystems [36–38]. Specific leaf water content (SLWC), a mechanistic trait most consistently correlated with climate, exhibits a significant relationship with leaf tissue density and leaf thickness [39]. SLWC is strongly influenced by temperature and generally exhibits a positive correlation with it under optimal environmental conditions [40]. Leaf thickness (LT) is a mechanistic leaf trait that is also related to a species' resource acquisition and utilization strategies and affects both light absorption and carbon dioxide diffusion rate [41,42]. Negative relationships have been observed between LT and both photosynthetic rate [43,44] and growth rate [45,46], making LT a useful screening tool for assessing a species' productivity [47] or overall ecological performance [48]. Under the influence of global warming, LT is expected to decrease as higher temperatures cause increased transpiration rates and reduced cell expansion, resulting in thinner leaves [49]. Stomatal density (SD), a mechanistic leaf trait, is widely recognized as an indicator of a plant's gas exchange and transpiration potential [50–54]. Environmental factors, especially temperature, play a crucial role in influencing stomatal density. Higher temperatures can lead to an increase in SD, which can contribute to increased transpiration and leaf cooling as plants adjust to the changing conditions. Nevertheless, it is important to acknowledge that stomatal density alone is not sufficient to draw definitive conclusions about transpiration and stomatal regulation, as the dynamic responsiveness of stomata to environmental factors often plays a more decisive role [55,56]. Chlorophyll content plays a central role in photosynthesis and serves as an important physiological marker for assessing plant stress, particularly in relation to temperature fluctuations [57,58]. High temperatures can accelerate chlorophyll degradation, which generally reduces photosynthetic efficiency [58,59]. However, under unfavorable conditions, many plant species reduce chlorophyll content to mitigate the harmful effects of high temperatures, which helps maintain stomatal conductance and consequently supports photosynthetic activity and growth [60].

Environmental manipulation studies are a widely applied approach for assessing the potential biological impacts of future climate change. In this study, such an experiment was conducted in an experimentally controlled growth room, where environmental conditions such as temperature were precisely regulated to simulate and evaluate plants' responses to predicted climate scenarios. Unlike *in situ* experiments, these controlled environments provide consistent and reproducible conditions, enabling the precise regulation of variables to isolate specific plant responses.

This study aimed to investigate and compare the plastic responses of functional (SLA, LDMC) and mechanistic (SLWC, LT, SD, and chlorophyll content estimated with the RGB index I_1) leaf traits expressed by the same clonal genotypes of *Iris pumila*, which were exposed to optimal and elevated (by 1 °C) air temperatures in a controlled growth environment. Specifically, the following questions were addressed: (1) What is the general pattern of plasticity in response to temperature increases? (2) Do functional and mechanistic leaf traits differ in the magnitude and direction of their plastic responses? (3) Is the correlation pattern between traits affected by temperature conditions? (4) Which category of traits, functional or mechanistic, is more sensitive to temperature change? By responding to these questions, the study intended to provide insights into the adaptive strategies of plants under thermal stress.

2. Results

2.1. Phenotypic Responses of SLA, LDMC, SLWC, LT, SD, and I_1 to Temperature

To assess the phenotypic responses of leaf functional and mechanistic traits to temperature increases, the values of SLA, LDMC, SLWC, LT, SD, and I_1 were determined in the same genotypes of *I. pumila* under the two temperature growth conditions—ambient and elevated by 1 °C. The mean values with the corresponding standard errors and coefficients of variation for the analyzed leaf traits are presented in Table 1. As can be seen, the phenotypic values of all analyzed leaf traits changed significantly with temperature. Regarding leaf functional traits, the mean SLA value was higher (up to 9%) and the LDMC value was lower (about 3%) in the leaves developed under the experimentally elevated temperature compared to the leaves of the same genotype developed at the ambient temperature. As for the leaf mechanistic traits, SLWC and LT decreased with the increase in temperature (both by 6%), while SD and I_1 increased (up to 23% and even 67%, respectively).

Table 1. The means ($\overline{X}$), standard errors (SE), and coefficients of variation (CV%) for the functional and mechanistic leaf traits: specific leaf area (SLA, in $cm^2\ g^{-1}$), leaf dry matter content (LDMC, in $g\ g^{-1}$), specific leaf water content (SLWC, in $g\ cm^{-2}$), leaf thickness (LT), stomatal density (SD, no. stomata/mm^2), and RGB index I_1 of *Iris pumila* genotypes ($N = 40$) grown at ambient and elevated temperature (by 1 °C) conditions in an environmentally controlled growth room. The *F*-values of the temperature effects obtained using repeated ANOVA are presented for each trait as well: ns—non significant, * $p < 0.05$, **** $p < 0.0001$.

Leaf Trait	Ambient Temperature			Elevated Temperature			*F* for Comparison of Trait Means
	$\overline{X}$	SE	CV%	$\overline{X}$	SE	CV%	
Functional							
SLA	186.4	3.5	11.9	202.3	2.8	8.7	24.63 ****
LDMC	0.2066	0.0026	7.8	0.2014	0.0016	4.9	4.86 *
Mechanistic							
SLWC	0.0209	0.0003	9.2	0.0197	0.0003	9.4	18.95 ****
LT	0.0263	0.0004	9.2	0.0247	0.0004	8.9	25.37 ****
SD	72.8	1.2	10.5	89.5	1.6	11.6	83.89 ****
I_1	21.3	1.5	44.3	35.5	2.6	47.1	45.14 ****

2.2. Phenotypic Plasticity of SLA, LDMC, SLWC, LT, SD, and I_1 to Temperature

The reaction norm plots for the SLA, LDMC, SLWC, LT, SD, and I_1 of the *I. pumila* genotypes grown at ambient and elevated temperatures are depicted in Figure 1. As can be seen, the mean reaction norms are relatively steep, indicating the capability of *I. pumila* to adjust its leaf morphology and physiology to suit the prevailing temperature conditions. The pattern of reaction norms proved to be trait-specific. Thus, the reaction norms for SLA, SD, and I_1 exhibited an upward trend with the increase in temperature, in contrast to those for LDMC, SLWC, and LT, which showed a downward trend (Figure 1). The reaction norms of individual genotypes exhibited a complex pattern, with some genotypes intersecting with each other, suggesting that besides their phenotypic values, their rank also changed at different temperatures. Additionally, a convergence trend was observed for the LDMC and I_1 reaction norms. The reaction norms for LDMC converged towards a single value at the elevated temperature, in contrast to the ambient temperature, where their phenotypic values appeared to diverge. Conversely, the reaction norms for I_1 converged towards a single value at the ambient temperature and diverged at the elevated temperature (Figure 1).

The individual variation of *I. pumila* genotypes, expressed as a coefficient of variation (CV%), changed in a trait- and temperature-specific manner (Table 1). Overall, the functional leaf traits showed less variation than the mechanistic traits. In addition, the functional traits exhibited a higher CV% at the ambient temperature, whereas the mechanistic traits exhibited a higher CV% at the elevated temperature. The highest CV% was observed for I_1, while the lowest was found for LDMC. For example, the CV% for I_1 was 44.3% and 47.1% at the ambient and elevated temperatures, respectively. Conversely, the CV% for LDMC was notably lower, amounting to 7.8% and 4.9% at the ambient and elevated temperatures, respectively. The CV% for SLA, LDMC, and LT was greater at the ambient temperature than at the elevated temperature, in contrast to the CV% for SLWC, SD, and I_1. The results of an *F*-test for equality of variance confirmed that the CV% for LDMC and I_1 differed between the alternative temperatures. However, these differences were only highly significant for I_1 ($F = 0.32$; $p < 0.001$), but marginal for LDMC ($F = 1.62$; $p = 0.058$).

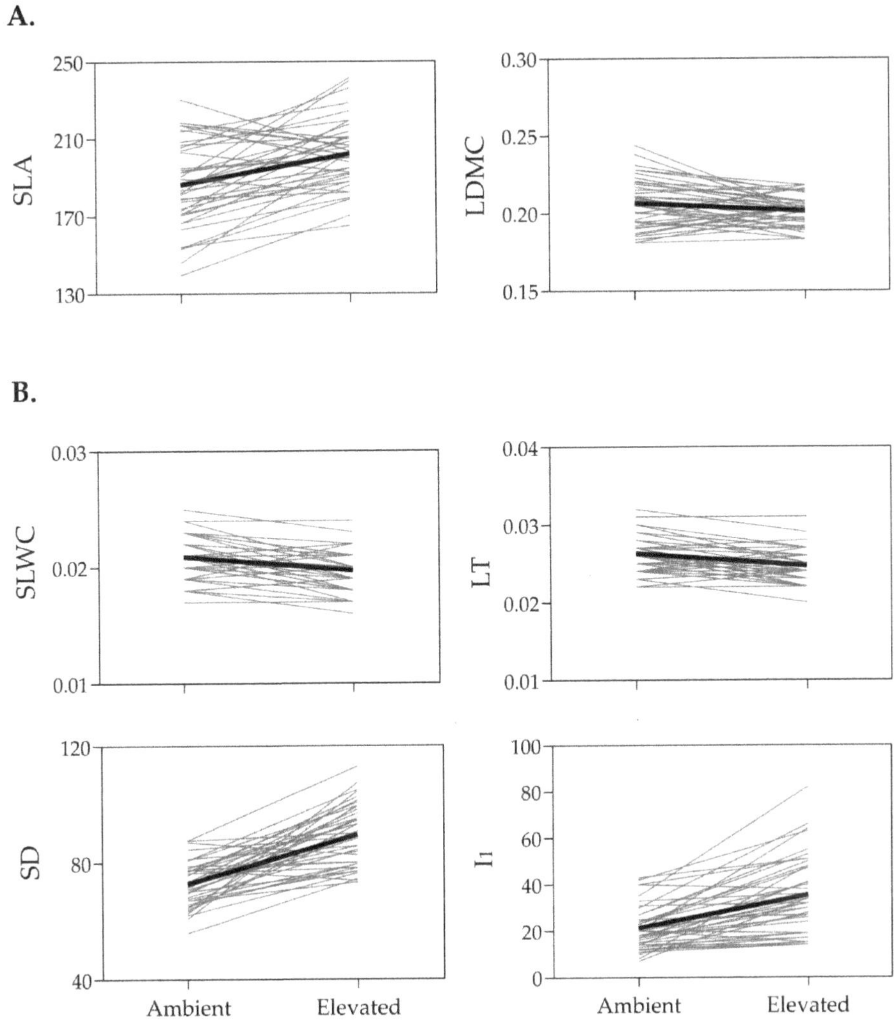

Figure 1. Reaction norm plots for the 40 *Iris* genotypes (fine lines) observed under two temperature treatments—ambient and elevated. The mean phenotypic plasticity is indicated by a heavy line. (**A**) Functional leaf traits: specific leaf area (SLA, in $cm^2\ g^{-1}$), leaf dry matter content (LDMC, in $g\ g^{-1}$). (**B**) Mechanistic leaf traits: specific leaf water content (SLWC, $g\ cm^{-2}$), leaf thickness (LT), stomatal density (SD, no. stomata/mm^2), and RGB index I_1.

The mean phenotypic plasticity of SLA, LDMC, SLWC, LT, SD, and I_1 under different temperature conditions was found to be trait-dependent (Figure 2). In general, the functional traits proved to be less plastic than the mechanistic ones ($PIv = 0.239$ vs. $PIv = 0.369$). Interestingly, the estimated plasticity index for the ratio of functional to mechanistic traits ($PIv = 0.463$) was found to be higher than the values considered for these traits individually. The most plastic traits were I_1 and SD ($PIv = 0.358$ and $PIv = 0.183$, respectively), whereas the least plastic one was LDMC ($PIv = 0.066$). An intermediate level of plasticity was detected for SLA, SLWC, and LT ($PIv = 0.104$, $PIv = 0.087$, and $PIv = 0.085$, respectively). The results of a Friedman ANOVA showed that the amount of plasticity differed for all leaf traits analyzed χ^2 $(5, N = 240) = 67.11$, $p < 0.0001$ (Figure 2).

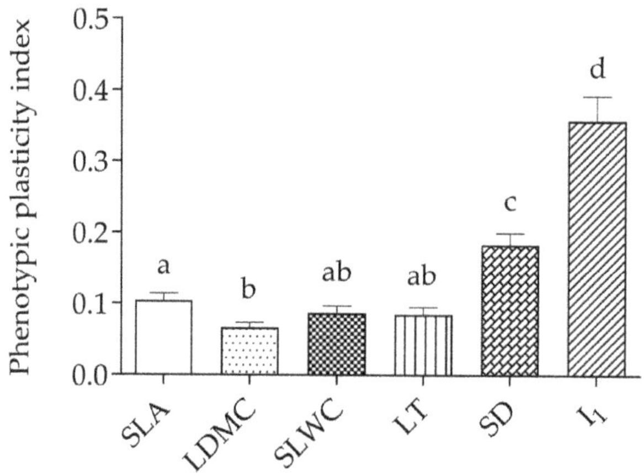

Figure 2. The phenotypic plasticity indices (mean ± SE) for the functional (SLA, LDMC) and mechanistic (SLWC, LT, SD, RGB index I_1) leaf traits of 40 distinct *I. pumila* genotypes expressed at ambient and elevated air temperatures. Means with the same letter are not significantly different at $p > 0.05$, according to the Wilcoxon rank-sum test.

2.3. Phenotypic Correlations Between Functional and Mechanistic Leaf Traits in Response to Temperature

The association between the phenotypic values of individuals for a pair of leaf traits is shown in Figure 3. Within each temperature treatment, a substantial number of significant correlations was observed, accounting for more than half of the total correlations. The positive correlation between SLWC and LT was the strongest among all correlations between pairs of leaf traits (r = 0.976 and r = 0.986, both $p < 0.0001$, in the ambient and elevated temperatures, respectively), whereas that between SLA and I_1 was moderate in strength (r = 0.525, $p < 0.001$ and r = 0.546, $p < 0.001$ in the ambient and elevated temperatures, respectively). The correlation between SLA and SLWC was negative in sign and moderate in strength (r = −0.604 and r = −0.568, both $p < 0.0001$, in the ambient and elevated temperatures, respectively), as was the correlation between SLA and LT (r = −0.760 and r = −0.667, both $p < 0.0001$, in the ambient and elevated temperatures, respectively). The weakest negative correlation was detected between LDMC and I_1 (r = −0.322, $p = 0.04$ and r = −0.353, $p = 0.02$ in the ambient and elevated temperatures, respectively).

Although the same total number of significant phenotypic correlations (two positive and six negative) was observed in both temperature treatments, the correlation structures suggest potential differences (Figure 3). For example, the significant correlation between SLA and LDMC (r = −0.663, $p < 0.0001$) as well as between SLWC and I_1 (r = −0.345, $p = 0.03$) was only detected at the ambient temperature. Conversely, at the elevated temperature, significant correlations between SLA and SD (r = −0.318, $p = 0.04$) and between SD and I_1 (r = −0.317, $p = 0.04$) were found. However, these differences do not appear to be strong enough for the correlation patterns between leaf traits expressed at ambient and elevated temperatures to differ significantly, as confirmed by the Mantel test (r = 0.902, $p < 0.01$).

Apart from the correlation analysis of the individual functional and mechanistic leaf traits, each trait category was also analyzed in its entirety. The first category, referred to as functional, showed a positive correlation between its phenotypic expression at ambient and elevated temperatures (r = 0.564, $p < 0.001$), as did the other category, referred to as mechanistic (r = 0.424, $p < 0.01$). In contrast, the functional and mechanistic trait categories were negatively correlated with each other during both temperature treatments (r = −0.794 and r = −0.819, both $p < 0.0001$, for ambient and elevated temperatures, respectively).

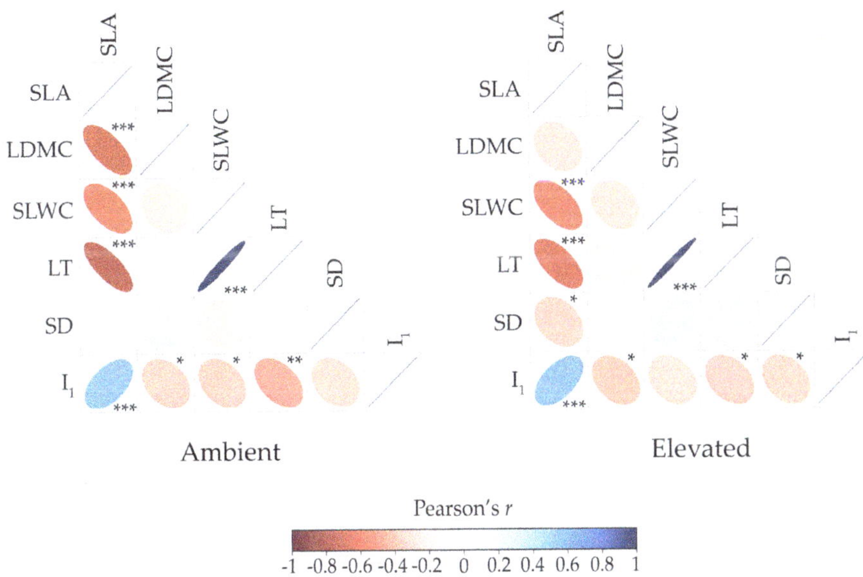

Figure 3. Heat maps of the correlations between the functional (SLA and LDMC) and mechanistic (SLWC, LT, SD, RGB index I_1) leaf traits under ambient and elevated temperature conditions. The color gradient and ellipse eccentricity represent the strength and direction of the correlations, with blue indicating stronger positive correlations and red indicating stronger negative correlations. Correlation values are identified as weak ($r < 0.4$), moderate ($0.4 \leq r \leq 0.8$), or strong ($r > 0.8$). Asterisks indicate the significance level of the correlations: * $p < 0.05$, ** $p < 0.01$, *** $p < 0.001$.

Furthermore, when correlating the plasticity indices, a strong and highly significant positive association was found between the SLWC and LT plasticity indices ($r = 0.936$, $p < 0.0001$). Additionally, a moderately positive association was discovered between the plasticity indices of SLA and I_1 ($r = 0.351$, $p = 0.03$), as well as between those of LDMC and I_1 ($r = 0.338$, $p = 0.03$). A marginally significant association was observed between the SLA and LT plasticity indices ($r = 0.317$, $p = 0.05$).

3. Discussion

3.1. Acclimation Responses to Increased Temperature

To withstand changing environments, plants have evolved diverse modifications in their morphology, physiology, phenology, and reproduction [61] or improved their capacity to tolerate and adjust to novel conditions through phenotypic plasticity [6,62]. The present study focuses on elucidating the role of leaf traits in response to elevated temperatures in *I. pumila*. By analyzing the phenotypic expression of its functional and mechanistic leaf traits at ambient and elevated temperatures, under controlled experimental warming conditions in a growth room, we aimed to assess the acclimation potential of *I. pumila* to adjust to new temperature environments. Despite *I. pumila* genotypes' remarkable ability to respond plastically to variations in temperature conditions, both the magnitude and pattern of plastic responses appear to be highly trait-specific [63]. However, the observed trait specificity does not refer to the functional and mechanistic traits as comprehensive categories, but rather to individual traits within this categorization. This suggests that different leaf traits can exhibit different plastic responses regardless of their functional categorization, emphasizing the complexity of trait–environment interactions [63].

The two most important functional leaf traits, specific leaf area (SLA) and leaf dry matter content (LDMC), proved to be reliable indicators of plant resource utilization strategies in the sand dune environment [20,63–66]. Furthermore, SLA is generally positively associated with both leaf nitrogen content per unit of dry mass and assimilation rate, but negatively with leaf lifespan [67–69]. This indicates that the accumulation of dry matter is related to efficient photosynthetic activity, which is due to both transpiration fluxes and Rubisco activity [28,70,71]. Although there is substantial evidence that SLA increases with temperature [34,72], the relationship between SLA and temperature is not generally positive and can vary depending on plant species, ecosystems, and environmental conditions [73,74]. In the present study, the plants that were subjected to experimental warming generally exhibited greater SLA than the plants that were grown at the ambient temperature. This result is fully consistent with the findings of our previous study conducted *in situ* in natural populations of *I. pumila* [63]. Since only temperature was manipulated in the current experiment, it confirms that the change in SLA is a direct consequence of increased temperature [33]. Since increased temperatures can accelerate metabolic processes, such an increase in SLA could be due to increased resource allocation towards leaf area expansion, thereby optimizing photosynthetic efficiency under stressful conditions [66]. In addition, larger leaves enhance transpiration efficiency, which helps to cool the plant and maintain an optimal internal temperature [75]. Since gas exchange was not measured in this study, we can only assume that an increase in SLA serves as an acclimation mechanism to cope with heat stress by enabling more efficient transpiration and photosynthesis [76]. Future studies using direct measurements of stomatal conductance and photosynthetic rates would provide a more comprehensive understanding and further strengthen this hypothesis. LDMC reflects a plant's investment in leaf structure, with higher values generally indicating tougher, more durable leaves that are resistant to environmental stresses [31,77,78]. A lower LDMC value is often associated with a strategy that favors resource acquisition and faster growth, as it corresponds to thinner, softer leaves that allow for faster carbon assimilation and water turnover. The relationship between LDMC and global warming is complex and context-dependent, implying that the response of LDMC may vary across plant species and ecosystems [36–38]. The decrease in LDMC at elevated temperatures observed in both this and our previous study [63] may indicate that plants are allocating fewer resources to the structural components of leaves, possibly favoring faster growth and increased metabolic activity [79]. In the context of the observed inverse relationship with SLA, this could indicate a trade-off between maximizing leaf area for photosynthesis (high SLA) and maintaining structural integrity or water retention (low LDMC), which could represent an adjustment mechanism to cope with higher temperatures.

Regarding the mechanistic leaf trait category, SLWC is closely responsive to global warming, as fluctuations in temperature and precipitation patterns affect water availability, which ultimately influences plants' resilience to these challenges [40,80,81]. The observed reduction in SLWC may indicate various issues, such as increased transpiration, reduced water retention in leaf tissue, or a disturbed balance between water uptake and loss [63,82,83]. Overall, the decrease in SLWC with increasing temperatures can serve as a potential indicator of physiological stress and reflect the challenges the plant faces in coping with thermal stress. Conversely, the increased SD observed at elevated temperatures suggests that plants may adjust to these conditions by enhancing their gas exchange capacity and improving cooling through transpiration [84]. This response could serve as a compensatory mechanism to maintain photosynthesis under heat stress. Although SD provides a structural foundation for gas exchange, transpiration is primarily regulated by stomatal conductance and dynamic control of stomatal aperture, rather than density alone [50]. Studies have shown that stomatal conductance may increase with rising temperatures, thereby enhancing evaporative cooling and optimizing gas exchange efficiency [85]. A combined analysis of stomatal conductance and stomatal density would have provided a more comprehensive understanding of the physiological mechanisms underlying these responses, distinguishing structural adjustments from functional regulation in *I. pumila*'s acclimation to elevated temperatures. Given that even small changes in SD can significantly affect gas exchange and transpiration [86], the observed 20% increase with just a 1 °C rise in temperature suggests a remarkable acclimation response in *I.* pumila, improving water regulation. However, SD alone is insufficient for drawing definitive conclusions about transpiration, as stomatal function—particularly stomata's ability to dynamically respond to environmental factors—plays a more critical role in regulating water loss. This underscores the need to consider both structural and functional aspects when studying plant transpiration mechanisms [50]. Leaf thickness (LT) is significantly affected by temperature, with higher temperatures generally leading to thinner leaves due to increased transpiration rates and reduced cell expansion [49,87]. The decrease in LT observed in this study probably reflects an acclimation strategy intended to protect the leaf tissue from overheating. Chlorophyll content, a key player in leaf photosynthesis, was significantly reduced under elevated temperature conditions, considering that the temperature increase was only one degree Celsius. This finding underlines the delicate balance that plants maintain with their environment and the potential implications of climate change on plant physiology. Even though some species do indeed reduce their chlorophyll content as a protective mechanism against heat stress [60,88], this study focused primarily on heat-induced damage to chlorophyll molecules and the photosynthetic machinery, which is a well-documented response in many species [89–91]. Unfortunately, no measurements of gas exchange were performed; these would have provided further insight into stomatal conductance and its role in photosynthetic activity under elevated temperatures.

3.2. *Thermal Plasticity of Functional and Mechanistic Leaf Traits*

Phenotypic plasticity has long been recognized as a key strategy that enables plants to adapt to variable environmental conditions [92]. As temperature is one of the most dynamic environmental factors and significantly affects plant metabolism, the plasticity of leaf functional traits is essential for maintaining physiological homeostasis and enhancing plant resilience to temperature fluctuations [93,94]. Our results confirm that the mean plastic responses of functional and mechanistic leaf traits, SLA, LDMC, SLWC, SD, I_1, and LT, changed with the increase in temperature. Of the traits analyzed, the highest plasticity was observed in chlorophyll content (namely its proxy, I_1), which suggests a strong capability of the plants to modulate their photosynthetic machinery in response to temperature fluctuations. The temperature sensitivity of photosynthesis is a crucial factor in predicting how plants will respond to warming [95]. By adjusting their photosynthetic apparatus, such as enzyme activity and membrane properties, to optimize carbon fixation under different environmental temperatures, plants will be able to maintain their productivity in changing climates [96].

Stomatal density (SD) was the second most plastic trait, reflecting the plant's ability to optimize gas exchange under varying environmental conditions, such as humidity or CO_2 concentration [55]. Modulating stomatal density helps the plant maintain a crucial balance between water loss and carbon dioxide uptake in response to fluctuating conditions. SLA, SLWC, and LT exhibited comparable levels of plasticity, which were relatively low in magnitude. This implies that these leaf traits remain relatively stable and show consistent expression across different thermal environments. The stability of these traits could be a result of physiological or ecological constraints, where maintaining consistent functionality is prioritized over flexibility in response to environmental changes [97]. In contrast, LDMC exhibited the lowest plasticity, suggesting that it represents a more conservative trait, being less responsive to short-term environmental changes and more reflective of a species' long-term adaptation to its habitat. Low plasticity in LDMC is often associated with stress-tolerant species, as it reflects a resource allocation strategy that favors structural stability and resistance to abiotic stresses, such as drought [31]. The converging reaction norms discovered for LDMC at elevated temperatures further emphasize its limited plasticity.

In general, leaf traits related to light absorption and gas exchange, such as chlorophyll content and stomatal density, are more flexible and can rapidly adjust to environmental changes. In contrast, traits like LDMC, which reflect a plant's resource use strategy and structural investment, are less plastic and may be more important for survival in stable or resource-limited environments.

In addition, our study provides evidence that the mean plastic response of mechanistic leaf traits, when considered collectively, is greater than that of functional leaf traits, indicating that mechanistic leaf traits have high sensitivity to temperature variation. This result is consistent with the perspective that mechanistic traits serve as physiological strategies that are specifically adjusted to the prevailing environmental factors and enable responses over short periods of time [98]. Higher plasticity allows plants to rapidly adjust to environmental changes, such as temperature fluctuations, enhancing survival in the short term. Conversely, functional leaf traits represent broader ecological strategies that respond to multiple environmental factors simultaneously and over longer periods of time [98]. Their relative stability and lower plasticity may result from their role in maintaining overall resource use efficiency and ecological balance across varying conditions. The observed higher plasticity of mechanistic traits emphasizes their role in fine-tuning physiological processes for acute stress management, whereas functional traits contribute to more generalized ecological strategies that prioritize consistency over flexibility. This dichotomy highlights the trade-off between rapid response and long-term stability in plant adaptation strategies [99–101].

3.3. Temperature-Driven Correlation Patterns of Leaf Traits

Since different traits often exhibit different plastic responses, it is likely that the correlations between them may also be shaped by environmental conditions [92,102]. Determining phenotypic correlations can provide important insights into the integration and stability of traits under varying environmental conditions [103]. It is well established that integrated phenotypes, that is, correlated suites of traits, can affect both ecological and evolutionary processes [104]. The high phenotypic correlations observed between leaf traits across different temperature conditions indicate a high degree of integration [105], suggesting that these traits function together in a coordinated manner. At the same time, the consistent phenotypic correlations across temperature treatments emphasize the stability of these leaf traits. Stable traits are likely to maintain their functional role despite environmental fluctuations and are, therefore, valuable targets for breeding resilient plant varieties [106,107].

To evaluate the integration and stability of leaf traits under varying temperature conditions, we assessed the phenotypic correlations between individual leaf trait expressions in each temperature treatment. The strongest positive correlation, observed between SLWC and LT at both temperature conditions, suggests a robust relationship between these two traits and may reflect an evolutionary adaptation in which thicker leaves improve water retention and overall structural integrity, contributing to higher leaf water content [108]. Moreover, the positive correlation of chlorophyll content with LDMC and LT, coupled with its negative correlation with SLA, suggests a trade-off in which thicker leaves invest more in chlorophyll to enhance photosynthetic efficiency, while thinner leaves prioritize maximizing light capture through their greater surface area [59]. In addition, we observed that different sets of leaf traits are correlated under distinct temperature conditions, highlighting the plants' ecological plasticity and potential adaptability to different environmental pressures [109–112].

Apart from determining the correlation patterns of the individual functional and mechanistic leaf traits, the relationship between the two trait categories as a whole was also assessed. The strong inverse relationship between the mechanistic and functional trait categories observed at both temperature treatments suggests a trade-off in which a higher investment in one set of traits may come at the expense of the other. Some of the potential scenarios reflecting this balance between mechanistic and functional trait categories may include, for instance, photosynthetic efficiency vs. protection, growth vs. stress resistance, or efficiency vs. longevity [113].

It is believed that correlations between the plasticity of different traits arise when the given traits exhibit the same sensitivity to environmental conditions, share the same function, and/or have a genetic basis [92,102]. The existence of similar plastic responses allows for the maintenance of an integrated phenotype despite the influence of variable environmental conditions. Traits that exhibit significantly correlated plasticity indices appear to respond jointly to environmental changes, suggesting a coordinated adaptive response [114,115]. The positive correlation observed between the plasticity indices for SLA and LT, as well as between SLWC and LT, suggests that these traits respond jointly to environmental changes. The former implies that the plants may simultaneously adjust their leaf area and thickness to optimize light capture and structural support under varying environmental conditions. The latter may indicate that the plants employ coordinated adjustment mechanisms to optimize water storage and photosynthesis under changing conditions to ensure that they manage resources efficiently, for example by altering leaf structure or physiological processes. In this way, they can improve water use efficiency and maintain photosynthetic activity, ensuring survival and growth even under stressful conditions such as drought or high temperatures [116]. Conversely, the negative correlation between the plasticity indices of SLA and chlorophyll content, as well as between LDMC and chlorophyll content, indicates a trade-off in the plants' adaptive strategies. If a plant adjusts its leaf area or dry matter content to optimize functions such as light capture or structural integrity, it may reduce its ability to modulate chlorophyll content and vice versa. This trade-off can help the plant to balance its resources and maintain its overall functionality under varying environmental conditions [117,118].

4. Materials and Methods

4.1. The Study Species

Iris pumila L. (Iridaceae) is a perennial, rhizomatous plant native to the Eurasian steppe with a wide geographical distribution. It extends from Austria in the west across central and southeastern Europe to the easternmost parts of western Siberia [119–125]. In Serbia, the species is native to the Deliblato Sands (44°90′23″ N, 21°11′32″ E), an isolated sand dune complex in the southern part of Banat [126], between the Danube and the western Carpathian slopes. Natural populations of *I. pumila* are widespread at exposed dune sites but can also be found in shaded habitats, usually under the canopy of *Pinus nigra*. Individual plants form rounded clones consisting of compact horizontally growing rhizome segments that extend from the center of a clone towards its margin. In such a way, the species establishes a rhizome system that can persist in an integrated state for many years [126]. In the Deliblato Sands, average annual temperatures have increased by 0.52 °C per decade, resulting in more arid conditions [127]. To reflect this trend, a 1 °C temperature increment was chosen in the present study to simulate conditions likely to occur in the near future.

4.2. Experimental Setup, Leaf Sampling, and Leaf Trait Measurement

The plants used for this study derived from a common garden experiment established back in 1986 from a natural population of *I. pumila* which inhabited a sun-exposed dune in the Deliblato Sands. In 2017, clonal genotypes from the experimental garden were individually planted into 500 cm^3 plastic pots containing a 2:1 (v/v) mixture of soil substrate and sand. The potted plants were placed in randomized positions on shelves in a growth room with an automated temperature control system and grown under a 16 h photoperiod, with 110.5 $\mu molm^2 s^{-1}$ of photosynthetically active radiation (PAR) provided by Philips TLD 36-W/54 fluorescent lamps (Philips Lighting (Signify N.V.), Eindhoven, Netherlands). Plants were watered regularly, once a week, with a nutrient solution. The growth temperature was maintained at 23/19 °C (day/night) over multiple generations; this is, therefore, considered the ambient temperature. To minimize the effects of position, the pots were rotated twice a week.

In 2019, leaf samples from 40 randomly selected *I. pumila* genotypes were collected. The ambient temperature was then increased by 1 °C (24/20 °C day/night), and after approximately one month the newly formed leaves from the same genotypes were collected. Since the part of the leaf that is sampled (top, middle, or base) and its developmental stage influence the traits analyzed, we only sampled 2/3 of the upper part of the last fully developed leaf. At both sampling times, the experimental procedure included cutting off the last fully developed leaf from one ramet per genotype and immediately measuring its fresh mass. Afterwards, a digital image of the leaf area was captured by an optical scanner (Epson perfection V600, Epson, Suwa, Japan) and the projected leaf area was determined using the program ImageJ (v1.51j8) [128]. Subsequently, the leaf samples were oven-dried at 60 °C for 72 h to a constant mass and their dry mass was weighed.

The phenotypic values of the studied leaf traits were determined using standardized protocols for plant trait measurements [129]. SLA (in cm^2 g^{-1}) was estimated as the ratio of leaf area to leaf dry mass [77], LDMC (in g g^{-1}) as the ratio of leaf dry mass to fresh mass [130], and SLWC as the difference between fresh and dry mass divided by leaf area [80]. LT was calculated according to the formula $1/SLA \times LDMC$ [131].

Stomatal density (SD) was estimated using the impression method [132]. Briefly, an area in the middle section of the adaxial leaf surface was coated with a transparent nail polish. Once the polish was dry, it was peeled off with the adhesive tape and fixed on a microscope slide. The number of stomata was counted in 10 randomly selected microscopic fields (0.161 mm^2 at $20 \times 3.2 \times$ magnification). *I. pumila* has amphistomatous, vertically oriented leaves, with no clear distinction between adaxial and abaxial leaf surfaces [133]. Since stomatal density does not differ between the two leaf surfaces (Figure S1), we arbitrarily selected the 'adaxial' surface—i.e., the side facing the rhizome—to ensure consistency across samples and to facilitate comparisons with our previous [53] and future studies.

The leaf chlorophyll content was estimated non-destructively using an image-based traits analysis [134]. Digital images of scanned leaves were analyzed for their green pixel intensity by extracting color information using an RGB model and ImageJ [128]. Previously, a pilot experiment (Table S1) had shown that the RGB index I_1 proposed by Sánchez-Sastre et al. [134] exhibits a moderately strong, but highly significant, linear negative correlation with the chlorophyll concentration, quantified by the DMSO method [135]. Furthermore, based on the regression analysis, which showed that all points are within the 99% prediction interval (Figure S2), it can be concluded that the non-destructively determined RGB index I_1 provides reliable estimates of chlorophyll content, making it a valid alternative to the destructive method of chlorophyll quantification. Therefore, this particular index, I_1, was used in the present study.

4.3. Statistical Analysis

All statistical analyses were performed with R Statistical Software (v4.2.3; R Core Team, 2023) in RStudio (v 2023.3.0.386) [136]: The *tidyr* [137], *dplyr* [138], and *reshape2* [139] packages were used to facilitate the data manipulation and reshaping. The *rstatix* package [140] was used to estimate the basic parameters of descriptive statistics (e.g., mean, standard error, coefficient of variation). The assumption of normality was checked and confirmed using the Shapiro–Wilk normality test from the base *stats* package [136]. Since each of the leaf trait measurements was completed on the same plant across two temperature regimes, a repeated measures ANOVA was performed using the *stats* package [136]. A comparison of the coefficients of variation (CV%) between the groups was performed using the *F*-test for equality of variances from the *stats* package [136]. The relationships between pairs of traits within a temperature regime were estimated using Pearson's correlation analysis. The correlation coefficient matrices for leaf traits expressed at ambient and elevated temperatures, along with the corresponding correlation heatmaps, were created using the *corrplot* package [141]. Afterwards, the correlation matrices were compared using the Mantel test from the *vegan* package [142]. The temperature-induced plasticity in the analyzed leaf traits was determined by calculating the plasticity index, PI_V [143]:

$$PI_V = |X_A - X_E| / X_E$$

where X_A is the value of a given leaf trait expressed at the ambient temperature, while X_E is the value of the same leaf trait expressed at an elevated temperature. PI_V denotes the measurement of the change in a trait from an elevated to the ambient temperature. A comparison of the plasticity indices of all analyzed leaf traits within the same plants grown under different temperature conditions was performed using Friedman's ANOVA, as the data were not normally distributed. The significance of the difference between each two plasticity indices was assessed using the post hoc Wilcoxon rank-sum test from the *stats* package [136].

To perform statistical analyses on the individual trait categories—functional and mechanistic leaf traits—rather than on individual traits within each category, the data were first normalized due to the differing ranges of absolute values across the traits. Normalization is essential when integrating variables with different units or scales, as it ensures that each variable contributes equally to the analysis [144]. By normalizing the data, the influence of large-scale variables is reduced, allowing for more meaningful comparisons between traits [145]. Once normalized, the mean values for each trait category were calculated and used in correlation and plasticity analyses.

5. Conclusions

This study explores the role of phenotypic plasticity in enabling *Iris pumila* to adjust its functional and mechanistic leaf traits in elevated temperature conditions. By applying an experimental temperature manipulation design, our study confirmed that a temperature increase of 1 °C—a change comparable to that caused by global warming—is sufficient to induce a plastic response in functional and mechanistic leaf traits. The observed variation in the plastic responses of different leaf traits demonstrates the complex and trait specific nature of plant acclimation strategies. Functional leaf traits, such as SLA and LDMC, and mechanistic traits, such as SD and chlorophyll content, exhibit different amounts of plasticity, reflecting the plant's ability to adjust its resource utilization and physiological functions in response to thermal stress. Notably, mechanistic traits exhibit higher plasticity than functional traits, highlighting the rapid physiological adjustments that plants make in the face of immediate environmental stress. These results suggest that while plants are capable of responding to temperature changes, such adjustments may involve a trade-off between maximizing resource acquisition and maintaining structural integrity. Understanding the interplay between functional and mechanistic traits and their plastic responses provides valuable insights into plants' resilience to climate warming, with implications for the prediction of ecological and evolutionary responses to future climate change.

Supplementary Materials: The following supporting information can be downloaded at: https://www.mdpi.com/article/10.3390/plants14060960/s1, Figure S1: Representative micrographs of *Iris pumila* leaf epidermal impressions. The images display the adaxial (A) and abaxial (B) surfaces of the same leaf. Figure S2: The relationship between RGB index I_1 (calculated according to Sánchez-Sastre et al. 2020 [134]) and chlorophyll concentration (in μgcm^{-2}) quantified by the DMSO method [135]. Table S1: The correlation coefficients for RGB index I_1 and chlorophyll concentration (in μgcm^{-2}) quantified by the DMSO method.

Author Contributions: Conceptualization, S.M.J.; methodology, S.M.J.; validation, K.H., A.V. and S.M.J.; formal analysis, K.H. and S.M.J.; investigation, K.H., A.V. and S.M.J.; writing—original draft preparation, S.M.J.; writing—review and editing, K.H., A.V. and S.M.J.; visualization, K.H. and S.M.J.; supervision, S.M.J. All authors have read and agreed to the published version of the manuscript.

Funding: This research was funded by the Ministry of Science, Technological Development and Innovation of the Republic of Serbia, grant number 451-03-136/2025-03/200007. The results presented in this manuscript are in line with Sustainable Development Goal 13 (Climate Action) of the United Nations 2030 Agenda.

Data Availability Statement: The original contributions presented in this study are included in the article/Supplementary Materials. Further inquiries can be directed to the corresponding author.

Acknowledgments: We express our gratitude to the three anonymous reviewers and the Academic Editors for their valuable and constructive comments and suggestions on a previous version of this manuscript.

Conflicts of Interest: The authors declare no conflicts of interest. The funders had no role in the design of the study; in the collection, analysis, or interpretation of data; in the writing of the manuscript; or in the decision to publish the results.

References

1. IPCC. Summary for Policymakers. In *Climate Change 2023: Synthesis Report. Contribution of Working Groups I, II and III to the Sixth Assessment Report of the Intergovernmental Panel on Climate Change*; Core Writing Team, Lee, H., Romero, J., Eds.; IPCC: Geneva, Switzerland, 2023; pp. 1–34. [CrossRef]
2. Alatalo, J.M.; Little, C.J.; Jägerbrand, A.K.; Molau, U. Vascular plant abundance and diversity in an alpine heath under observed and simulated global change. *Sci. Rep.* **2015**, *5*, 10197. [CrossRef] [PubMed]
3. Elmendorf, S.C.; Henry, G.H.; Hollister, R.D.; Fosaae, A.M.; Gould, W.A.; Hermanutz, L.; Hofgaard, A.; Jónsdóttir, I.S.; Jorgenson, J.C.; Lévesquel, E.; et al. Experiment, monitoring, and gradient methods used to infer climate change effects on plant communities yield consistent patterns. *Proc. Natl. Acad. Sci. USA* **2015**, *112*, 448–452. [CrossRef] [PubMed]
4. Catullo, R.A.; Liewelyn, J.; Phillips, B.L.; Moritz, C.C. The potential for rapid evolution under anthropogenic climate change. *Curr. Biol.* **2019**, *29*, R996–R1007. [CrossRef] [PubMed]
5. Greenspoon, P.B.; Spencer, H.G. Avoiding extinction under nonlinear environmental change: Models of evolutionary rescue with plasticity. *Biol. Lett.* **2021**, *17*, 20210459. [CrossRef]
6. Nicotra, A.B.; Atkin, O.K.; Bonser, S.P.; Davidson, A.M.; Finnegan, E.J.; Mathesius, U.; Poot, P.; Purugganan, M.D.; Richards, C.L.; Valladares, F.; et al. Plant phenotypic plasticity in a changing climate. *Trends Plant Sci.* **2010**, *15*, 684–692. [CrossRef]
7. Chevin, L.M.; Collins, S.; Lefevre, F. Phenotypic plasticity and evolutionary demographic responses to climate change: Taking theory out to the field. *Funct. Ecol.* **2013**, *27*, 967–979. [CrossRef]
8. Merilä, J.; Hendry, A.P. Climate change, adaptation, and phenotypic plasticity: The problem and the evidence. *Evol. Appl.* **2014**, *7*, 1–14. [CrossRef]
9. Scheiner, S.M.; Barfield, M.; Holt, R.D. The genetics of phenotypic plasticity. XVII. Response to climate change. *Evol. Appl.* **2020**, *13*, 388–399. [CrossRef]

10. Gienapp, P.; Teplitsky, C.; Alho, J.S.; Mills, J.A.; Merilä, J. Climate change and evolution: Disentangling environmental and genetic responses. *Mol. Ecol.* **2008**, *17*, 167–178. [CrossRef]
11. Hoffmann, A.A.; Sgrò, C.M. Climate change and evolutionary adaptation. *Nature* **2011**, *470*, 479–485. [CrossRef]
12. Nogués-Bravo, D.; Rodríguez-Sánchez, F.; Orsini, L.; de Boer, E.; Jansson, R.; Morlon, H.; Damien, A.; Jackson, S.T. Cracking the code of biodiversity responses to past climate change. *Trends Ecol. Evol.* **2018**, *33*, 765–776. [CrossRef] [PubMed]
13. Ghalambor, C.K.; McKay, J.K.; Carroll, S.P.; Reznick, D.N. Adaptive versus non-adaptive phenotypic plasticity and the potential for contemporary adaptation in new environments. *Funct. Ecol.* **2007**, *21*, 394–407. [CrossRef]
14. Niinemets, Ü. Responses of forest trees to single and multiple environmental stresses from seedlings to mature plants: Past stress history, stress interactions, tolerance and acclimation. *For. Ecol. Manag.* **2010**, *260*, 1623–1639. [CrossRef]
15. Davidson, A.M.; Jennions, M.; Nicotra, A.B. Do invasive species show higher phenotypic plasticity than native species and, if so, is it adaptive? A meta-analysis. *Ecol. Lett.* **2011**, *14*, 419–431. [CrossRef]
16. Wang, S.P.; Althoff, D.M. Phenotypic plasticity facilitates initial colonization of a novel environment. *Evolution* **2019**, *73*, 303–316. [CrossRef]
17. Snell-Rood, E.C.; Kobiela, M.E.; Sikkink, K.L.; Shephard, A.M. Mechanisms of plastic rescue in novel environments. *Annu. Rev. Ecol. Evol. Syst.* **2018**, *49*, 331–354. [CrossRef]
18. Westoby, M. A leaf-height-seed (LHS) plant ecology strategy scheme. *Plant Soil* **1998**, *199*, 213–227. [CrossRef]
19. Vendramini, F.; Diaz, S.; Gurvich, D.E.; Wilson, P.J.; Thomson, K.; Hodgson, J.G. Leaf traits as indicator of resource-use strategy in floras with succulent species. *New Phytol.* **2002**, *154*, 147–157. [CrossRef]
20. Madani, N.; Kimball, J.S.; Ballantyne, A.P.; Affleck, D.L.; Van Bodegom, P.M.; Reich, P.B.; Kattge, J.; Sala, A.; Nazeri, M.; Jones, M.O.; et al. Future global productivity will be affected by plant trait response to climate. *Sci. Rep.* **2018**, *8*, 2870. [CrossRef]
21. Liu, Y.; Tang, L.; Qiu, X.; Liu, B.; Chang, X.; Liu, L.; Zhang, X.; Cao, W.; Zhu, Y. Impacts of 1.5 and 2.0 °C global warming on rice production across China. *Agric. For. Meteorol.* **2020**, *284*, 107900. [CrossRef]
22. Gao, J.; Zhao, P.; Shen, W.; Rao, X.; Hu, Y. Physiological homeostasis and morphological plasticity of two tree species subjected to precipitation seasonal distribution changes. *Perspect. Plant Ecol. Evol. Syst.* **2017**, *25*, 1–19. [CrossRef]
23. Kassout, J.; Terral, J.F.; Hodgson, J.G.; Ater, M. Trait-based plant ecology a flawed tool in climate studies? The leaf traits of wild olive that pattern with climate are not those routinely measured. *PLoS ONE* **2019**, *14*, e0219908. [CrossRef] [PubMed]
24. Lavorel, S.; Garnier, E. Predicting changes in community composition and ecosystem functioning from plant traits: Revisiting the Holy Grail. *Funct. Ecol.* **2002**, *16*, 545–556. [CrossRef]
25. Atkin, O.K.; Scheurwater, I.; Pons, T.L. High thermal acclimation potential of both photosynthesis and respiration in two lowland Plantago species in contrast to an alpine congeneric. *Glob. Change Biol.* **2006**, *12*, 500–515. [CrossRef]
26. Meziane, D.; Shipley, B. Direct and indirect relationships between specific leaf area, leaf nitrogen and leaf gas exchange. Effects of irradiance and nutrient supply. *Ann. Bot.* **2001**, *88*, 915–927. [CrossRef]
27. Vuleta, A.; Manitašević-Jovanović, S.; Tucić, B. Pattern of plasticity to irradiance levels and genotypic correlations between structural and physiological leaf traits in *Iris pumila*. *Arch. Biol. Sci.* **2011**, *63*, 655–660. [CrossRef]
28. Poorter, H.; Niinemets, Ü.; Poorter, L.; Wright, I.J.; Villar, R. Causes and consequences of variation in leaf mass per area (LMA): A meta-analysis. *New Phytol.* **2009**, *182*, 565–588. [CrossRef]
29. Larigauderie, A.; Hilbert, D.W.; Oechel, W.C. Effect of CO_2 enrichment and nitrogen availability on resource acquisition and resource allocation in a grass, *Bromus mollis*. *Oecologia* **1988**, *77*, 544–549. [CrossRef]
30. Bussotti, F. Functional leaf traits, plant communities and acclimation processes in relation to oxidative stress in trees: A critical overview. *Glob. Change Biol.* **2008**, *14*, 2727–2739. [CrossRef]
31. Poorter, H.; De Jong, R.O.B. A comparison of specific leaf area, chemical composition and leaf construction costs of field plants from 15 habitats differing in productivity. *New Phytol.* **1999**, *143*, 163–176. [CrossRef]
32. Garnier, E.; Cortez, J.; Billè, G.; Navas, M.L.; Roumet, C.; Debussche, M.; Laurent, G.; Blanchard, A.B.; Aubry, D.; Bellmann, A.; et al. Plant functional markers capture ecosystem properties during secondary succession. *Ecology* **2004**, *85*, 2630–2637. [CrossRef]
33. Manitašević-Jovanović, S.; Vuleta, A.; Tucić, B. Does physiological integration among intraclonal ramets of *Iris pumila* enhance stress tolerance in heterogeneous environments? *Arch. Biol. Sci.* **2014**, *66*, 713–720. [CrossRef]
34. Rosbakh, S.; Römermann, C.; Poschlod, P. Specific leaf area correlates with temperature: New evidence of trait variation at the population, species and community levels. *Alp. Bot.* **2015**, *125*, 79–86. [CrossRef]
35. Shipley, B.; Vu, T.T. Dry matter content as a measure of dry matter concentration in plants and their parts. *New Phytol.* **2002**, *153*, 359–364. [CrossRef]
36. Duru, M.; Al Haj Khaled, R.; Ducourtieux, C.; Theau, J.P.; De Quadros, F.L.F.; Cruz, P. Do plant functional types based on leaf dry matter content allow characterizing native grass species and grasslands for herbage growth pattern? In *Herbaceous Plant Ecology*; Van Der Valk, A.G., Ed.; Springer: Dordrecht, The Netherlands, 2008; pp. 57–69, ISBN 978-90-481-2797-9.

37. Liancourt, P.; Boldgiv, B.; Song, D.S.; Spence, L.A.; Helliker, B.R.; Petraitis, P.S.; Casper, B.B. Leaf-trait plasticity and specie vulnerability to climate change in a Mongolian Steppe. *Glob. Change Biol.* **2015**, *21*, 3489–3498. [CrossRef]
38. Niklas, K.J.; Shi, P.; Gielis, J.; Schrader, J.; Niinemets, Ü. Leaf functional traits: Ecological and evolutionary implications. *Fron Plant Sci.* **2023**, *14*, 1169558. [CrossRef]
39. Garnier, E.; Laurent, G. Leaf anatomy, specific mass and water content in congeneric annual and perennial grass species. *Ne Phytol.* **1994**, *128*, 725–736. [CrossRef]
40. Wang, Z.; Huang, H.; Wang, H.; Peñuelas, J.; Sardans, J.; Niinemets, Ü.; Niklas, K.J.; Li, Y.; Xie, J.; Wright, I.J. Leaf water conter contributes to global leaf trait relationships. *Nat. Commun.* **2022**, *13*, 5525. [CrossRef]
41. Agusti, S.; Enriquez, S.; Frost-Christensen, H.; Sand-Jensen, K.; Duarte, C.M. Light harvesting among photosynthetic organism *Funct. Ecol.* **1994**, *8*, 273–279. [CrossRef]
42. Syvertsen, J.P.; Lloyd, J.; McConchie, C.; Kriedemann, P.E.; Farquhar, G.D. On the relationship between leaf anatomy and CO diffusion through the mesophyll of hypostomatous leaves. *Plant Cell Environ.* **1995**, *18*, 149–157. [CrossRef]
43. Garnier, E.; Salager, J.L.; Laurent, G.; Sonié, L. Relationship between photosynthesis, nitrogen and leaf structure in 14 gras species and their dependence on the basis of expression. *New Phytol.* **1999**, *143*, 119–149. [CrossRef]
44. Enríquez, S.; Sand-Jensen, K. Variation in light absorption properties of *Mentha aquatica* L. as a function of leaf form: Implication for plant growth. *Int. J. Plant Sci.* **2003**, *164*, 125–136. [CrossRef]
45. Poorter, H.; Remkes, C.; Lambers, H. Carbon and nitrogen economy of 24 wild species differing in relative growth rate. *Plan Physiol.* **1990**, *94*, 621–627. [CrossRef]
46. Nielsen, S.L.; Enríquez, S.; Duarte, C.M.; Sand-Jensen, K. Scaling maximum growth rates across photosynthetic organisms. *Func Ecol.* **1996**, *10*, 167–175. [CrossRef]
47. White, J.W.; Montes-R, C. Variation in parameters related to leaf thickness in common bean (*Phaseolus vulgaris* L.). *Field Crops Res* **2005**, *91*, 7–21. [CrossRef]
48. Díaz, S.; Hodgson, J.G.; Thompson, K.; Cabido, M.; Cornelissen, J.H.; Jalili, A.; Montserrat-Martí, G.; Grime, J.P.; Zarrinkamar, F Asri, Y.; et al. The plant traits that drive ecosystems: Evidence from three continents. *J. Veg. Sci.* **2004**, *15*, 295–304. [CrossRef]
49. Wright, I.J.; Dong, N.; Maire, V.; Prentice, I.C.; Westoby, M.; Díaz, S.; Gallagher, R.V.; Jacobs, B.F.; Kooyman, R.; Law, E.A.; et a Global climatic drivers of leaf size. *Science* **2017**, *357*, 917–921. [CrossRef]
50. Lawson, T.; Blatt, M.R. Stomatal size, speed, and responsiveness impact on photosynthesis and water use efficiency. *Plant Physio* **2014**, *164*, 1556–1570. [CrossRef]
51. Beerling, D.J.; Woodward, F.I. Changes in land plant function over the Phanerozoic: Reconstructions based on the fossil record *Bot. J. Linn.* **1997**, *124*, 137–153. [CrossRef]
52. Casson, S.; Gray, J.E. Influence of environmental factors on stomatal development. *New Phytol.* **2008**, *178*, 9–23. [CrossRef]
53. Vuleta, A.; Manitašević-Jovanović, S.; Tucić, B. Light intensity influences variations in the structural and physiological traits in the leaves of *Iris pumila* L. *Arch. Biol. Sci.* **2011**, *63*, 1099–1110. [CrossRef]
54. Zhu, Y.; Zheng, J.; Kang, H.; Hui, N.; Yin, S.; Chen, Z.; Du, B.; Liu, C. Spatial variations in leaf trichomes and their coordinatior with stomata in *Quercus variabilis* across Eastern Asia. *J. Plant Ecol.* **2024**, *17*, rtae023. [CrossRef]
55. Hetherington, A.M.; Woodward, F.I. The role of stomata in sensing and driving environmental change. *Nature* **2003**, *424*, 901–908 [CrossRef]
56. Drake, P.L.; Froend, R.H.; Franks, P.J. Smaller, faster stomata: Scaling of stomatal size, rate of response, and stomatal conductance *J. Exp. Bot.* **2013**, *64*, 495–505. [CrossRef]
57. Croft, H.; Chen, J.M.; Luo, X.; Bartlett, P.; Chen, B.; Staebler, R.M. Leaf chlorophyll content as a proxy for leaf photosynthetic capacity. *Glob. Change Biol.* **2017**, *23*, 3513–3524. [CrossRef]
58. Peñuelas, J.; Filella, I.; Llusia, J.; Siscart, D.; Piñol, J. Comparative field study of spring and summer leaf gas exchange and photobiology of the Mediterranean trees *Quercus ilex* and *Phillyrea latifolia*. *J. Exp. Bot.* **1998**, *49*, 229–238. [CrossRef]
59. Li, Y.; Liu, C.; Zhang, J.; Yang, H.; Xu, L.; Wang, Q.; Sack, L.; Wu, X.; Hou, J.; He, N. Variation in leaf chlorophyll concentration from tropical to cold-temperate forests: Association with gross primary productivity. *Ecol. Indic.* **2018**, *85*, 383–389. [CrossRef]
60. Cutolo, E.A.; Guardini, Z.; Dall'Osto, L.; Bassi, R. A paler shade of green: Engineering cellular chlorophyll content to enhance photosynthesis in crowded environments. *New Phytol.* **2023**, *239*, 1567–1583. [CrossRef]
61. Gratani, L. Plant phenotypic plasticity in response to environmental factors. *Adv. Bot.* **2014**, *2014*, 208747. [CrossRef]
62. Matesanz, S.; Gianoli, E.; Valladares, F. Global change and the evolution of phenotypic plasticity in plants. *Ann. N. Y. Acad. Sci.* **2010**, *1206*, 35–55. [CrossRef]
63. Manitašević Jovanović, S.; Hočevar, K.; Vuleta, A.; Tucić, B. Predicting the responses of functional leaf traits to global warming: An *in situ* temperature manipulation design using *Iris pumila* L. *Plants* **2023**, *12*, 3114. [CrossRef] [PubMed]

64. Luo, W.; Zuo, X.; Griffin-Nolan, R.J.; Xu, C.; Ma, W.; Song, L.; Helsen, K.; Lin, Y.; Cai, J.; Yu, Q.; et al. Long term experimental drought alters community plant trait variation, not trait means, across three semiarid grasslands. *Plant Soil* **2019**, *442*, 343–353. [CrossRef]
65. Manitašević, S.; Dunđerski, J.; Matić, G.; Tucić, B. Seasonal variation in heat shock proteins Hsp70 and Hsp90 expression in an exposed and a shaded habitat of *Iris pumila*. *Plant Cell Environ.* **2007**, *30*, 1–11. [CrossRef] [PubMed]
66. Liu, Z.; Dong, N.; Zhang, H.; Zhao, M.; Ren, T.; Liu, C.; Westerband, A.; He, N. Divergent long- and short-term responses to environmental gradients in specific leaf area of grassland species. *Ecol. Indic.* **2021**, *130*, 108058. [CrossRef]
67. Cunningham, S.A.; Summerhayes, B.; Westoby, M. Evolutionary divergences in leaf structure and chemistry, comparing rainfall and soil nutrient gradients. *Ecol. Monogr.* **1999**, *69*, 569–588. [CrossRef]
68. McDonald, P.G.; Fonseca, C.R.; Overton, J.M.C.; Westoby, M. Leaf-size divergence along rainfall and soil-nutrient gradients: Is the method of size reduction common among clades? *Funct. Ecol.* **2003**, *17*, 50–57. [CrossRef]
69. Ackerly, D.; Knight, C.; Weiss, S.; Barton, K.; Starmer, K. Leaf size, specific leaf area and microhabitat distribution of chaparral woody plants: Contrasting patterns in species level and community level analyses. *Oecologia* **2002**, *130*, 449–457. [CrossRef]
70. Correia, P.M.; da Silva, A.B.; Vaz, M.; Carmo-Silva, E.; Marques da Silva, J. Efficient regulation of CO_2 assimilation enables greater resilience to high temperature and drought in maize. *Front. Plant Sci.* **2021**, *12*, 675546. [CrossRef]
71. Allahverdiyev, T.; Huseynova, I. Influence of water deficit on photosynthetic activity, dry matter partitioning and grain yield of different durum and bread wheat genotypes. *Cereal Res. Commun.* **2017**, *45*, 432–441. [CrossRef]
72. Atkin, O.K.; Atkinson, L.J.; Fisher, R.A.; Campbell, C.D.; Zaragoza-Castells, J.; Pitchford, J.W.; Woodward, F.I.; Hurry, V. Using temperature-dependent changes in leaf scaling relationships to quantitatively account for thermal acclimation of respiration in a coupled global climate–vegetation model. *Glob. Change Biol.* **2008**, *14*, 2709–2726. [CrossRef]
73. Kühn, N.; Tovar, C.; Carretero, J.; Vandvik, V.; Enquist, B.J.; Willis, K.J. Globally important plant functional craits for coping with climate change. *Front. Biogeogr.* **2021**, *13*, e53774. [CrossRef]
74. Henn, J.J.; Buzzard, V.; Enquist, B.J.; Halbritter, A.H.; Klanderud, K.; Maitner, B.S.; Michaletz, S.T.; Pötsch, C.; Seltzer, L.; Telford, R.J.; et al. Intraspecific trait variation and phenotypic plasticity mediate alpine plant species response to climate change. *Front. Plant Sci.* **2018**, *9*, 1548. [CrossRef] [PubMed]
75. Michaletz, S.T.; Weiser, M.D.; Zhou, J.; Kaspari, M.; Helliker, B.R.; Enquist, B.J. Plant thermoregulation: Energetics, trait–environment interactions, and carbon economics. *Trends Ecol. Evol.* **2015**, *30*, 714–724. [CrossRef]
76. Gong, H.; Gao, J. Soil and climatic drivers of plant SLA (Specific Leaf Area). *Glob. Ecol. Conserv.* **2019**, *20*, e00696. [CrossRef]
77. Wilson, P.J.; Thompson, K.; Hodgson, J.G. Specific leaf area and leaf dry matter content as alternative predictors of plant strategies. *New Phytol.* **1999**, *143*, 155–162. [CrossRef]
78. Pontes, L.D.S.; Soussana, J.F.; Louault, F.; Andueza, D.; Carrère, P. Leaf traits affect the above-ground productivity and quality of pasture grasses. *Funct. Ecol.* **2007**, *21*, 844–853. [CrossRef]
79. Gorné, L.D.; Díaz, S.; Minden, V.; Onoda, Y.; Kramer, K.; Muir, C.; Michaletz, S.T.; Lavorel, S.; Sharpe, J.; Jansen, S.; et al. The acquisitive–conservative axis of leaf trait variation emerges even in homogeneous environments. *Ann. Bot.* **2022**, *129*, 709–722. [CrossRef]
80. Hulshof, C.M.; Swenson, N.G. Variation in leaf functional trait values within and across individuals and species: An example from a Costa Rican dry forest. *Funct. Ecol.* **2010**, *24*, 217–223. [CrossRef]
81. Wang, R.; He, N.; Li, S.; Xu, L.; Li, M. Variation and adaptation of leaf water content among species, communities, and biomes. *Environ. Res. Lett.* **2021**, *16*, 124038. [CrossRef]
82. Yoo, C.Y.; Pence, H.E.; Hasegawa, P.M.; Mickelbart, M.V. Regulation of transpiration to improve crop water use. *Crit. Rev. Plant Sci.* **2009**, *28*, 410–431. [CrossRef]
83. Sikorska, D.; Papierowska, E.; Szatyłowicz, J.; Sikorski, P.; Suprun, K.; Hopkins, R.J. Variation in leaf surface hydrophobicity of wetland plants: The role of plant traits in water retention. *Wetlands* **2017**, *37*, 997–1002. [CrossRef]
84. Xu, Z.; Zhou, G. Responses of leaf stomatal density to water status and its relationship with photosynthesis in a grass. *J. Exp. Bot.* **2008**, *59*, 3317–3325. [CrossRef] [PubMed]
85. Urban, J.; Ingwers, M.; McGuire, M.A.; Teskey, R.O. Stomatal Conductance Increases with Rising Temperature. *Plant Signal. Behav.* **2017**, *12*, e1356534. [CrossRef] [PubMed]
86. Brodribb, T.J.; McAdam, S.A.M.; Jordan, G.J.; Feild, T.S. Evolution of stomatal responsiveness to CO_2 and optimization of water-use efficiency among land plants. *New Phytol.* **2009**, *183*, 839–847. [CrossRef]
87. Leigh, A.; Sevanto, S.; Close, J.D.; Nicotra, A.B. The influence of leaf size and shape on leaf thermal dynamics: Does theory hold up under natural conditions? *Plant Cell Environ.* **2017**, *40*, 237–248. [CrossRef]
88. Tardy, F.; Créach, A.; Havaux, M. Photosynthetic pigment concentration, organization and interconversions in a pale green syrian landrace of barley (*Hordeum vulgare* L., Tadmor) adapted to harsh climatic conditions. *Plant Cell Environ.* **1998**, *21*, 479–489. [CrossRef]

89. Silla, F.; González-Gil, A.; González-Molina, M.E.; Mediavilla, S.; Escudero, A. Estimation of chlorophyll in Quercus leaves usin a portable chlorophyll meter: Effects of species and leaf age. *Ann. For. Sci.* **2010**, *67*, 108. [CrossRef]
90. Li, Y.; He, N.; Hou, J.; Xu, L.; Liu, C.; Zhang, J.; Wang, Q.; Zhang, X.; Wu, X. Factors influencing leaf chlorophyll content in natura forests at the biome scale. *Front. Ecol. Evol.* **2018**, *6*, 64. [CrossRef]
91. Hu, X.; Gu, T.; Khan, I.; Zada, A.; Jia, T. Research progress in the interconversion, turnover and degradation of chlorophyll. *Cel* **2021**, *10*, 3134. [CrossRef]
92. Bradshaw, A.D. Evolutionary significance of phenotypic plasticity in plants. In *Advances in Genetics*; Elsevier: Amsterdan The Netherlands, 1965; Volume 13, pp. 115–155, ISBN 978-0-12-017613-7.
93. Knight, C.A.; Ackerly, D.D. Evolution and plasticity of photosynthetic thermal tolerance, specific leaf area and leaf size: Congeneri species from desert and coastal environments. *New Phytol.* **2003**, *160*, 337–347. [CrossRef]
94. Huang, W.; Gielis, J.; Shi, P. The adaptation, plasticity and extinction of forest plants to climate change: Mechanisms behind th morphological, physiological, phenological and ecological traits. *Front. Ecol. Evol.* **2024**, *12*, 1488465. [CrossRef]
95. Kumarathunge, D.P.; Medlyn, B.E.; Drake, J.E.; Tjoelker, M.G.; Aspinwall, M.J.; Battaglia, M.; Cano, F.J.; Carter, K.R.; Cavaler M.A.; Cernusak, L.A.; et al. Acclimation and adaptation components of the temperature dependence of plant photosynthesis a the global scale. *New Phytol.* **2019**, *222*, 768–784. [CrossRef] [PubMed]
96. Kaiser, E.; Morales, A.; Harbinson, J.; Kromdijk, J.; Heuvelink, E.; Marcelis, L.F.M. Dynamic photosynthesis in differen environmental conditions. *J. Exp. Bot.* **2015**, *66*, 2415–2426. [CrossRef] [PubMed]
97. Poorter, L.; Bongers, F. Leaf traits are good predictors of plant performance across 53 rain forest species. *Ecology* **2006**, *87* 1733–1743. [CrossRef]
98. Volaire, F.; Gleason, S.M.; Delzon, S. What do you mean "functional" in ecology? Patterns versus processes. *Ecol. Evol.* **2020**, *1C* 11875–11885. [CrossRef]
99. Reich, P.B.; Wright, I.J.; Cavender-Bares, J.; Craine, J.M.; Oleksyn, J.; Westoby, M.; Walters, M.B. The Evolution of plant functiona variation: Traits, spectra, and strategies. *Int. J. Plant Sci.* **2003**, *164*, S143–S164. [CrossRef]
100. Nicotra, A.B.; Davidson, A. Adaptive phenotypic plasticity and plant water use. *Funct. Plant Biol.* **2010**, *37*, 117. [CrossRef]
101. Burns, J.H.; Strauss, S.Y. Effects of competition on phylogenetic signal and phenotypic plasticity in plant functional traits. *Ecolog* **2012**, *93*, S126–S137. [CrossRef]
102. Schlichting, C.D. Phenotypic integration and environmental change. *BioScience* **1989**, *39*, 460–464. [CrossRef]
103. Shaar-Moshe, L.; Hayouka, R.; Roessner, U.; Peleg, Z. Phenotypic and metabolic plasticity shapes life-history strategies unde combinations of abiotic stresses. *Plant Direct* **2019**, *3*, e00113. [CrossRef]
104. Sheehy, K.A.; Laskowski, K.L. Correlated behavioural plasticities: Insights from plasticity evolution, the integrated phenotype and behavioural syndromes. *Anim. Behav.* **2023**, *200*, 263–271. [CrossRef]
105. Gong, H.; Yang, M.; Wang, C.; Tian, C. Leaf phenotypic variation and its response to environmental factors in natural populations of *Eucommia ulmoides*. *BMC Plant Biol.* **2023**, *23*, 562. [CrossRef] [PubMed]
106. Chaudhary, D.; Pal, N.; Arora, A.; Prashant, B.D.; Venadan, S. Plant functional traits in crop breeding: Advancement and challenges. In *Plant Functional Traits for Improving Productivity*; Kumar, N., Singh, H., Eds.; Springer Nature: Singapore, 2024 pp. 169–202, ISBN 978-981-97-1509-1.
107. Franks, S.J.; Sim, S.; Weis, A.E. Rapid evolution of flowering time by an annual plant in response to a climate fluctuation. *Proc. Natl. Acad. Sci. USA* **2007**, *104*, 1278–1282. [CrossRef]
108. Mitchell, P.J.; Veneklaas, E.J.; Lambers, H.; Burgess, S.S.O. Leaf water relations during summer water deficit: Differential responses in turgor maintenance and variation in leaf structure among different plant communities in south-western Australia. *Plant Cell Environ.* **2008**, *31*, 1791–1802. [CrossRef]
109. Napier, J.D.; Heckman, R.W.; Juenger, T.E. Gene-by-environment interactions in plants: Molecular mechanisms, environmental drivers, and adaptive plasticity. *Plant Cell* **2023**, *35*, 109–124. [CrossRef]
110. Yang, J.; Gao, Y.; Zhao, C.; Chen, H. Leaf phenotypic plasticity and integration balance plant adaptation to water table decline: A mesocosm experiment. *Plant Soil* **2024**, *497*, 611–627. [CrossRef]
111. Wang, X.; Ji, M.; Zhang, Y.; Zhang, L.; Akram, M.A.; Dong, L.; Hu, W.; Xiong, J.; Sun, Y.; Li, H.; et al. Plant trait networks reveal adaptation strategies in the drylands of China. *BMC Plant Biol.* **2023**, *23*, 266. [CrossRef]
112. Bita, C.E.; Gerats, T. Plant tolerance to high temperature in a changing environment: Scientific fundamentals and production of heat stress-tolerant crops. *Front. Plant Sci.* **2013**, *4*, 273. [CrossRef]
113. Kazakou, E.; Violle, C.; Roumet, C.; Navas, M.; Vile, D.; Kattge, J.; Garnier, E. Are trait-based species rankings consistent across data sets and spatial scales? *J. Veg. Sci.* **2014**, *25*, 235–247. [CrossRef]
114. Wright, I.J.; Reich, P.B.; Westoby, M.; Ackerly, D.D.; Baruch, Z.; Bongers, F.; Cavender-Bares, J.; Chapin, T.; Cornelissen, J.H.C.; Diemer, M.; et al. The worldwide leaf economics spectrum. *Nature* **2004**, *428*, 821–827. [CrossRef]

15. Arnold, P.A.; Wang, S.; Catling, A.A.; Kruuk, L.E.B.; Nicotra, A.B. Patterns of phenotypic plasticity along a thermal gradient differ by trait type in an alpine plant. *Funct. Ecol.* **2022**, *36*, 2412–2428. [CrossRef]
16. Chaves, M.M.; Flexas, J.; Pinheiro, C. Photosynthesis under drought and salt stress: Regulation mechanisms from whole plant to cell. *Ann. Bot.* **2009**, *103*, 551–560. [CrossRef] [PubMed]
17. Liu, Y.; Dawson, W.; Prati, D.; Haeuser, E.; Feng, Y.; Van Kleunen, M. Does greater specific leaf area plasticity help plants to maintain a high performance when shaded? *Ann. Bot.* **2016**, *118*, 1329–1336. [CrossRef]
18. Pierce, S.; Maffi, D.; Faoro, F.; Cerabolini, B.E.L.; Spada, A. The leaf anatomical trade-offs associated with plant ecological strategy variation. *Plant Ecol.* **2022**, *223*, 1233–1246. [CrossRef]
19. Randolph, L.F.; Mitra, J. Karyotypes of *Iris pumila* and related species. *Am. J. Bot.* **1959**, *46*, 93–102. [CrossRef]
20. Randolph, L.F. The geographic distribution of European and eastern Mediterranean species of bearded Iris. In *Iris Year Book*; British Iris Society: Tunbridge Wells, UK, 1955; pp. 35–46.
21. Dalmady, J. Iris species on Slovenski Kras area. *Nat. Prot* **1972**, *3*, 64–65.
22. Randuška, D.; Križo, M. *Chránené Rastliny ["Protected Plants"]*; Príroda: Bratislava, Slovakia, 1986; pp. 1–430. (In Slovak)
23. Teixeira da Silva, J.A. *Floriculture, Ornamental and Plant Biotechnology: Advances and Topical Issues*, 1st ed.; Global Science Books: London, UK, 2006; Volume I–IV, p. 2506.
24. Eliáš, P.; Dítě, D.; Kliment, J.; Hrivnák, R.; Feráková, V. Red list of ferns and flowering plants of Slovakia, 5th Edition (October 2014). *Biologia* **2015**, *70*, 218–228. [CrossRef]
25. Parnikoza, I.Y.; Andreev, I.O.; Bublyk, O.M.; Spiridonova, K.V.; Gołębiewska, J.; Kubiak, M.; Kuczyńska, A.; Mystkowska, K.; Olędrzyńska, N.; Urasińska, B.; et al. The current state of steppe perennial plants populations: A case study on *Iris pumila*. *Biologia* **2017**, *72*, 24–35. [CrossRef]
26. Tucić, B.; Milojković, S.; Vujčić, S.; Tarasjev, A. Clonal diversity and dispersion in *Iris pumila*. *Acta Oecol.* **1988**, *9*, 211–219.
27. Kadović, R.; Spasov, P.; Bohajar, Y.; Belanovic-Simić, S.; Košanin, O. Analysis of aridity indicators in the Deliblato Sands. *Glas. Sumar. Fak.* **2014**, *109*, 97–112. [CrossRef]
28. Schneider, C.A.; Rasband, W.S.; Eliceiri, K.W. NIH Image to ImageJ: 25 years of image analysis. *Nat. Methods* **2012**, *9*, 671–675. [CrossRef] [PubMed]
29. Pérez-Harguindeguy, N.; Díaz, S.; Garnier, E.; Lavorel, S.; Poorter, H.; Jaureguiberry, P.; Bret-Harte, M.S.; Cornwell, W.K.; Craine, J.M.; Gurvich, D.E.; et al. Corrigendum to: New handbook for standardised measurement of plant functional traits worldwide. *Aust. J. Bot.* **2016**, *64*, 715–716. [CrossRef]
30. Yulin, L.I.; Johnson, D.A.; Yongzhong, S.U.; Jianyuan, C.U.I.; Zhang, T. Specific leaf area and leaf dry matter content of plants growing in sand dunes. *Bot. Bull. Acad. Sin.* **2005**, *46*, 127–134. [CrossRef]
31. Vile, D.; Garnier, É.; Shipley, B.; Laurent, G.; Navas, M.-L.; Roumet, C.; Lavorel, S.; Díaz, S.; Hodgson, J.G.; Lloret, F.; et al. Specific leaf area and dry matter content estimate thickness in laminar leaves. *Ann. Bot.* **2005**, *96*, 1129–1136. [CrossRef]
32. Hilu, K.W.; Randall, J.L. Convenient method for studying grass leaf epidermis. *TAXON* **1984**, *33*, 413–415. [CrossRef]
33. Wu, Q.G.; Cutler, D.F. Taxonomic, evolutionary and ecological implications of the leaf anatomy of rhizomatous Iris species. *Bot. J. Linn.* **1985**, *90*, 253–303. [CrossRef]
34. Sánchez-Sastre, L.F.; Alte Da Veiga, N.M.S.; Ruiz-Potosme, N.M.; Carrión-Prieto, P.; Marcos-Robles, J.L.; Navas-Gracia, L.M.; Martín-Ramos, P. Assessment of RGB vegetation indices to estimate chlorophyll content in sugar beet leaves in the final cultivation stage. *AgriEngineering* **2020**, *2*, 128–149. [CrossRef]
35. Wellburn, A.R. The spectral determination of chlorophylls a and b, as well as total carotenoids, using various solvents with spectrophotometers of different resolution. *J. Plant Physiol.* **1994**, *144*, 307–313. [CrossRef]
36. R Core Team. *R: A Language and Environment for Statistical Computing*; R Foundation for Statistical Computing: Vienna, Austria, 2024. Available online: https://www.R-project.org/ (accessed on 25 December 2024).
37. Wickham, H.; Vaughan, D.; Girlich, M. Tidyr: Tidy Messy Data. R package Version 1.3.0. 2023. Available online: https://CRAN.R-project.org/package=tidyr (accessed on 25 December 2024).
38. Wickham, H.; François, R.; Henry, L.; Müller, K.; Vaughan, D. Dplyr: A Grammar of Data Manipulation. R package Version 1.1.4. 2023. Available online: https://CRAN.R-project.org/package=dplyr (accessed on 25 December 2024).
39. Wickham, H. Reshaping Data with the reshape Package. *J. Stat. Softw.* **2007**, *21*, 1–20. Available online: http://www.jstatsoft.org/v21/i12/ (accessed on 25 December 2024). [CrossRef]
40. Kassambara, A. Rstatix: Pipe-Friendly Framework for Basic Statistical Tests. R Package Version 0.7.2. 2023. Available online: https://CRAN.R-project.org/package=rstatix (accessed on 25 December 2024).
41. Wei, T.; Simko, V. Corrplot: Visualization of a Correlation Matrix. R Package Version 0.92. 2021. Available online: https://github.com/taiyun/corrplot (accessed on 25 December 2024).

142. Oksanen, J.; Simpson, G.; Blanchet, F.; Kindt, R.; Legendre, P.; Minchin, P.; O'Hara, R.; Solymos, P.; Stevens, M.; Szoecs, E.; et a Vegan: Community Ecology Package. R Package Version 2.6-4. 2022. Available online: https://CRAN.R-project.org/package vegan (accessed on 25 December 2024).
143. Valladares, F.; Sanchez-Gomez, D.; Zavala, M.A. Quantitative estimation of phenotypic plasticity: Bridging the gap between th evolutionary concept and its ecological applications. *J. Ecol.* **2006**, *94*, 1103–1116. [CrossRef]
144. Ali, P.J.M.; Faraj, R.H.; Koya, E.; Ali, P.J.M.; Faraj, R.H. Data normalization and standardization: A technical report. *Mach. Lear Tech. Rep.* **2014**, *1*, 1–6.
145. Singh, D.; Singh, B. Investigating the impact of data normalization on classification performance. *Appl. Soft Comput.* **202** *97*, 105524. [CrossRef]

plants

rticle

Changes in the Stress Response and Fitness of Hybrids Between Transgenic Soybean and Wild-Type Plants Under Heat Stress

i Zhang [1], Qi Yu [1], Xin Yin [1], Laipan Liu [1], Zhentao Ren [1], Zhixiang Fang [1], Wenjing Shen [1], Shengnan Liu [2,*]
nd Biao Liu [1,*]

1 Key Laboratory on Biodiversity and Biosafety, Nanjing Institute of Environmental Sciences, Ministry of Ecology and Environment, Nanjing 210042, China; feiniao8897@126.com (L.Z.); 15968821722@163.com (Q.Y.); njfu_shin@163.com (X.Y.); liulaipan@163.com (L.L.); rztkkk@163.com (Z.R.); zxfang23@126.com (Z.F.); swj@nies.org (W.S.)

2 Institute of Plant Protection, Sichuan Academy of Agricultural Science, Chengdu 610066, China

* Correspondence: hnulshn2006@163.com (S.L.); 85287064@163.com (B.L.)

Abstract: Understanding the ability of hybrids of genetically modified (GM) soybean and wild soybean to survive and reproduce under unfavorable conditions is critical for answering questions regarding risk assessment and the existence of transgenes in the environment. To investigate the effects of high-temperature stress on soybean growth and competitive ability, the GM soybean DBN8002, which expresses the VIP3Aa and PAT proteins, and F_2 generations derived from a cross between GM soybean and NJW (wild soybean) were placed in a greenhouse with an elevated temperature (38/32 °C) for 14 days, and the plant agronomic performance and foreign protein levels of hybrid soybean were evaluated to observe their responses to high temperature. The results revealed that the VIP3Aa and PAT protein levels in F_2 and GM were not influenced by high-temperature stress. In contrast, the pollen germination, pod number, hundred-seed weight, and seed vigor of the F_2 hybrid and parent soybean plants decreased after high-temperature stress. However, except for the number of fully filled seeds per plant, the above parameters of the F_2 hybrid were similar to or slightly lower than those of wild soybean, and no significant difference in fitness was observed between the F_2 hybrid and wild soybean, indicating that the growth and competitive ability of the hybrid were similar to those of its female parent under heat stress conditions, resulting in the transgenes persisting and spreading within agricultural ecosystems. Our results enhance the understanding of the GM soybean plant's response to heat stress, lay the foundation for breeding heat-resistant soybean varieties, and provide new insights and advanced information on the ecological risks arising from the escape of transgenes.

Keywords: genetically modified soybean; transgene escape; F_2 hybrid; high temperature; competition

1. Introduction

Genetically modified (GM) soybeans were one of the earliest introduced GM crops for commercial cultivation and rank first in terms of planting area worldwide [1]. By 2019, GM soybean plants with beneficial traits, such as insect resistance, herbicide tolerance, and salt tolerance, were extensively planted in 29 countries, resulting in important economic and social benefits [1,2]. However, GM soybean plants have not been officially commercialized to date in East Asian countries such as China, Japan, and Korea. There are two limiting factors for the application and commercialization of GM soybean cultivation in these countries. The first is poor public acceptance of transgenic technology, mainly because of concerns about the potential adverse effects of GM soybeans on human and animal health [3–6]. Another significant problem is gene flow via seeds during transportation or the pollen-mediated gene transfer of transgenic soybean plants to their wild relatives, which could result in resistant weeds and a series of environmental issues [7–9].

The wild resource of cultivated soybeans, *Glycine soja*, is naturally distributed in East Asia, and China is a center for wild soybean cultivation [10–12]. Wild soybeans are widely distributed in 31 provinces and the autonomous regions of China [13]. Since cultivated soybeans and wild soybeans both belong to the genus *Soja*, they can cross freely and produce fertile hybrid offspring [14]; therefore, wild soybean is commonly used as an important genetic resource for soybean breeding. After GM soybean application, GM soybean and its transgenes can naturally spread via seeds during transportation and through pollen-mediated gene transfer to wild soybean plants [15,16]; if this occurs, the genetic introgression of foreign genes could contribute to the loss of genetic integrity in wild soybean. However, hybrids often show higher growth rates and greater fecundity than the parental lines due to heterosis [17,18], and they can rapidly accumulate and disseminate in weedy and wild populations, which may lead to a decline in or even the extinction of wild soybean populations [19,20].

Recently, some researchers have focused on the ecological impact of gene flow on wild soybean populations. Yook et al. [21] reported that hybrids between glufosinate-resistant (GR) soybean and wild soybean (IT182932) had similar numbers of pods and seeds to wild soybean, indicating that transgenes of the GR soybean might disperse into wild populations and persist in the agroecosystem due to the relatively high fitness of the hybrid progeny. Pot-based experimental studies conducted by Zhang et al. [8] revealed that F_2 and F_3 hybrids between the Roundup Ready soybean and wild soybean exhibited lower seed germination rates and higher seed productivity than the GM soybean, indicating that the adaptability of hybrids may increase the possibility of dispersal of transgenes in wild soybean relatives. Guan et al. [22] and Liu et al. [23] reported no obvious decline in the fitness of hybrids and wild-type soybeans. The above studies were conducted mainly under suitable growing conditions and well-resourced cultivation practices. However, soybean plants are usually challenged by various abiotic stressors during their growth cycles, such as cold, drought, and salt stress, which can directly or indirectly influence plant fitness via their effects on survival, growth, and reproduction. Nevertheless, few studies have been conducted to evaluate the competitive ability of hybrids under stressful environmental conditions.

In recent years, extreme high-temperature events have occurred increasingly frequently worldwide [24–26], severely affecting normal crop growth and production. Many studies have shown that the high-temperature stress caused by global warming has led to pollen abortion, kernel shrinkage, and reduced pod setting and seed weight [27–29], which has become an increasingly serious problem and a limiting factor in soybean production [30–33]. Compared with cultivated soybean plants, wild soybean plants are generally more resistant to abiotic stressors such as extreme temperatures, drought, and salt stress [34–36]. For example, Veremeichik et al. [37] reported that 35 days of cold (16/12 °C) and heat (36/34 °C) treatments did not affect the growth and biomass accumulation of wild soybean compared with cultivated soybean. Thus, the vegetative and reproductive growth abilities of crop-wild hybrid soybean plants under relatively high temperatures cannot be effectively predicted. If hybrids have low tolerance to high-temperature stress, their growth and reproductive abilities will be adversely affected by relatively high temperatures. In contrast, if the tolerance of the hybrid to high temperatures is similar to that of the female parent or if the hybrid is even more resistant to high temperatures (i.e., hybrid vigor), this would indicate that these heat-resistant genes in wild soybean plants are effectively utilized to enhance resistance through introgression breeding, which may considerably increase the competitive ability of the hybrid in the agroecosystem.

GM soybeans have consistently been the focus of research and development related to GM technology in China [38]. Since 2008, the government has implemented and successfully conducted the GMO Special Project [39,40], and independent innovation capacity in this field has significantly increased in recent years. By 2024, numerous GM soybean varieties were developed and showed substantial application prospects in agriculture, and three GM soybean varieties were issued with biosafety certificates by the Chinese government. One of these certificates was issued for DBN8002 soybean, which is highly resistant to lepidopteran pests and tolerant to glyphosate [41], demonstrating strong potential for commercialization. In 2023, DBN8002 was approved for future application in the Huang-Huai-Hai region of China. The region has a vast land area and abundant resources, with rain and high temperatures in summer. In recent years, extremely high temperatures have occurred frequently in the region during summer (July–August) [42,43], which severely affects the yield and quality of soybean [30–32]. However, little is known about the competitive ability of GM soybean and hybrid offspring under heat stress. In this study, we analyzed the seed vigor, foreign gene expression, and plant growth performance of GM, wild, and hybrid populations under heat stress conditions, to evaluate how different soybean accessions respond to heat stress, provide a theoretical basis for the breeding of heat-resistant GM soybean crop varieties in the future, and offer further insight to understand the risk of transgene escape from GM soybean to its wild relative.

2. Results

2.1. In Vitro Pollen Germination Analysis

Compared with the control, high temperatures during the flowering and pod formation periods significantly decreased the pollen germination rates of GM soybean, $F_2(+)$, and $F_2(-)$ plants by 35.57%, 35.45%, and 31.03%, respectively (Table 1). In addition, the pollen germination rates were reduced in the wild soybean groups, but the germination rate did not significantly differ from that of the control pollen.

Table 1. Pollen germination rates of GM, wild, and hybrid soybean plants under normal conditions and under high-temperature stress.

Material	Pollen Germination Rate (%)	
	Control	High Temperature
GM	74.5 ± 3.77 a	48.0 ± 4.69 a*
NJW	73.5 ± 3.59 a	59.0 ± 1.73 a*
$F_2(+)$	72.5 ± 1.71 a	50.0 ± 3.65 a*
$F_2(-)$	77.5 ± 2.63 a	50.0 ± 3.36 a*

Notes: Asterisks indicate significant differences between control and high-temperature treatment materials (* for $p < 0.05$). Different lowercase letters in a column represent statistically significant differences between different materials ($p < 0.05$). GM, genetically modified; NJW, Nanjing wild soybean; $F_2(+)$, F_2 plants with foreign genes; $F_2(-)$, F_2 plants negative for foreign genes.

Additionally, under the same conditions, no significant difference in pollen viability was detected among the different soybean materials.

2.2. Aboveground Biomass

Under control conditions, the F_2 hybrid plants had the greatest biomass, at 134.87 and 130.02 g for $F_2(-)$ and $F_2(+)$, respectively, followed by the wild-type soybeans, and the genetically modified soybeans had the lowest biomass. There was a significant difference among the three groups (Figure 1).

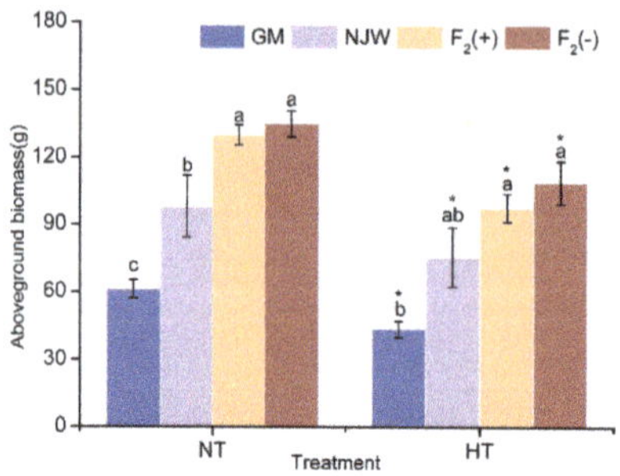

Figure 1. Mean biomasses of GM, wild, and hybrid soybean plants under normal and temperature-stress conditions. The values are presented as the means ± standard errors of the means (SEs) (n = 10 in each group). The same letter indicates a nonsignificant difference and different letters indicate a significant difference ($p < 0.05$). Asterisks indicate significant differences between control and high-temperature treatment specimens (* for $p < 0.05$). NT, normal temperature treatment; HT, high-temperature treatment; GM, genetically modified; NJW, Nanjing wild soybean; $F_2(+)$, F_2 plants with foreign genes; $F_2(-)$, F_2 plants negative for foreign genes.

After heat stress treatment, the aboveground biomass of the F_2 generation (108.81 g and 97.62 g) was slightly greater than that of the wild-type soybeans (75.55 g) and significantly greater than that of the GM-type soybeans (43.71 g), and there were no significant differences in aboveground biomass between the F_2 population and the wild-type soybeans. In addition, there were varying degrees of aboveground biomass decrease in all four soybean varieties after heat stress. Among them, significant reductions in biomass were found for GM and $F_2(+)$, with decreases of 28.54% (from 61.17 to 43.71 g) and 24.92% (from 130.02 to 97.62 g), respectively. Compared with that of the control, the biomass of NJW and $F_2(-)$ decreased by 18.78% and 19.32%, respectively, with no significant differences.

2.3. *Fecundity*

2.3.1. Number of Pods per Plant

Compared with the control treatment, heat stress significantly reduced the number of pods per plant. The number of pods of $F_2(+)$ and $F_2(-)$ decreased from 645.60 to 380.90 and from 619.90 to 401.40, significantly decreasing by 41% and 35.25%, respectively (Table 2). Similarly, the number of pods on NJW and GM soybean plants decreased after high-temperature treatment, with decreases of 22.30% and 21.94%, respectively, this being significantly different from those on the control.

Table 2. Effects of high temperature on the number of pods per plant of GM, wild, and hybrid soybeans.

Material	Number of Pods per Plant	
	Control	High Temperature
GM	43.30 ± 2.35 b	33.80 ± 2.28 b*
NJW	630.50 ± 33.45 a	489.90 ± 56.07 a*
$F_2(+)$	645.60 ± 39.17 a	380.90 ± 30.16 a*
$F_2(-)$	619.90 ± 31.27 a	401.40 ± 20.67 a*

Notes: The values are presented as the mean ± standard errors of the means (SEs). Asterisks indicate significant differences between control and high-temperature treatment materials (* for $p < 0.05$). Different lowercase letters in a column represent statistically significant differences between different materials ($p < 0.05$). GM, genetically modified; NJW, Nanjing wild soybean; $F_2(+)$, F_2 plants with foreign genes; $F_2(-)$, F_2 plants negative for foreign genes.

Analysis of variance revealed that there was no significant difference in the number of pods per plant between the hybrid F_2 plants and the wild soybean parent under control conditions or heat-stress conditions.

2.3.2. Seed Number and Full Seed Number per Plant

Table 3 shows that the seed number and the full seed number per plant decreased significantly in GM, F_2(+), and F_2(−) hybrid soybean plants after 14 days of heat stress. The seed numbers decreased from 81.30, 944.3, and 919.50 to 46.40, 520.20, and 447.50, respectively. In particular, F_2 (+) showed the greatest decrease (52.56%). Moreover, the seed number and full seed number in wild soybean plants also decreased but not significantly.

Table 3. Effects of high temperatures on the number of seeds and full seeds per plant of GM, wild, and hybrid soybeans.

Material	Number of Seeds		Full Seed Number	
	Control	High Temperature	Control	High Temperature
GM	81.30 ± 5.31 b	46.40 ± 6.32 c*	73.30 ± 16.30 b	41.40 ± 11.32 c*
NJW	1096.5 ± 72.54 a	838.60 ± 112.09 a	919.70 ± 231.50 a	629.20 ± 307.67 a
F_2(+)	944.30 ± 95.85 a	447.50 ± 43.23 b*	684.6 ± 278.65 a	157.2 ± 99.45 b*
F_2(−)	919.50 ± 36.36 a	520.20 ± 36.38 a*	724.80 ± 87.42 a	264.90 ± 229.30 b*

Notes: The values are presented as the mean ± standard errors of the means (SEs). Asterisks indicate significant differences between control and high-temperature treatment materials (* for $p < 0.05$). Different lowercase letters in a column represent statistically significant differences between different materials ($p < 0.05$). GM, genetically modified; NJW, Nanjing wild soybean; F_2(+), F_2 plants with foreign genes; F_2(−), F_2 plants negative for foreign genes.

Under both control and heat-stress conditions, both the seed number and full seed number were highest in the wild soybean seeds and lowest in the GM soybean seeds. The seed number and full seed number of the F_2 hybrid seeds were similar to those of the wild-type soybean plants in the control group; however, significant differences were observed in the seed number and full seed number between the F_2(+) and wild-type soybean plants under heat stress.

2.3.3. One-Hundred-Seed Weight and Seed Weight per Plant

In both the control and heat-stress treatments, the 100-seed weights were highest in the GM soybean seeds and lowest in the wild-type soybean seeds, and the F_2(+) and F_2(−) hybrids presented higher 100-seed weights than their wild-type soybean counterparts did; no parameters significantly differed between the F_2(+) and F_2(−) populations (Table 4). After 14 days of heat-stress treatment, the 100-seed weight per plant decreased significantly in GM and F_2(+) and F_2(−) hybrid soybean plants but did not significantly decrease in wild soybean plants.

Table 4. Effects of high temperature on the hundred-seed weight and seed weight per plant of GM, wild, and hybrid soybeans.

Material	100-Seed Weight (g)		Seed Weight (g)	
	Control	High Temperature	Control	High Temperature
GM	19.11 ± 0.58 a	12.61 ± 0.37 a*	18.48 ± 1.42 c	9.55 ± 0.66 b*
NJW	1.91 ± 0.02 c	1.68 ± 0.04 c*	31.27 ± 2.91 b	17.47 ± 2.01 ab*
F_2(+)	5.47 ± 0.11 b	3.05 ± 0.17 b*	45.32 ± 2.74 a	22.52 ± 1.56 a*
F_2(−)	5.21 ± 0.06 b	3.11 ± 0.26 b*	47.66 ± 5.98 a	22.46 ± 2.80 a*

Note: The values are presented as the mean ± standard errors of the means (SEs). Asterisks indicate significant differences between control and high-temperature treatment materials (* for $p < 0.05$). Different lowercase letters in a column represent statistically significant differences between different materials ($p < 0.05$). GM, genetically modified; NJW, Nanjing wild soybean; F_2(+), F_2 plants with foreign genes; F_2(−), F_2 plants negative for foreign genes.

The F_2 hybrids presented greater seed weights than the wild-type soybean plants under both normal and high-temperature conditions, and the difference was significant

($p < 0.05$) under normal conditions. After high-temperature treatment, seed weight also significantly decreased in all the soybean varieties, with significant decreases of 52.88%, 50.32%, 48.33%, and 44.13% for $F_2(-)$, $F_2(+)$, GM, and wild soybean, respectively, as shown in Table 4.

2.4. Vip3Aa and PAT Protein Contents in Soybean Leaves

An ELISA was performed to quantify the levels of PAT and Vip3Aa proteins in different soybean leaf samples. The results revealed that the wild soybean, $F_2(-)$, and control samples were negative for PAT and Vip3Aa expression, whereas foreign proteins were detected in all the GM and $F_2(+)$ leaf samples.

In both the control and high-temperature treatment samples, the PAT and Vip3Aa contents in the leaves of the GM soybean plants were, on average, greater than those in the leaves of the $F_2(+)$ soybean plants (Figure 2). There were no significant differences in the foreign protein contents between the treatments, and both the PAT and Vip3Aa contents were similar in the plants grown under control and high-temperature conditions.

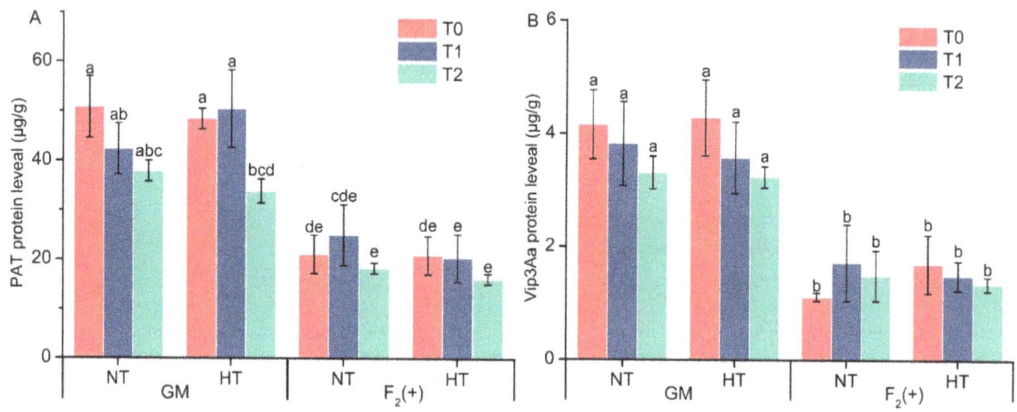

Figure 2. Changes in the expression levels of PAT (**A**) and Vip3Aa (**B**) in the leaves of two soybean varieties before treatment (T_0), after treatment (T_1), and 7 days after treatment (T_2) under normal conditions and high-temperature conditions. The values are presented as the mean ± standard errors of the means (SEs). The same letter indicates a nonsignificant difference and different letters indicate a significant difference ($p < 0.05$). NT, normal temperature treatment; HT, high-temperature treatment; GM, genetically modified; NJW, Nanjing wild soybean; $F_2(+)$, F_2 plants with foreign genes; $F_2(-)$, F_2 plants negative for foreign genes.

2.5. Seed Vigor

In both the control and high-temperature treatments, the total germination rate of the wild-type soybean plants was the highest, followed by that of the F_2 hybrids and, finally, the GM soybean plants (Table 5). The seed germination rates of the NJW and $F_2(+)$ hybrid seeds were slightly higher than those of the GM and $F_2(-)$ hybrid seeds in the control samples, and the seed germination rates of the NJW and $F_2(-)$ hybrid seeds were significantly higher than those of the GM and $F_2(+)$ hybrid seeds under high-temperature treatment. In addition, no significant differences were detected between $F_2(+)$ and $F_2(-)$ hybrid seeds.

Table 5. Effects of high temperature on the viability of GM, wild, and hybrid soybean seeds.

Material	Seed Germination Rates (%)		Dead Seeds (%)	
	Control	High Temperature	Control	High Temperature
GM	86.5 ± 1.26 b	46.50 ± 7.14 c*	13.50 ± 1.26 a	53.50 ± 7.14 a*
NJW	94.5 ± 0.96 a	84.50 ± 2.75 a*	5.50 ± 0.96 b	15.5 ± 2.75 c*
F_2(+)	89.00 ± 1.73 ab	58.00 ± 6.32 bc*	11.00 ± 1.73 ab	42.00 ± 6.32 ab*
F_2(−)	86.50 ± 2.50 b	72.50 ± 4.35 ab*	13.50 ± 2.50 a	27.5 ± 4.35 bc*

Notes: Data are presented as the mean ± standard errors of the means (SEs). Asterisks indicate significant differences between control and high-temperature treatments (* for $p < 0.05$). Different lowercase letters in a column represent statistically significant differences between different materials ($p < 0.05$). GM, genetically modified; NJW, Nanjing wild soybean; F_2(+), F_2 plants with foreign genes; F_2(−), F_2 plants negative for foreign genes.

After 14 days of heat treatment, the seed germination rates greatly varied among the soybean materials. The germination rates of the NJW and F_2(−) seeds were 84.50% and 72.50%, respectively, which were greater than those of the GM and F_2(+) seeds (46.50% and 58.00%, respectively). Moreover, a significant decrease in seed germination was recorded in the NJW, F_2(−), F_2(+), and GM soybean seeds after high-temperature treatment compared with the control seeds; however, the decrease was not large in the NJW and F_2(−) samples, at 10.58% and 16.18%, but was greater for the GM and F_2(+) samples at 46.24% and 34.83%, respectively. Additionally, the percentage of dead seeds increased significantly in all soybean materials after heat-temperature treatment, especially in the GM and F_2(+) hybrid seeds; approximately 53.5% and 42% of the GM soybeans and F_2(+) hybrid seeds failed to germinate (they had generally rotted or mildewed), respectively, during the test.

2.6. Relative Composite Fitness

Under normal growth conditions, the fitness values of the F_2(+) and F_2(−) hybrids were 1.07 and 1.15, respectively, both of which values were greater than those of the wild-type soybean parent (Figure 3). In contrast, the fitness of F_2(+) and F_2(−) hybrids was lower than that of their wild soybean counterparts. However, no differences in relative composite fitness were detected between the hybrid and the wild soybean, either in the control samples or under high-temperature conditions.

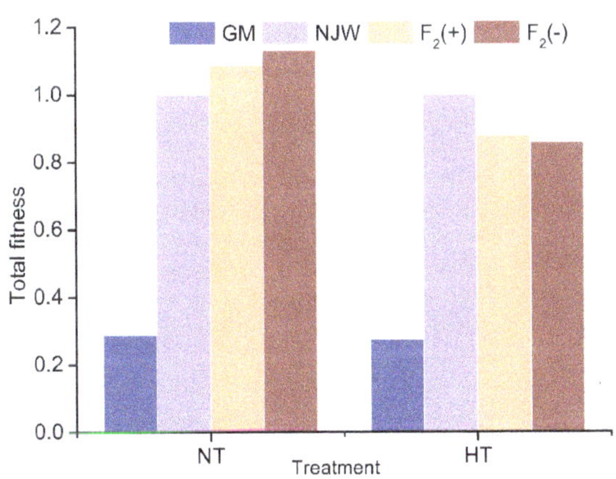

Figure 3. Composite fitness of F_2 hybrids and wild-type soybeans under normal and high-temperature conditions. NT, normal temperature treatment; HT, high-temperature treatment; GM, genetically modified; NJW, Nanjing wild soybean; F_2(+), F_2 plants with foreign genes; F_2(−), F_2 plants that were negative for foreign genes.

3. Discussion

Whether the escape of foreign genes from GM soybeans poses potential ecological risks largely depends on the fitness of the hybrid offspring [44–46]. Generally, the higher the level of hybrid fitness is, the greater the risk of transgene escape and transmission [47,48]. In addition to the parental genome, environmental stress related to soil conditions, insect pressure, and climatic conditions can affect plant fitness [49–51]. Thus, we evaluated the ecological risk of GM and its hybrid soybean under heat-stress conditions to gain a comprehensive understanding of the ecological risk of planting GM soybeans, supplementing previous research in this field of study.

3.1. *Effects of High-Temperature Stress on the Aboveground Biomass of Soybean*

Aboveground biomass is not only an important predictor of crop growth status and yield but is also an indicator of plant competitive ability and biotic resistance [52–55]. In the present study, biomass decreased in all three soybean varieties; however, although there was a significant decrease in biomass in the F_2 soybean, its biomass was still greater than that of wild soybean after high-temperature stress, suggesting that the F_2 hybrid still had high vegetative growth at high temperatures. In a two-year agricultural field study, Yook et al. [21] found that hybrids showed similar characteristics to wild soybeans in terms of above-ground biomass. Zhang et al. [8] and Liu et al. [23] also obtained similar results under controlled conditions. The increase in F_2 soybean biomass may be beneficial for individual competitiveness and organizational performance. Notably, Veremeichik et al. [37] reported that long-term (35 days) heat (36/34 °C) treatments did not affect growth and biomass accumulation in wild soybeans, whereas a significant decrease in biomass was observed in two cultivated soybeans. These findings are consistent with the results of this study, indicating that wild soybeans may be more tolerant to heat stress conditions than GM cultivated soybeans.

3.2. *Effects of High-Temperature Stress on Pollen Viability and Soybean Fecundity*

During the period of anthesis, heat stress causes pollen deformity and anther indehiscence, inhibits pollen-tube growth, decreases pollen viability, and ultimately leads to a reduction in soybean yield [56,57]. Similar phenomena were observed in this study; a decrease in pollen viability was observed in all soybean materials after heat stress, with the greatest decrease occurring in the $F_2(-)$ plants, which was close to 35.5%, whereas the $F_2(+)$ plants presented the lowest pollen viability. Nevertheless, no significant differences in pollen germination were detected among the four plant materials, suggesting that the influences of high-temperature stress on the pollen from different soybean materials were roughly the same. To date, only a few researchers have evaluated changes in the viability of GM soybean or hybrid pollen under stress conditions. Liang et al. [58] reported that weed competition or barren soil can reduce the pollen germination rate of wild soybeans and the two hybrid offspring, but no difference was found in the pollen germination rate between hybrid and wild parents under the same planting conditions. In the present study, similar conclusions were reached.

After heat stress, both parent plants and hybrid plants presented significant decreases in the numbers of pods and seeds, but these two parameters did not differ between the F_2 hybrid and the wild-type parent. Like in our study, in a two-year pot-based experiment, Liang et al. [58] reported that F_2 hybrids between herbicide-tolerant transgenic soybeans (T14R1251-70) and wild soybeans produced more pods and full seeds than did wild soybeans under weed-stress conditions. Our findings and those of previous studies suggest that hybrids possess greater reproductive ability, even under unfavorable conditions. In addition, previous results have shown that F_2 and F_3 hybrids between GM and wild soybeans have larger and heavier seeds [8,21,46]. In the present study, the F_2 hybrid plants presented larger seeds and greater 100-seed weights than did the wild-type plants, which may also explain why the seed weight was greater in the F_2 hybrid soybean than in the wild-type soybean plants in this study. To date, several studies have been conducted to assess the effects of heat stress on soybean field performance, with a particular focus on cultivated soybeans. These studies showed that a few days of plant exposure to high temperatures (35 °C or above) during the growth period can cause large yield losses through delays in flowering, flower and pod abortion, and a reduction in the 100-seed weight [29,41,59–61]; their findings are similar to the conclusions of this study.

3.3. Effects of High-Temperature Stress on Vip3Aa and PAT Protein Contents in Soybean Leaves

After two weeks of heat stress, the foreign genes were expressed normally in the GM- and F_2-positive soybean plants and their expression levels differed little from those in the control plants, suggesting that high temperatures at the R3 stage had little or no effect on foreign gene expression levels. To date, few studies have investigated changes in foreign proteins in GM crops and hybrid plants under heat stress. Trtikova et al. [62] investigated the expression levels of the Bt protein in the leaves of two transgenic maize varieties, namely, *Bt* PAN 6Q-321B (white Bt maize) and *Bt* PAN 6Q-308B (yellow Bt maize), under high-temperature stress (21–30 °C, with a maximum of 45 °C at noon), and their results revealed that high temperature had little effect on the Bt protein level. A similar result was reported by Chen et al. [63], who reported that the Cry1Ac contents in the leaves of Sikang1 and Sikang3 cotton plants under short-term high-temperature stress (37 °C for 24 h) were not significantly different from those under optimal conditions. However, Zhang et al. [64] reached the opposite conclusion; they reported a significant decrease in the Cry1Ac protein content after *Bt* cotton was exposed to 38 °C continuously for 24 h. Our findings and the abovementioned results suggest that foreign gene content is not only influenced by the genetic background of a given plant variety but also by stress-inducing temperatures and exposure time. Under stressful conditions, the concentration of foreign protein is highly variable and is, therefore, difficult to predict [62].

3.4. Effects of High-Temperature Stress on Seed Vigor

Seed vigor is an important indicator of seed germination, seedling growth potential, plant stress resistance, and production potential [65]. The results of the seed germination tests demonstrated that both the GM, wild, and F_2 hybrid seed germination rates significantly decreased after high-temperature treatment. In the present research, visual evidence of seed damage, such as incomplete filling, wrinkling, cracking, and sunken lesions was more common in the high-temperature treatment group than in the control group (Figure S2). Shu et al. [66] and Liu et al. [67] reported that high temperatures result in damage to seeds, including stress cracks, shrinkage, and decreased seed germination [68,69], which is consistent with our observations. Such damage caused by heat stress may increase the risk of seed mold infection, leading to a decrease in seed germination and emergence rates.

The germination rates of both parent and hybrid seeds decreased significantly under heat-stress conditions. However, the decreases in the germination rates of the F_2(+) plants and GM soybean seeds were much greater than those of the wild-type and F_2(−) plants. A sharp decrease in germination was observed for the F_2(+) plants and GM soybean seeds, for which the germination rates decreased by 58.0% and 46.5%, respectively, whereas the germination rates of the NJW and F_2(−) plants were 16.18% and 10.58%, respectively, compared with those of the control plants. In addition, the germination rates of the wild-type soybean and F_2(−) plants were much greater than those of the GM and F_2(+) plants.

Our results indicate that high temperature may have a greater effect on F_2(+) and GM soybean seeds than on NJW and F_2(−) seeds.

In this study, GM soybean plants produced large yellow seeds, whereas F_2(+) and F_2(−) had medium-sized seeds that were colored light green and black, respectively (Figure S2). In contrast, wild soybean plants produced small, blackish seeds. Some studies have shown that the seed coat's color, thickness, and structure both directly and indirectly affect seed germination [70]. Zhang et al. [46] reported that the physical characteristics of the seed coats of F_3 plants were closely related to seed germination; the darker-colored (brown and black) seeds of the F_3 hybrid between GM soybean and wild soybean presented a large area of honeycomb-like surface deposits and a thick palisade layer and had a low germination rate and greater overwintering ability than did lighter-colored (yellow-green) F_3 seeds. Ma et al. [71] and Santos [72] also suggested that soybean seeds with dark seed coats (black and brown) have more agronomic advantages than those with yellow seed coats, such as greater coat thickness, lower permeability, and greater resistance to mechanical damage. In addition, previous studies have demonstrated that dark seed coats have relatively high concentrations of anthocyanins [73], phenolics [74,75], and flavonoids [76,77], and that these substances may increase the capacity of seeds to cope with abiotic stressors [78]. Therefore, the higher germination rates of NJW and F_2(−) seeds after high-temperature stress compared to those of GM and F_2(+) soybeans may be related to the coat structure and chemical composition of the seeds, but the specific reasons need to be studied further.

To our knowledge, this is the first report of fitness changes in F_2 hybrids between genetically modified soybeans and wild soybeans after high-temperature stress. The findings described in this study indicate that high-temperature stress ultimately resulted in reduced fitness in all soybean materials, especially the GM soybean. However, although the average fitness of the hybrids was lower than the fitness of the wild-type relative, no difference was observed in total fitness between the two, indicating that the survival and competitive ability of hybrids are similar to those of their female parents under high-temperature stress conditions. Moreover, it was found that high temperatures did not seem to affect the expression levels of foreign proteins, which could contribute to the ability of hybrid plants to cope with pest stressors. Gompert et al. [79] and Lu et al. [47] suggested that plants with heterosis/hybrid vigor are more able to adapt to more extreme conditions than their parental genotypes, especially in later generations. Thus, continuing to perform studies on fitness changes in later-generation hybrids under stress conditions might contribute to a more comprehensive and systematic understanding of the ecological risks of hybridization.

In addition, in this study, as the hybrids were generated using wild soybean as the pollen recipient, we speculate that the significantly increased fitness of the F_2 hybrid compared with the GM was mostly due to maternal inheritance. Therefore, searching and mining the related heat tolerance genes from wild soybean plants will be critical to soybean cultivation and food production in the future under global climate warming conditions.

4. Materials and Methods

4.1. Plant Materials

The GM soybean DBN8002, which is highly toxic to Lepidopteran pests such as *Spodoptera frugiperda* and *Spodoptera exigua* and is tolerant to herbicides [41], can be transformed with the *Vip3Aa* and *PAT* genes [80]. Wild soybean plants were collected from Jiangpu County, Nanjing, Jiangsu Province, China. The seeds of both soybean types were provided by Beijing DaBeiNong Science & Technology Group Co., Ltd. (Beijing, China).

To obtain F_1 hybrid offspring, the GM soybean DBN8002 was used as a male parent (pollen donor), and wild soybean was used as a female parent (pollen recipient). Artificial pollination was conducted during the flowering season in 2019. In the next year, F_1 seeds from the cross were collected and selfed to generate the F_2 generation.

As F_2 lines are a segregating population, two traits segregate the F_2 population: the transgenic-positive plants contain foreign genes, whereas the nontransgenic plants do not. To confirm the expression of the foreign gene, the presence or absence of the foreign gene in F_2 soybean plants was identified using a QuickStix Kit for PAT (EnviroLogix, Portland, ME, USA) and VIP3Aa (EnviroLogix, Portland, ME, USA). Among the 176 plants in the F_2 population, 42 plants were negative for foreign genes, marked as $F_2(-)$, whereas 134 plants were positive for foreign genes, marked as $F_2(+)$. The test results are shown in Supplementary Figure S1.

4.2. Plant Growth and Heat-Stress Treatment

In 2022, Nanjing city in China experienced abnormally high temperatures; the average maximum and minimum temperatures in August were 34 °C and 27 °C, respectively, and extremely high temperatures (38 °C or above) lasted for 12 days, with an extremely high temperature of 41 °C. High temperatures in August at the flowering and pod-setting stages (mid-July to mid- to late August) strongly affect the normal growth and development of soybean plants and lead to decreased yield and quality in the field. Thus, to investigate the sensitivity of GM soybean plants, especially hybrid soybean plants, to heat stress, we designed and conducted the current study, and the treatment temperatures were set at 38/32 °C (day/night); 38 °C and 32 °C were the average day and night temperatures during high-temperature weather in August 2022, respectively. In addition, the treatment time and soybean growth period during which these high temperatures occurred were set according to the actual field conditions in Nanjing in the summer of 2022. Because field trials are currently limited by policies in China, the study was conducted under controlled conditions.

Pot-based experiments were conducted at the Nanjing Institute of Environmental Sciences. GM, wild soybean plants, and F_2 hybrids were given the high-temperature treatment. In 2023, a total of 200 F_2 seeds, together with 80 seeds each from the parent plants, were sown in seedling trays and incubated in a greenhouse under controlled conditions (25–30 °C, with a day length of 12 h). After 40 days of growth, 50 parent plants and 176 F_2 seedlings were transplanted into cultivation pots containing soil or nutrient-enriched soil (1:1, *v:v*), with 1 seedling per pot. When both flowers and young raised pods (0.5 to 2 cm in length) were present in the soybean plants (the R3 growth stage, according to Fehr et al. [81], 20 plants were randomly selected from each soybean material and moved to a greenhouse for heat stress treatment at 38/32 °C (day/night) for 2 weeks (high-temperature treatment, or HT), while the remaining plants were kept in greenhouse chambers as the controls (normal temperature treatment, or NT). The NT treatment plants were grown in a 16 h light/8 h dark cycle at temperatures of 28 °C during light periods and 18 °C during darkness, with 35–70% humidity. After heat treatment, all the plants were allowed to grow under normal conditions until harvest. In addition, to prevent seed loss due to pod shattering, 1 mm nylon fabric was used to cover each plant when the color of the pods changed from green to brown, and manual weeding and pesticide application were performed as necessary during the experiment.

4.3. *Effects of High-Temperature Stress on the Growth and Reproduction of Soybean Plants*

4.3.1. In Vitro Pollen Germination Rate

Pollen germination was determined using an optimized solid medium, as described previously [23]. At the end of the heat stress treatment, mature pollen from 5 plants was randomly selected and scattered on solid media, and pollen germination was observed under a microscope after 30 min of incubation in a chamber at 28 °C and 100% relative humidity in the dark. For each Petri dish, four replicates were performed in three random visual fields (with approximately 100 pollen grains in each field). Pollen was recorded as having germinated when the pollen tube length exceeded the diameter of the pollen grain [82], and the pollen germination rate was calculated as the percentage of germinated pollen grains among the total number of pollen grains.

4.3.2. Aboveground Biomass and Fecundity

At harvest, 10 plants of each soybean material were randomly chosen, cut at a point 5 cm above the ground, air-dried, and weighed via a weighing balance (PB602-N, Mettler Toledo, Greifensee, Zürich, Switzerland). After the biomass was recorded, fecundity traits such as the number of pods, number of seeds, number of full seeds, 100-seed weight, and seed weight were recorded per plant.

4.3.3. Vip3Aa and PAT Protein Contents in Soybean Leaves

To assess whether heat stress might affect the expression of Vip3Aa and PAT proteins in GM and F_2 plants, a commercial double-antibody sandwich ELISA kit (enzyme-linked immunosorbent assay) was used to determine the concentration of proteins in soybean leaves. Leaf samples of the same plants from GM, wild soybean, and F_2 offspring were collected before treatment (T_0), after treatment (T_1), and 7 days after treatment (T_2). The collected leaves were quickly frozen in liquid nitrogen and stored at −20 °C until analysis.

For foreign protein determination, 10~12 mg of leaf tissue was ground in liquid nitrogen, extracted in 1 mL PBS buffer (provided with the kit), and centrifuged for 15 min at 3000× *g*. The supernatant was analyzed for protein content using quantitative ELISA kits (QuantiPlate™ Kit for PAT/bar, Envirologix, Inc., Portland, ME, USA; QuantiPlate™ Kit for Vip_3A, Youlong Biological Engineering Co., Ltd., Shanghai, China). The subsequent steps were performed according to the instructions provided in the kit manual. The optical density values of the samples were obtained via a microplate reader (EnviroLogix, Portland, ME, USA) at 450 nm, and all the samples were measured in triplicate.

4.3.4. Seed Vigor

To investigate the effects of extremely high temperatures on seed viability, seeds harvested from heat-treated/control GM, wild-type, F_2(+), and F_2(−) plants were subjected to germination tests. Prior to the germination test, the seeds of wild and F_2 soybean plants were carefully scarified by nicking the seed coat with a scalpel to break seed dormancy. For each soybean material, four replicate groups with a total number of 200 seeds (4 × 50 seeds) were prepared. Seeds were placed in Petri dishes containing two layers of moist filter paper and incubated in an incubator (Binder KBF720, Binder Instrument Co., Ltd., Tuttlingen, Baden-Wuerttemberg, Germany) at 25 °C with a 12 h photoperiod and 55% relative humidity for 14 days. The number of germinated seeds was counted daily, and the seeds were subsequently removed from the Petri dishes. Seeds were considered germinated when the radicle length exceeded 2 mm.

4.4. Statistical Analysis

SPSS 20.0 statistical software was used for data analysis. The data are expressed as the mean ± standard deviation (x ± SE). A one-way analysis of variance (ANOVA) and Tukey's multiple comparison test were used for comparisons, and a *t*-test was used to compare the high-temperature treatments with the controls. For all statistical tests, $p < 0.05$ was considered to indicate statistically significant differences.

Supplementary Materials: The following supporting information can be downloaded at: https://www.mdpi.com/article/10.3390/plants14040622/s1, Figure S1: Analysis of PAT (A,B) and Vip3Aa (C,D) proteins in soybean plants using colloidal gold immunochromatographic assay test strips (N, negative control; P, positive control; 1-176, F_2 progenies; C, control line; T, test line); Figure S2: Effects of high temperatures on seed morphology. Range of phenotypes observed in mature seeds: (A) fully filled seeds, (B) partially filled seeds, and (C) seed coat cracks.

Author Contributions: Conceptualization, L.Z. and B.L.; methodology, L.Z., Q.Y. and B.L.; data curation, L.Z., X.Y. and B.L.; writing—original draft preparation, L.Z.; writing—review and editing, L.Z., Z.F., L.L., W.S., S.L. and B.L.; visualization, L.Z. and Z.R.; funding acquisition, B.L. All authors have read and agreed to the published version of the manuscript.

Funding: The work was funded by the National Natural Science Foundation of China (32171656) and the Biological Breeding-National Science and Technology Major Project (2022ZD04021).

Data Availability Statement: The data collected in this study are available upon request. The data are not publicly available due to institutional restrictions.

Conflicts of Interest: The authors declare that they have no known competing financial interests or personal relationships that could have appeared to influence the work reported in this paper.

References

1. ISAAA (International Agricultural Biotechnology Application Service Organization). Global Biotechnology/GM Crop Commercialization Development Trend in 2019. *China Bioeng. J.* **2021**, *41*, 114–119.
2. Machado, E.P.; Rodrigues Junior, G.L.S.; Somavilla, J.C.; Führ, F.M.; Zago, S.L.; Marques, L.H.; Santos, A.C.; Nowatzki, T.; Dahmer, M.L.; Omoto, C.; et al. Survival and Development of *Spodoptera eridania, Spodoptera cosmioides* and *Spodoptera albula* (Lepidoptera: Noctuidae) on Genetically Modified Soybean Expressing Cry1Ac and Cry1F Proteins. *Pest. Manag. Sci.* **2020**, *76*, 4029–4035. [CrossRef]
3. Globus, R.; Qimron, U. A Technological and Regulatory Outlook on CRISPR Crop Editing. *J. Cell Biochem.* **2018**, *119*, 1291–1298. [CrossRef] [PubMed]
4. Pental, D. When Scientists Turn Against Science: Exceptionally Flawed Analysis of Plant Breeding Technologies. *Curr. Sci.* **2019**, *117*, 932–939. [CrossRef]
5. Zhang, L.; Shen, W.J.; Fang, Z.X.; Liu, B. Effects of Genetically Modified Maize Expressing Cry1Ab and EPSPS Proteins on Japanese Quail. *Poultry Sci.* **2021**, *100*, 1068–1075. [CrossRef] [PubMed]
6. Fei, X.; Huang, X.; Li, Z.; Li, X.; He, C.; Xiao, S.; Li, Y.; Zhang, X.; Deng, X. Effect of Marker-Free Transgenic Chlamydomonas on the Control of *Aedes* Mosquito Population and on Plankton. *Parasit. Vectors* **2023**, *16*, 18. [CrossRef] [PubMed]
7. García, M.J.; Palma-Bautista, C.; Vazquez-Garcia, J.G.; Rojano-Delgado, A.M.; Osuna, M.D.; Torra, J.; De, P.R. Multiple Mutations in the EPSPS and ALS Genes of Amaranthus Hybridus Underlie Resistance to Glyphosate and ALS Inhibitors. *Sci. Rep.* **2020**, *10*, 17681. [CrossRef] [PubMed]
8. Zhang, L.; Liu, L.; Fang, Z.; Shen, W.; Dai, Y.; Jia, R.Z.; Liang, J.G.; Liu, B. Fitness Changes in Wild Soybean Caused by Gene Flow from Genetically Modified Soybean. *BMC Plant Biol.* **2023**, *23*, 424. [CrossRef] [PubMed]
9. Liu, L.; Zhang, L.; Fang, Z.; Shen, W.J.; Yin, X.; Ren, Z.T.; Yu, Q.; Liang, J.G., Liu, B. Glyphosate Resistance and No Fitness Cost in Backcross Offspring of Wild Soybean and Transgenic Soybean with *Epsps* Gene. *BMC Plant Biol.* **2024**, *24*, 849. [CrossRef] [PubMed]
10. Tian, B.; Talukder, S.K.; Fu, J.; Fritz, A.K.; Trick, H.N. Expression of a Rice Soluble Starch Synthase Gene in Transgenic Wheat Improves the Grain Yield under Heat Stress Conditions. *Vitr. Cell Dev. Biol. Plant* **2018**, *54*, 216–227. [CrossRef]
11. Hailemariam, H.M. Adaptability and Stability for Soybean Yield by AMMI and GGE Models in Ethiopia. *Front. Plant Sci.* **2022**, *13*, 950992. [CrossRef]

12. Li, W.; Liu, M.; Lai, Y.C.; Liu, J.X.; Fan, C.; Yang, G.; Wang, L.; Liang, W.W.; Di, S.F.; Yu, D.Y.; et al. Genome-Wide Associatio Study of Partial Resistance to *P. sojae* in Wild Soybeans from Heilongjiang Province, China. *Curr. Issues Mol. Biol.* **202** *44*, 3194–3207. [CrossRef]
13. Wang, K.J.; Li, X.H. Genetic Diversity and Gene Flow Dynamics Revealed in the Rare Mixed Populations of Wild Soybean (*Glycir soja*) and Semi-Wild Type (*Glycine gracilis*) in China. *Genet. Resour. Crop Evol.* **2013**, *60*, 2303–2318. [CrossRef]
14. Zeng, Q.Y.; Yang, C.Y.; Ma, Q.B.; Li, X.P.; Dong, W.W.; Nian, H. Identification of Wild Soybean Mirnas and Their Target Gene Responsive to Aluminum Stress. *BMC Plant Biol.* **2012**, *12*, 182. [CrossRef]
15. Clark, M.; Maselko, M. Transgene Biocontainment Strategies for Molecular Farming. *Front. Plant Sci.* **2020**, *11*, 210. [CrossRef] [PubMed]
16. Sohn, S.I.; Pandian, S.; Oh, Y.J.; Kang, H.J.; Ryu, T.H.; Cho, W.S.; Shin, E.K.; Shin, K.S. A Review of the Unintentional Release c Feral Genetically Modified Rapeseed into the Environment. *Biology* **2021**, *10*, 1264. [CrossRef] [PubMed]
17. Zhou, Z.; Zhang, C.; Lu, X.; Wang, L.; Hao, Z.; Li, M.; Zhang, D.; Yong, H.; Zhu, H.; Weng, J.; et al. Dissecting the Genetic Basi Underlying Combining Ability of Plant Height Related Traits in Maize. *Front. Plant Sci.* **2018**, *9*, 1117. [CrossRef] [PubMed]
18. Xu, B.; Wu, R.; Shi, F.; Gao, C.; Wang, J. Transcriptome Profiling of Flower Buds of Male-Sterile Lines Provides New Insights int Male Sterility Mechanism in Alfalfa. *BMC Plant Biol.* **2022**, *22*, 199. [CrossRef] [PubMed]
19. Lu, B.R.; Xia, H. Environmental Biosafety of Transgenic Plants: Research and Assessment of Transgene Escape and Its Potentia Ecological Impacts. *Chin. Sci. Bull.* **2011**, *23*, 186–194. [CrossRef]
20. Sohn, S.I.; Thamilarasan, S.K.; Pandian, S.; Oh, Y.J.; Ryu, T.H.; Lee, G.S.; Shin, E.K. Interspecific Hybridization of Transgeni *Brassica napus* and *Brassica rapa*—An Overview. *Genes* **2022**, *13*, 1442. [CrossRef] [PubMed]
21. Yook, M.J.; Park, H.R.; Zhang, C.J.; Lim, S.H.; Kim, D.S. Environmental Risk Assessment of Glufosinate-Resistant Soybean by Pollen-Mediated Gene Flow under Field Conditions in the Region of the Genetic Origin. *Sci. Total Environ.* **2020**, *762*, 143073 [CrossRef] [PubMed]
22. Guan, Z.J.; Zhang, P.; Wei, W.; Mi, X.C.; Kang, D.M.; Liu, B. Performance of Hybrid Progeny Formed Between Genetically Modified Herbicide-Tolerant Soybean and Its Wild Ancestor. *AoB Plants* **2015**, *7*, plv121. [CrossRef] [PubMed]
23. Liu, L.L.; Zhang, L.; Fu, J.M.; Shen, W.J.; Fang, Z.X.; Dai, Y.; Jia, R.Z.; Liu, B.; Liang, J.G. Fitness and Ecological Risk of Hybrid Progenies of Wild and Herbicide-Tolerant Soybeans with *Epsps* Gene. *Front. Plant Sci.* **2022**, *13*, 922215. [CrossRef] [PubMed]
24. Mimić, G.; Brdar, S.; Brkić, M.; Panić, M.; Marko, O.; Crnojević, V. Engineering Meteorological Features to Select Stress Tolerant Hybrids in Maize. *Sci. Rep.* **2020**, *10*, 3421. [CrossRef]
25. Zandalinas, S.I.; Fritschi, F.B.; Mittler, R. Global Warming, Climate Change, and Environmental Pollution: Recipe for a Multifactorial Stress Combination Disaster. *Trends Plant Sci.* **2021**, *26*, 588–599. [CrossRef]
26. Wang, Q.; Li, X.; Qiu, Z.; Yang, S.; Zhou, W.; Zhao, J. Depth-Keeping Control for a Deep-Sea Self-Holding Intelligent Buoy System Based on Inversion Time Constraint Stability Strategy Optimization. *Sensors* **2022**, *22*, 1096. [CrossRef] [PubMed]
27. Wheeler, T.R.; Craufurd, P.Q.; Ellis, R.H.; Porter, J.R.; Prasad, P.V.V. Temperature Variability and The Yield of Annual Crops. *Agric Ecosyst. Environ.* **2000**, *82*, 159–167. [CrossRef]
28. Allen, L.H.; Zhang, L.; Boote, K.J.; Hauser, B.A. Elevated Temperature Intensity, Timing, and Duration of Exposure Affect Soybean Internode Elongation, Mainstem Node Number, and Pod Number Per Plant. *Crop J.* **2018**, *6*, 148–161. [CrossRef]
29. Djanaguiraman, M.; Schapaugh, W.; Fritschi, F.; Nguyen, H.; Vara, P.P.V. Reproductive Success of Soybean (*Glycine max* L. Merril) Cultivars and Exotic Lines under High Daytime Temperature. *Plant Cell Environ.* **2019**, *42*, 321–336. [CrossRef]
30. Das, A.; Rushton, P.J.; Rohila, J.S. Metabolomic Profiling of Soybeans (*Glycine max* L.) Reveals the Importance of Sugar and Nitrogen Metabolism under Drought and Heat Stress. *Plants* **2017**, *6*, 21. [CrossRef]
31. Zhang, G.; Bahn, S.C.; Wang, G.; Zhang, Y.; Chen, B.; Zhang, Y.; Wang, X.; Zhao, J. PLDα1-knockdown Soybean Seeds Display Higher Unsaturated Glycerolipid Contents and Seed Vigor in High Temperature and Humidity Environments. *Biotechnol. Biofuels* **2019**, *12*, 9. [CrossRef] [PubMed]
32. Narayanan, S.; Zoong-Lwe, Z.S.; Gandhi, N.; Welti, R.; Fallen, B.; Smith, J.R.; Rustgi, S. Comparative Lipidomic Analysis Reveals Heat Stress Responses of Two Soybean Genotypes Differing in Temperature Sensitivity. *Plants* **2020**, *9*, 457. [CrossRef]
33. Lippmann, R.; Babben, S.; Menger, A.; Delker, C.; Quint, M. Development of Wild and Cultivated Plants under Global Warming Conditions. *Curr. Biol.* **2022**, *29*, R1326–R1338. [CrossRef]
34. Zhang, J.L.; Wang, J.X.; Wei, J.; Liu, J.G.; Yang, S.N.; Gai, J.Y.; Li, Y. Identification and Analysis of $NaHCO_3$ Stress Responsive Genes in Wild Soybean (*Glycine soja*) Roots by RNA-seq. *Front. Plant Sci.* **2016**, *7*, 1842. [CrossRef] [PubMed]
35. Zhang, J.; Yang, D.; Li, M.; Shi, L.X. Metabolic Profiles Reveal Changes in Wild and Cultivated Soybean Seedling Leaves under Salt Stress. *PLoS ONE* **2016**, *11*, e0159622. [CrossRef]
36. Fu, H.; Guo, R.; Shen, W.Y.; Li, M.X.; Shi, L.X. Changes in the Metabolome of Two Soybean Genotypes under Drought Stress. *Russian J. Plant Physl.* **2020**, *67*, 472–481. [CrossRef]
37. Veremeichik, G.N.; Brodovskaya, E.V.; Grigorchuk, V.P.; Butovets, E.S.; Lukyanchuk, L.M.; Bulgakov, V.P. ABA-Dependent Regulation of Calcium-Dependent Protein Kinase Gene *GmCDPK5* in Cultivated and Wild Soybeans. *Life* **2022**, *12*, 1576. [CrossRef]

28. Zhao, Y.; Deng, H.; Yu, C.; Hu, R. The Chinese Public's Awareness and Attitudes Toward Genetically Modified Foods with Different Labeling. *NPJ Sci. Food* **2019**, *3*, 17. [CrossRef] [PubMed]
29. Cai, J.; Hu, R.; Huang, J.; Wang, X. Investment, Research Ability and Progress of Agricultural Biotechnology in China. *Chin. Rural. Econ.* **2016**, *6*, 61–70.
30. Jin, Y.; Schaub, S.; Tosun, J.; Wesseler, J. Does China Have a Public Debate on Genetically Modified Organisms? A Discourse Network Analysis of Public Debate on Weibo. *Public. Underst. Sci.* **2022**, *31*, 732–750. [CrossRef] [PubMed]
31. Xiang, S.; Wang, S.; Xu, M.; Wang, W.; Liu, W. YOLO POD: A Fast and Accurate Multi-Task Model for Dense Soybean Pod Counting. *Plant Methods* **2023**, *19*, 8. [CrossRef]
32. Guo, J.; Wang, Z.; Qu, L.; Hu, Y.; Lu, D. Transcriptomic and Alternative Splicing Analyses Provide Insights into the Roles of Exogenous Salicylic Acid Ameliorating Waxy Maize Seedling Growth under Heat Stress. *BMC Plant Biol.* **2022**, *22*, 432. [CrossRef] [PubMed]
33. Hu, X.; Wei, Q.; Wu, H.; Huang, Y.; Peng, X.; Han, G.; Ma, Q.; Zhao, Y. Identification and Characterization of Heat-Responsive Lncrnas in Maize Inbred Line CM1. *BMC Genom.* **2022**, *23*, 208. [CrossRef] [PubMed]
34. Huang, Y.; Wang, Y.Y.; Qiang, S.; Song, X.L.; Dai, W.M. Fitness of F_1 hybrids Between Stacked Transgenic Rice T1c-19 With Cry1c*/Bar Genes and Weedy Rice. *J. Integr. Agric.* **2019**, *18*, 2793–2805. [CrossRef]
35. Liu, J.Y.; Sheng, Z.W.; Hu, Y.Q.; Liu, Q.; Qiang, S.; Song, X.L.; Liu, B. Fitness of F_1 Hybrids Between 10 Maternal Wild Soybean Populations and Transgenic Soybean. *Transgenic Res.* **2021**, *30*, 105–119. [CrossRef] [PubMed]
36. Zhang, L.; Jia, R.Z.; Liu, L.; Shen, W.; Fang, Z.; Zhou, B.; Liu, B. Seed Coat Colour and Structure Are Related to the Seed Dormancy and Overwintering Ability of Crop-to-Wild Hybrid Soybean. *AoB Plants* **2023**, *15*, plad081. [CrossRef] [PubMed]
37. Lu, B.R. Transgene Escape from GM Crops and Potential Biosafety Consequences: An Environmental Perspective. *Int. Centre. Genet. Eng. Biotechnol. Collect. Biosaf. Rev.* **2008**, *4*, 66–141.
38. Liu, B.; Xue, K.; Liu, L.P.; Shen, W.J.; Guo, H. Research on the Gene Flow from Transgenic EPSPS + PAT Soybean S4003.14 to Non-Transgenic Soybeans. *J. Ecol. Rural. Environ.* **2020**, *36*, 367–373.
39. Hegland, S.J.; Nielsen, A.; Lázaro, A.; Bjerknes, A.L.; Totland, Ø. How Does Climate Warming Affect Plant-Pollinator Interactions? *Ecol. Lett.* **2009**, *12*, 184–195. [CrossRef] [PubMed]
40. Wang, C.H.; Li, B. Salinity and Disturbance Mediate Direct and Indirect Plant-Plant Interactions in An Assembled Marsh Community. *Oecologia* **2016**, *182*, 139–152. [CrossRef] [PubMed]
41. Du, L.; Qu, M.; Jiang, X.; Li, X.; Ju, Q.; Lu, X.; Wang, J. Fitness Costs Associated with Acetyl-Coenzyme A Carboxylase Mutations Endowing Herbicide Resistance in *American sloughgrass* (*Beckmannia syzigachne* Steud.). *Ecol. Evol.* **2019**, *9*, 2220–2230. [CrossRef] [PubMed]
42. Snow, A.A.; Pilson, D.; Rieseberg, L.H.; Paulsen, M.J.; Pleskac, N.; Reagon, M.R.; Wolf, D.E.; Selbo, S.M. A Bt Transgene Reduces Herbivory and Enhances Fecundity in Wild Sunflowers. *Ecol. Appl.* **2003**, *13*, 279–286. [CrossRef]
43. Rinella, M.J.; Pokorny, M.L.; Rekaya, R. Grassland Invader Responses to Realistic Changes in Native Species Richness. *Ecol. Appl.* **2007**, *17*, 1824–1831. [CrossRef] [PubMed]
44. Di, K.; Stewart, C.N.; Wei, W.; Shen, B.C.; Tang, Z.X.; Ma, K.P. Fitness and Maternal Effects in Hybrids Formed Between Transgenic Oilseed Rape (*Brassica napus* L.) and Wild Brown Mustard [*B. juncea* (L.) Czern et Coss.] in the Field. *Pest. Manag. Sci.* **2009**, *65*, 753–760. [CrossRef] [PubMed]
45. Song, X.; Wang, Z.; Qiang, S. Agronomic Performance of F_1, F_2 And F_3 Hybrids Between Weedy Rice and Transgenic Glufosinate-Resistant Rice. *Pest. Manag. Sci.* **2011**, *67*, 921–931. [CrossRef]
46. Hinojosa, L.; Matanguihan, J.B.; Murphy, K.M. Effect of High Temperature on Pollen Morphology, Plant Growth and Seed Yield in Quinoa (*Chenopodium quinoa* Willd). *J. Agron. Crop Sci.* **2019**, *205*, 33–45. [CrossRef]
47. Raja, M.M.; Vijayalakshmi, G.; Naik, M.L.; Basha, P.O.; Sergeant, K.; Hausman, J.F.; Khan, P.S.S.V. Pollen Development and Function under Heat Stress: From Effects to Responses. *Acta Physiol. Plant.* **2019**, *41*, 47. [CrossRef]
48. Liang, R.; Ji, X.; Sheng, Z.; Liu, J.; Qiang, S.; Song, X. Fitness and Rhizobacteria of F_2, F_3 Hybrids of Herbicide-Tolerant Transgenic Soybean and Wild Soybean. *Plants* **2022**, *11*, 3184. [CrossRef]
49. Sehgal, A.; Sita, K.; Kumar, J.; Kumar, S.; Singh, S.; Siddique, K.H.M.; Nayyar, H. Effects of Drought, Heat and Their Interaction on the Growth, Yield and Photosynthetic Function of Lentil (*Lens culinaris* Medikus) Genotypes Varying in Heat and Drought Sensitivity. *Front. Plant Sci.* **2017**, *8*, 1776. [CrossRef] [PubMed]
50. Adeyemi, N.O.; Atayese, M.O.; Olubode, A.A.; Akan, M.E. Effect of Commercial Arbuscular Mycorrhizal Fungi Inoculant on Growth and Yield of Soybean under Controlled and Natural Field Conditions. *J. Plant Nutr.* **2020**, *43*, 487–499. [CrossRef]
51. Tang, Y.; Lu, S.; Fang, C.; Liu, H.; Dong, L.; Li, H.; Su, T.; Li, S.; Wang, L.; Cheng, Q.; et al. Diverse Flowering Responses Subjecting to Ambient High Temperature in Soybean under Short-Day Conditions. *Plant Biotechnol. J.* **2023**, *21*, 782–791. [CrossRef] [PubMed]
52. Trtikova, M.; Wikmark, O.G.; Zemp, N.; Widmer, A.; Hilbeck, A. Transgene Expression and Bt Protein Content in Transgenic Bt maize (MON810) under Optimal and Stressful Environmental Conditions. *PLoS ONE* **2015**, *10*, e0123011. [CrossRef]

63. Chen, Y.; Chen, Y.J.; Chen, Y.; Zhang, X.; Wang, Y.H.; Chen, D.H. The Recovery of Bt Toxin Content After Temperature Stres Termination in Transgenic Cotton. *Span. J. Agric. Res.* **2013**, *11*, 438–446. [CrossRef]
64. Zhang, X.; Rui, Q.Z.; Liang, P.P.; Wei, C.H.; Deng, G.Q.; Chen, Y.; Dong, Z.D.; Chen, D.H. Dynamics of Bt Cotton Cry1Ac protei Content under an Alternating High Temperature Regime and Effects on Nitrogen Metabolism. *J. Integr. Agric.* **2018**, *17*, [CrossRef]
65. Feng, L.; Zhu, S.; Liu, F.; He, Y.; Bao, Y.; Zhang, C. Hyperspectral Imaging for Seed Quality and Safety Inspection: A Review. *Plan Methods* **2019**, *15*, 91. [CrossRef] [PubMed]
66. Shu, Y.; Zhou, Y.; Mu, K.; Hu, H.; Chen, M.; He, Q.; Huang, S.; Ma, H.; Yu, X. A Transcriptomic Analysis Reveals Soybea Seed Pre-Harvest Deterioration Resistance Pathways under High Temperature and Humidity Stress. *Genome* **2020**, *63*, 115–12 [CrossRef]
67. Liu, S.; Liu, Y.; Liu, C.; Li, Y.; Zhang, F.; Ma, H. Isolation and Characterization of the GmMT-II Gene and Its Role in Response t High Temperature and Humidity Stress in *Glycine max*. *Plants* **2022**, *11*, 1503. [CrossRef] [PubMed]
68. Musielak, G. Influence of the Drying Medium Parameters on Drying Induced Stresses. *Dry. Technol.* **2000**, *18*, 561–581. [CrossRef
69. Igathinathane, C.; Chattopadhyay, P.K.; Pordesimo, L.O. Moisture Diffusion Modeling of Parboiled Paddy Accelerated Temperin Process with Extended Application to Multi-Pass Drying Simulation. *J. Food Eng.* **2008**, *88*, 239–253. [CrossRef]
70. Liu, F.; Li, N.; Yu, Y.; Chen, W.; Yu, S.; He, H. Insights into the Regulation of Rice Seed Storability by Seed Tissue-Specifi Transcriptomic and Metabolic Profiling. *Plants* **2022**, *11*, 1570. [CrossRef]
71. Ma, F.; Ewa, C.; Tasneem, M.; Carol, A.P.; Mark, G. Cracks in The Palisade Cuticle of Soybean Seed Coats Correlate with Thei Permeability to Water. *Ann. Bot-London* **2004**, *94*, 213–228. [CrossRef] [PubMed]
72. Santos, E.L.D.; José Nivaldo, P.; Barros, A.S.D.R.; Prete, C.E.C. Soybean Seed Coat Variation and Its Influence on the Physiologica Quality and Chemical Composition. *Rev. Bras. De Sementes* **2007**, *29*, 20–26. [CrossRef]
73. Nurzyńska-Wierdak, R.; Łabuda, H.; Buczkowska, H.; Sałata, A. Pericarp of Colored-Seeded Common Bean (*Phaseolus vulgaris* L. Varieties a Potential Source of Polyphenolic Compounds. *Agron. Res.* **2019**, *17*, 2005–2015. [CrossRef]
74. Zhou, S.; Sekizaki, H.; Yang, Z.; Sawa, S.; Pan, J. Phenolics in the Seed Coat of Wild Soybean (*Glycine soja*) and Their Significance for Seed Hardness and Seed Germination. *J. Agric. Food Chem.* **2010**, *58*, 10972. [CrossRef] [PubMed]
75. Malenčić, D.; Cvejić, J.; Miladinović, J. Polyphenol Content and Antioxidant Properties of Colored Soybean Seeds from Centra Europe. *J. Med. Food* **2012**, *15*, 89–95. [CrossRef]
76. Wang, M.; Gillaspie, A.; Morris, J.; Pittman, R.; Davis, J.; Pederson, G. Flavonoid Content in Different Legume Germplasm Seeds Quantified by HPLC. *Plant Genet. Resour.-C* **2008**, *6*, 62–69. [CrossRef]
77. Ren, Z.; Yin, X.; Liu, L.; Zhang, L.; Shen, W.; Fang, Z.; Yu, Q.; Qin, L.; Chen, L.; Jia, R.; et al. Flavonoid Localization in Soybear Seeds: Comparative Analysis of Wild (*Glycine soja*) and Cultivated (*Glycine max*) Varieties. *Food Chem.* **2024**, *456*, 139883. [CrossRef
78. Hinman, E.D.; Fridley, J.D.; Parry, D. Plant Defense Against Generalist Herbivores in the Forest Understory: A Phylogeneti Comparison of Native and Invasive Species. *Biol. Invasions* **2019**, *21*, 1269–1281. [CrossRef]
79. Gompert, Z.; Fordyce, J.A.; Forister, M.L.; Shapiro, A.M.; Nice, C.C. Homoploid Hybrid Speciation in An Extreme Habitat. *Science* **2006**, *314*, 1923–1925. [CrossRef] [PubMed]
80. Xiang, D.; Luo, M.; Jiang, F.; Wen, Z.; Chen, X.; Wang, X.; Xu, X.; Wei, W.; Xu, J. Safety Assessment of Subchronic Feeding of Insect-Resistant and Herbicide-Resistant Transgenic Soybeans to Juvenile Channel Catfish (*Ictalurus punctatus*). *Sci. Rep.* **2023** *13*, 5445. [CrossRef] [PubMed]
81. Fehr, W.R.; Caviness, C.E.; Burmood, D.T.; Pennington, J.S. Stage of Development Description for Soybeans, *Glycine max* (L.) Merrill1. *Crop Sci.* **1971**, *11*, 929–931. [CrossRef]
82. Djanaguiraman, M.; Prasad, P.V.V.; Boyle, D.L.; Schapaugh, W.T. Soybean Pollen Anatomy, Viability and Pod Set under High Temperature Stress. *J. Agron. Crop Sci.* **2013**, *199*, 171–177. [CrossRef]

plants

MDPI

Article

Changes in the Suitable Habitat of the Smoke Tree (*Cotinus coggygria* Scop.), a Species with an East Asian–Tethyan Disjunction

Zichen Zhang [1], Xin Yan [1], Chang Guo [1], Wenpan Dong [1], Liangcheng Zhao [2,*] and Dan Liu [3,*]

[1] School of Ecology and Nature Conservation, Beijing Forestry University, Beijing 100083, China; zhangzichen@bjfu.edu.cn (Z.Z.); Yanxin999@bjfu.edu.cn (X.Y.); guochang@bjfu.edu.cn (C.G.); wpdong@bjfu.edu.cn (W.D.)

[2] Museum of Beijing Forestry University, Beijing Forestry University, Beijing 100083, China

[3] Shandong Provincial Center of Forest and Grass Germplasm Resources, Ji'nan 250102, China

* Correspondence: lczhao@bjfu.edu.cn (L.Z.); 1821618@163.com (D.L.)

Abstract: The smoke tree (*Cotinus coggygria* Scop.) is a woody species mainly distributed in the Mediterranean region and East Asia, known for its high ecological and ornamental value. Investigation of changes in suitable habitats under different conditions can provide valuable insights with implications for predicting the distribution of *C. coggygria*. In this study, we employed a MaxEnt model to simulate the current, historical, and future suitable habitat of *C. coggygria* using distribution records and environmental variables. The results indicated that climatic variables had a much stronger impact on the suitable habitat of this species compared with soil and topographic variables, and bio11 (mean temperature of the coldest quarter) and bio12 (annual precipitation) played particularly important roles in determining the suitable habitat. The core distribution of *C. coggygria* exhibited an East Asian–Tethyan disjunction. During the glacial period (Last Glacial Maximum), *C. coggygria* in Europe was concentrated in the glacial refugia in southern Europe; its range was substantially smaller during the glacial period than during interglacial periods (mid-Holocene). In contrast, *C. coggygria* in East Asia survived in regions similar to those of the interglacial period. Future climate change led to a gradual northward expansion of suitable habitats for *C. coggygria*, and the area of suitable habitat was substantially larger in Europe than in East Asia. There were significant differences among the four climate scenarios in Europe, with minimal variation in East Asia. Our findings provide valuable insights into the contrasting effects of climate change on European and East Asian populations of *C. coggygria*, which enhances our understanding of Eurasian species with discontinuous distributions.

Keywords: *Cotinus coggygria*; species distribution models; environment variables; suitable habitat; glacial refugia

1. Introduction

The East Asian–Tethyan disjunction refers to the discontinuous distribution of species between East Asia and the Mediterranean region, including Mediterranean Europe and adjacent Africa [1]. This distributional pattern, which is often examined in biogeographical and phylogenetic studies, has been documented in many plant and animal species [2–5]. Previous studies have suggested that the cause of the East Asian–Tethyan disjunction is related to climate change associated with geological changes, which includes the drought caused by the gradual retreat of the Tethys Sea during the Eocene [6] and the monsoon circulation associated with the violent uplift of the Qinghai–Tibetan Plateau [7]. Given that climate change has had a significant influence on the East Asian–Tethyan disjunction, studies of changes in distributions under the climatic background of different historical glacial periods and future climatic scenarios are needed to clarify the distributions of species.

Numerous studies have demonstrated that many variables affect species distributions [8]. For example, temperature and precipitation within the tolerance range of a species can affect its geographic range [9]. In addition, soil, which serves as a crucial medium for plant growth, provides water and nutrients to plants, which affects the ecological niche of species [10]. Soil properties can serve as indicators of habitat conditions and human activities at small scales and thus provide valuable information for simulating species distributions [11]. Soil represents an additional environmental variable that drives changes in the distributions of plant species. Topography plays a significant role in shaping species distributions [12], and elevation, aspect, and slope have been identified as key topographic variables affecting species diversity [13,14].

Species distribution models (SDMs), also known as ecological niche models, are widely used in biogeography [15], ecology, conservation biology, and wildlife management [16]. SDMs describe the relationship between species and the environment and can be used to predict spatial distributions via statistical methods, machine learning models, species location information or diversity data, and environmental variables that affect habitat suitability [17,18]. Several species distribution models have been used to simulate species niches, including MaxEnt (Maximum Entropy modeling) [19], Random Forest [20], and Boosting Regression Tree [21]. MaxEnt is the most widely used SDM due to its wide range of applications and high accuracy of simulations based on small sample sizes [22]. SDMs have been widely applied in many fields, such as biogeography, species diversity, and global climate change [15].

The smoke tree (*Cotinus coggygria* Scop.) is the most extensively cultivated species, and it also has the widest natural distribution in the genus *Cotinus* (Anacardiaceae) [23]. *C. coggygria* spread to the Mediterranean from South-Central Europe; it then crossed the continent, including the Himalayas, and entered China. During this dispersal process, it has adapted to different habitats and diversified in the phenotype [23,24]. Previous phylogenetic studies indicate that *Cotinus* and its sister genus *Pistacia* diverged in the middle and late Eocene (38.41 Ma), and it was not until the early Miocene (17.91 Ma) that *Cotinus* species began to diverge [25] and subsequently occupy Eurasia [23], which is consistent with the biogeography of *Pistacia* [2]. Because of its high ecological value and ornamental value, *C. coggygria* has been widely planted in Asia, Europe, and North America [23,26]. *C. coggygria* is most commonly cultivated using stem cuttings [24]. Previous studies have confirmed that the differentiation between *C. coggygria* populations is related to environmental differences [27,28], and precipitation is a key driver of genetic differentiation [28]. The potential distribution of this species in China under current conditions and during the Last Glacial Maximum (LGM) was simulated using MaxEnt [27]. However, these studies have only investigated the population differentiation of *C. coggygria* in the Yellow River Basin in Central and Northern China; the suitability of other regions on the Eurasian continent for *C. coggygria* thus remains unclear. The effects of soil and topographic variables should also be considered in analyses of the potential distribution of *C. coggygria*.

In this study, MaxEnt models were used to simulate the suitable habitat for *C. coggygria* under current conditions and during historical and future periods based on occurrence records and soil, topographic, and climatic variables. Specifically, we aimed to (1) identify the most important environmental variables affecting the distribution of *C. coggygria*, (2) characterize the current suitable habitat of *C. coggygria* in Eurasia, and (3) determine changes in the East Asian–Tethyan disjunction of *C. coggygria* between historical and future periods.

2. Materials and Methods

2.1. Collection and Processing of Distribution Records

A total of 1771 occurrence records of *C. coggygria* were obtained after selecting preserved specimen records with coordinates from the Global Biodiversity Information Facility (GBIF, https://www.gbif.org/, accessed on 24 October 2023) [29] and field-collected data obtained in nature. Duplicate records and records not located in Eurasia were excluded. The SDM Toolbox v 2.5 in ArcGIS v 10.8 was used to filter the data. SDMToolbox v 2.5 is a Python-based ArcGIS toolkit, which can simplify the complex processing required for species distribution modeling and geospatial analysis [30]. With SDMToolbox, only one distribution record was retained in each 10 km × 10 km grid to reduce errors caused by spatial autocorrelation. Finally, 335 effective occurrence records were obtained, of which GBIF data accounted for 94.3% and field-collected data accounted for 5.7% (Figure S1). The distribution range of field-collected data was largely consistent with the descriptions in the Flora of China. It covers the core distribution range in China and provides support for the East Asian portion of the GBIF data. All the occurrence records covered the distribution range of *C. coggygria* in previous studies [23,27,31].

2.2. Environmental Variable Selection

Data on 39 environmental variables (climatic variables, topographic variables, and soil variables) were obtained in this study (Table S1). The 19 current (1970–2000) climatic variables (bio1–bio19) were downloaded from WorldClim (https://worldclim.org/, accessed on 19 October 2023); the data had a resolution of 2.5 arc-min (5 km spatial resolution at the equator). Global elevation data, which had a resolution of 2.5 arc-min, were downloaded from WorldClim (https://worldclim.org/, accessed on 15 December 2023), and the slope and aspect were extracted using ArcGIS v 10.8. The 17 soil variables were extracted from the soil data in the World Soil Database (HWSD, https://www.fao.org/soils-portal/data-hub/en/, accessed on 16 December 2023). To make these data consistent with the climatic and topographic data, the resolution of the soil data was transformed from 30 arc-seconds to 2.5 arc-min using the Resample tool in ArcGIS v 10.8 in the WGS 1984 geographic coordinate system.

Due to the complex correlations among environmental variables, the 39 environmental variables were screened according to their correlations and contributions to reduce the influence of collinearity between variables on the prediction accuracy of the model [32]. First, the contribution rate of 39 environmental variables to the distribution of *C. coggygria* was determined by MaxEnt, and the contribution rates ranged from 0 to 1. Environmental variables with a contribution rate of less than 0.25 were removed. Second, correlation coefficients (r) of 39 environmental variables were calculated using ENMTools v 1.4.4 [33]. If $|r| > 0.75$ for a pair of environmental variables, the one with the higher contribution rate was selected. Finally, sixteen environmental variables, including six climatic variables, seven soil variables, and three topographic variables, were selected for model analysis based on the contribution rate and correlation coefficient (Figure S2). The contribution rate of climatic variables was the highest (59.2%), followed by topographic variables (9.6%) and soil variables (8.1%) (Figure S2a).

2.3. MaxEnt Simulations and Model Accuracy Evaluation

The current suitable habitat was modeled by MaxEnt 3.4.1 using 335 *C. coggygria* occurrence records and 16 environmental variables. A jackknife test was performed to measure the importance of each environmental variable. Model training was performed using 75% of the distribution data, and the remaining 25% of the data were used for model testing [34]. Model values were output in logistic format [35]. The max number of background points was set to 10,000. The maximum iterations were set to 5000, which means that the training will stop after 5000 iterations of the optimization algorithm. The running process was repeated 10 times with bootstrap as the replicated run type and an average was taken as the result. Other settings were set to their default values.

The accuracy of the model was evaluated using the area under the receiver operating characteristic (ROC) curve (AUC), which ranges from 0 to 1. We used the average test AUC value of 10 repeated runs to evaluate the model accuracy in this study. Higher AUC values indicate higher performance [36], and the simulation results were considered highly accurate when AUC > 0.9.

2.4. Classification of Suitable Habitat

Habitat suitability is expressed by a value between 0 and 1 in the MaxEnt result; values closer to 1 indicate higher suitability. We used the Reclassify tool in ArcGIS v 10.8 to classify the whole research region into four suitability grades. For its equality, objectivity, and discriminability [37], the mean maximum test sensitivity plus specificity logistic threshold of 10 replicates in each simulation result was used to distinguish between suitable and unsuitable habitats. Areas with values greater than the mean maximum test sensitivity plus specificity logistic threshold were designated as suitable habitat, and areas with values less than the threshold were designated as unsuitable habitat. Suitable habitats were divided into lowly suitable habitats, moderately suitable habitats, and highly suitable habitats (Table S2). The Zonal Geometry as Table tool in ArcGIS v 10.8 was used to calculate the area of suitable habitat under the Goode Homolosine projection. The spatial distribution and area of habitat with each suitability level were obtained for comparative analysis.

2.5. Simulation of the Historical Suitable Habitat

For the historical period, including the Last Glacial Maximum and mid-Holocene (MH), we downloaded climatic data of the CCSM4 model from WorldClim (https://worldclim.org/, accessed on 19 October 2023) at a resolution of 2.5 arc-min. To reduce the influence of collinearity between variables on the prediction accuracy, we extracted the climatic variables from the 16 environmental variables that were retained according to their contribution rates and correlation coefficients in Section 2.2.

Historical climatic variables were used as projection layers based on the occurrence records and current climatic variables. This historical model was built based on the same MaxEnt parameters and classification method as the simulation of the current suitable habitat.

2.6. Simulation and Migration of the Future Suitable Habitat

For future periods (2050s, 2070s, and 2090s), we downloaded climatic variables extracted in 2.2 from WorldClim (https://worldclim.org/, accessed on 19 October 2023) with a resolution of 2.5 arc-min. Considering the uncertainty in future climate projections, we calculated the arithmetic averages of three general circulation models (GCMs) (ACCESS-CM2, BCC-CSM2-MR, and GISS-E2-1-G) to take into account various future models [32]. Given that emission scenarios are affected by various socioeconomic assumptions, the predicted values of climatic variables varied under the different scenarios. In each GCM, four scenarios (SSP126, SSP245, SSP370, and SSP585) were examined, which combine representative concentration pathways (RCPs) and shared socioeconomic pathways (SSPs) to drive Coupled Model Intercomparison Project (CMIP6) climate models [38,39]. SSPs show future development trajectories related to human society's ability to cope with climate change [39]. SSP1 represents a sustainable development society with a high degree of environmental friendliness, SSP2 represents a society with a more moderate development mode, SSP3 represents a society with slow economic growth and highly unbalanced development due to high emissions and rapid population growth, and SSP5 represents a society lacking climate strategies with high fuel consumption and few alternative energy sources [40]. RCPs refer to radiative forcing values, which range from 2.6 to 8.5 W/m^2 by 2100 and reflect land use and atmospheric emissions [41]. As in the simulation of the historical suitable habitat, future climatic variables were used as projection layers in MaxEnt.

To clarify spatial changes in the suitable habitat under different future climate scenarios, the Mean Center tool in ArcGIS [42] was used to calculate the distribution center of the suitable habitat in each climate scenario; the center point was then converted into a line to show the migration direction and distance.

3. Results

3.1. Environmental Variable Selection

The AUC values of the ROC curves for the MaxEnt models based on test data were all above 0.97 after 10 repetitions. The average of these 10 values was 0.9832 (Figure S3a), which indicates that the model was effective in simulating the suitable habitat for *C. coggygria*.

The jackknife test results showed that climatic variables were more important than soil variables and topographic variables. The mean temperature of the coldest quarter (bio11) was the most significant variable in the model, as it showed the highest test gain when used in isolation; the omission of this variable also resulted in the largest reduction in test gain among all variables. The four variables with the highest test gain when used in isolation were all climatic variables, including the mean temperature of the coldest quarter (bio11), temperature seasonality (standard deviation ×100) (bio4), annual precipitation (bio12), and mean diurnal range (bio2). The first two variables were temperature factors (Figure S3b). According to the response curves, the optimal values of bio11, bio4, bio12, and bio2 were 3.43 °C, 671.67 °C, 688.42 mm, and 8 °C, respectively (Figure S3c–f).

3.2. Suitable Habitat Under Current Conditions

The simulated results showed that the main suitable habitat for *C. coggygria* was widely distributed in East Asia and Europe, but a disjunction was present. Moderately suitable habitat and lowly suitable habitat were widely distributed from Southern to Central and even Northern Europe, as well as Southwest to North China, and highly suitable habitat was concentrated along the northern coast of the Mediterranean and North China (Figure 1a). The total area of suitable habitat was 8.47×10^6 km^2; the area of lowly suitable habitat, moderately suitable habitat, and highly suitable habitat accounted for 68.84%, 29.39%, and 1.77% of the total area of suitable habitat, respectively.

3.3. Changes in the Distribution of Historical Suitable Habitat

The suitable habitat of *C. coggygria* was mainly distributed in East Asia and Europe in all periods examined, and the suitable habitat at high latitudes increased from the Last Glacial Maximum to the mid-Holocene and current period (Figure 1b–g). During the Last Glacial Maximum, the suitable habitat was restricted to the Mediterranean coast around 40° N. By the mid-Holocene, the suitable habitat expanded to higher latitudes in Europe, reaching a maximum latitude of 70° N, which is similar to its current distribution (Figure 1b–d). The distribution of suitable habitats in East Asia did not change much, and this was in contrast to the patterns observed in Europe. During the Last Glacial Maximum, the distribution of suitable habitat was between 20° N and 40° N, and it spread north of 40° N in the mid-Holocene. The range of suitable habitat in the current period did not differ much from that in the mid-Holocene; only a slight shift to higher latitudes was observed (Figure 1e–g).

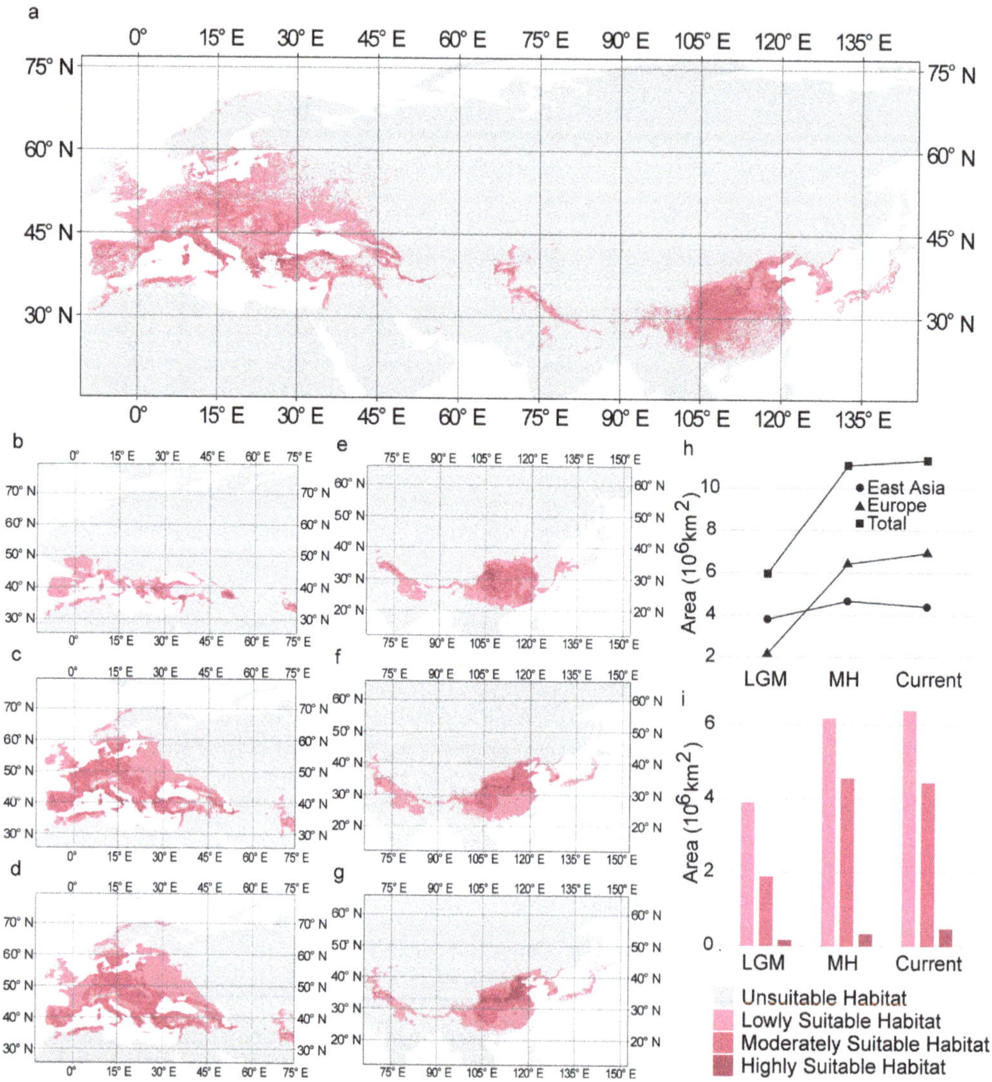

Figure 1. Historical changes in suitable habitat for *Cotinus coggygria*. (**a**) Current suitable habitat of *C. coggygria* modeled with 16 environmental variables. (**b–d**) The suitable habitat in Europe during the (**b**) Last Glacial Maximum, (**c**) mid-Holocene, and (**d**) current period. (**e–g**) The suitable habitat in East Asia during the (**e**) Last Glacial Maximum, (**f**) mid-Holocene, and (**g**) current period. (**h**) The area of suitable habitat during different periods. (**i**) The area of lowly suitable habitat, moderately suitable habitat, and highly suitable habitat during different periods.

Changes in the distribution of habitat in each suitability grade varied in Europe and East Asia. (Figure 1b–g). From the Last Glacial Maximum to the current period, the highly suitable habitat in Europe was consistently clustered in the southern margin of the entire suitable habitat at approximately 40° N. Moderately suitable habitat and lowly suitable habitat in Europe expanded to higher latitudes (Figure 1b–d). In East Asia, most of the highly suitable habitats gradually expanded to North China. The moderately suitable habitat was clustered around the highly suitable habitat from Southwest China to North China. As the northward migration of moderately suitable habitat expanded, the southern part of the moderately suitable habitat transformed into a lowly suitable habitat (Figure 1e–g).

3.4. Changes in the Area of Historical Suitable Habitat

The total area of suitable habitat in the Last Glacial Maximum was 5.95×10^6 km^2, and this increased to 11.09×10^6 km^2 in the mid-Holocene, which was 1.86 times higher than that in the Last Glacial Maximum. An increase in the area of suitable habitat of 0.24×10^6 km^2 was observed from the mid-Holocene to the current period (Figure 1h, Table S2). The area of suitable habitat in Europe was much smaller than that in East Asia during the Last Glacial Maximum. By the mid-Holocene, the area of suitable habitat in Europe increased by 197.95% and exceeded that in East Asia. From the mid-Holocene to the current period, the area of suitable habitat in Europe increased, and that in East Asia began to decrease (Figure 1h, Table S2).

Changes in the area of habitat in each suitability grade varied in Europe and East Asia. In both Europe and East Asia, the proportion of lowly suitable habitat was the largest, followed by moderately suitable habitat and highly suitable habitat, in all periods (Figure 1i, Table S2). The proportion of highly suitable habitat during the Last Glacial Maximum and the current period was lower in Europe than in East Asia, and it was higher in Europe than in East Asia only during the mid-Holocene. Although the total area of suitable habitat in Europe in the current period was large, only the proportion of moderately suitable habitats and lowly suitable habitats was larger; the proportion of highly suitable habitats was much smaller in Europe than in East Asia (Table S2).

3.5. Suitable Habitat of *C. coggygria* Under Future Climate Change

The distribution of suitable habitats for *C. coggygria* changed significantly over time. The distribution center of the suitable habitat in Europe and East Asia moved northward from the current period to the 2090s (Figure 2a–b). Changes in the distribution of suitable habitat were highly significant under SSP585, which is the scenario with the highest emissions. The suitable habitat in Europe rapidly expanded into most of Europe, and the southern boundary of the suitable habitat remained almost unchanged (Figure 3a–d). The main suitable habitat in East Asia gradually shifted to Northern China, and this caused wide areas of suitable habitat in Southern China to be transformed into unsuitable habitat (Figure 3e–h); this contrasted with the patterns of change observed over time in Europe. Regardless of the specific scenario, the highly suitable habitat in both Europe and East Asia was mainly distributed from 30° N to 40° N. The moderately suitable habitat and lowly suitable habitat below 30° N gradually decreased (Figure 3a–h). The trends observed in the other scenarios were similar to those observed under SSP585.

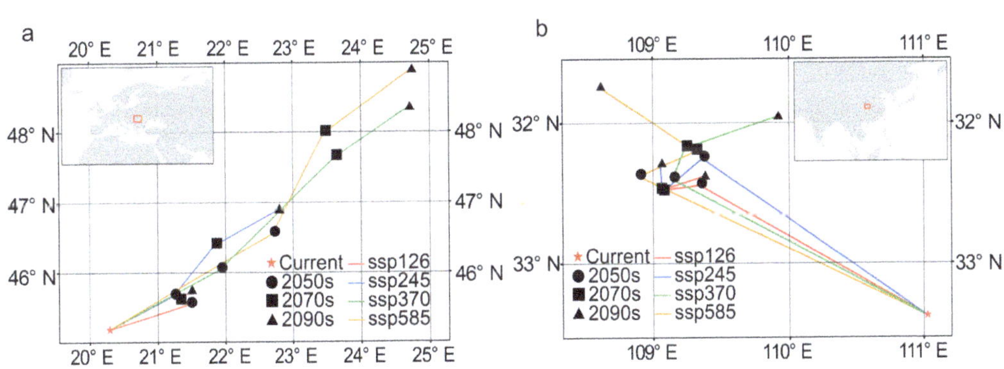

Figure 2. The migration of the center of suitable habitat in (**a**) Europe and (**b**) East Asia.

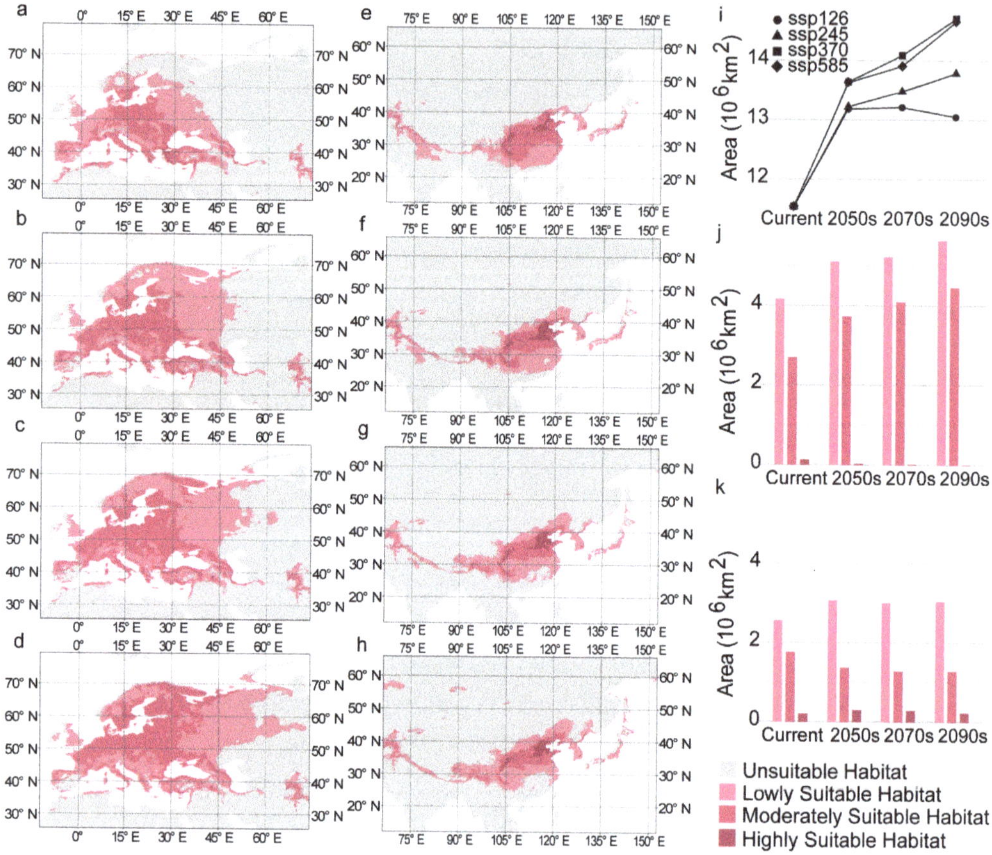

Figure 3. Suitable habitat of *C. coggygria* under the future climate. (**a**–**d**) The suitable habitat in Europe during the (**a**) current period, (**b**) 2050s, (**c**) 2070s, and (**d**) 2090s under SSP 585. (**e**–**h**) The suitable habitat in East Asia during the (**e**) current period, (**f**) 2050s, (**g**) 2070s, and (**h**) 2090s under SSP585. (**i**) Changes in the area of suitable habitat in different periods under SSP126, SSP245, SSP370, and SSP585. (**j**,**k**) The area of lowly suitable habitat, moderately suitable habitat, and highly suitable habitat in (**j**) Europe and (**k**) East Asia during different periods under SSP585.

The total area of suitable habitat under most scenarios gradually increased over time and was much larger than the area of suitable habitat under current conditions (Figure 3i). The average total area reached 14.05×10^6 km^2 in the future, which was 1.22 times higher than that under current conditions (Table S3). Under SSP585, the total area of suitable habitat in Europe, more than 99% of which comprised moderately suitable habitat and lowly suitable habitat, was always larger than that in East Asia (Table S3). The total area of suitable habitat in Europe substantially increased from the current period to the 2090s (increasing from 7.01×10^6 km^2 to 10.11×10^6 km^2), and the area of highly suitable habitat decreased to 0.00724×10^6 km^2, which only accounted for 0.07% of the area of suitable habitat in Europe (Figure 3j, Table S3). In East Asia, the total area of suitable habitat fluctuated steadily between 4.52×10^6 km^2 and 4.75×10^6 km^2 (Table S3). The area of highly suitable habitat was larger in East Asia than in Europe and was more than 0.2×10^6 km^2 in all periods, which accounted for 4.56–6.37% of the area of suitable habitat in East Asia (Figure 3k, Table S3). The trends under other scenarios were similar to those observed under SSP585 (Table S3).

3.6. Variation in Suitable Habitat Under Different Future Climate Scenarios

The suitability of habitat for *C. coggygria* in the same areas varied among scenarios. In the 2090s, the habitat suitability values for *C. coggygria* increased in the northern part of the entire range of suitable habitats and decreased in the southern part of the entire range of suitable habitats as expected emissions increased; this indicates that the suitable habitat increased in the north (Figure 4a–c). The trends in other periods were similar to those observed in the 2090s.

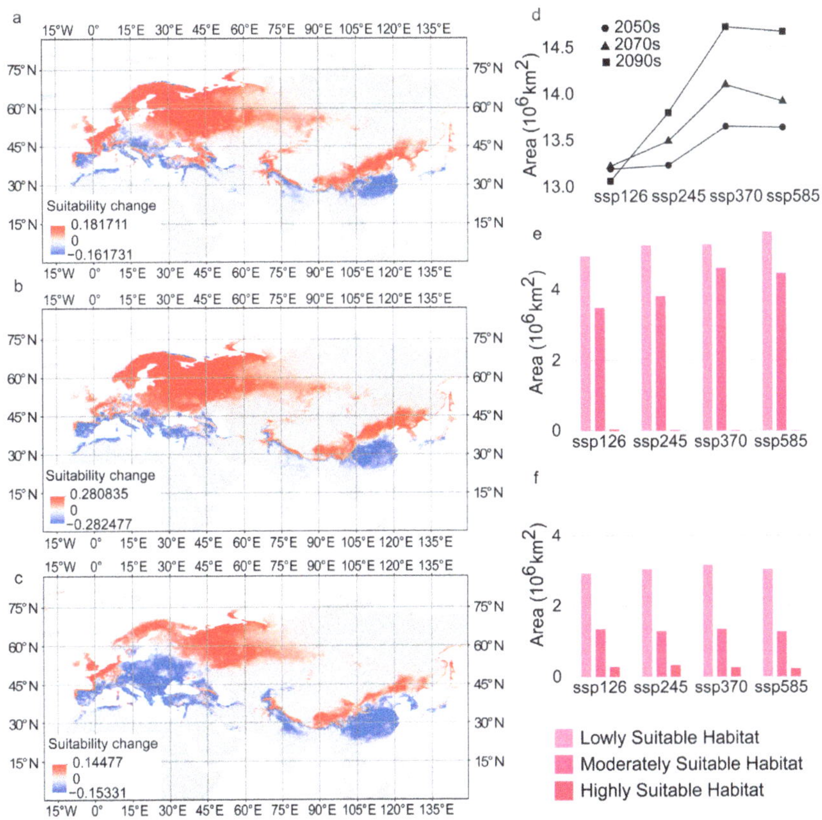

Figure 4. Suitable habitat of *C. coggygria* under different SSPs in the future. (**a**–**c**) Suitability changes (**a**) from SSP126 to SSP245, (**b**) from SSP245 to SSP370, and (**c**) from SSP370 to SSP585 in the 2090s. (**d**) Changes in the area of suitable habitat under different SSPs in the 2050s, 2070s, and 2090s. (**e**,**f**) Areas of lowly suitable habitat, moderately suitable habitat, and highly suitable habitat in (**e**) Europe and (**f**) East Asia under different SSPs.

The suitable habitat area was larger under each future climate scenario than under current conditions (Figure 4d). The area of suitable habitat and the proportion of different habitat suitability grades under different climate scenarios varied greatly in Europe and varied little in East Asia (Table S3). In the 2090s, changes in the distribution of habitat in each suitability grade under different scenarios ranged from 0.18×10^6 km^2 and 1.62×10^6 km^2 in Europe, but all were less than 0.24×10^6 km^2 in East Asia (Figure 4e–f, Table S3). The trends in other periods were similar to those observed in the 2090s (Table S3).

4. Discussion

4.1. Differences in the Effects of Environmental Variables on C. coggygria

Six of the sixteen climatic variables screened contributed nearly 60% to the model (Figure S2b). This result was similar to the results of the jackknife test, in which the four environmental variables with the greatest effect on the MaxEnt model were all climatic variables (Figure S3b). This indicates that climate conditions played the most important role in determining the distribution of *C. coggygria*. Many previous SDM studies have shown that the contribution of climatic variables was greater than that of soil variables and topographic variables [43–45].

Previous studies have shown that *C. coggygria* is a locally adapted species with strong responses to precipitation and temperature conditions [28,46]. In this study, our results suggested that bio11 (mean temperature of coldest quarter) is a key variable determining the suitability of *C. coggygria*. Bio12 (annual precipitation) is also an important variable affecting the suitability of *C. coggygria* (Figure S3b). In previous studies of adaptive genetic variation in *C. coggygria*, bio12 was found to be significantly correlated with two alleles (Cc 025D and Cc 025F) in *C. coggygria*, and Cc 025F was also significantly correlated with precipitation variables (bio5, bio6, and bio8) [28]. In the models established based on historical, current, and future climate data, bio11 and bio12 were the variables with the highest contribution rate, which further verified the important effects of these two variables on the distribution of *C. coggygria*. Therefore, bio11 and bio12 have key effects on the suitability of habitat for *C. coggygria*.

Environmental stress induces responses at the cellular and molecular level in plants [47]. Low-temperature stress can make the extracellular water and water in the tracheary element freeze, resulting in cell dehydration, blocking water transport, and thus affecting the physiological metabolic processes of plant cells [48,49]. The important effect of the mean temperature of the coldest quarter on the suitable habitat reflects the sensitivity of *C. coggygria* to low-temperature stress, which can also explain the northward shift in the suitable habitat of *C. coggygria* under climate warming caused by increasing emissions in the future. Similar to low temperatures, insufficient precipitation can also limit physiological processes such as plant growth, development, and reproduction [47]. In this study, the relationship between the suitable habitat of *C. coggygria* and annual precipitation indicates that precipitation has a major effect on the suitability of *C. coggygria*.

4.2. Changes in the Distribution of C. coggygria

The distribution of *C. coggygria* has been discontinuous between Europe and East Asia since the Last Glacial Maximum (2.1 Ma) (Figure 1). The distribution of *C. coggygria* was affected by the glacial period, and changes in its distribution in East Asia and Europe were not consistent. Specifically, East Asia provides a habitat for many Tertiary relict species since it was less affected by ice sheets during the glacial period [50]. The mountains of Southwest China and Northern Vietnam in particular are considered long-term stable refugia because of their mild climatic conditions [51]. In this study, the survival range of *C. coggygria* in East Asia during the Last Glacial Maximum is not much different from that in the interglacial (mid-Holocene) and current period (Figure 1e–g), which was comparable to the simulation results obtained using the distribution data in China [27]. These survival ranges were consistent with observed patterns in China's warm-temperate zone. In contrast to patterns observed in East Asia, during the Last Glacial Maximum, the area of suitable habitat in Europe was much lower than that during the interglacial period (mid-Holocene) and concentrated in Southern Europe (Figure 1b); it overlapped with the refugia of numerous temperate tree species during the Quaternary glacial period [52,53]. This result was consistent with the southern refugia hypothesis [54].

The suitable habitat of *C. coggygria* shifted to the north from the current period to the 2090s (Figure 2). This finding is consistent with the prediction of the future suitable habitat of many species located in Europe [55,56] and East Asia [57–59]. For example, the distribution of European Hop Hornbeam (*Ostrya virginiana*) in Europe will move northward over the next 60 years [55]. The potential habitat of *Ziziphus jujuba* in China will shift northeastward to adapt to global warming [59]. Future climate conditions are more suitable for the survival of *C. coggygria* compared with current conditions, and the suitable habitat is mainly located to the north of the current distribution. Specifically, the suitable habitat in Europe is rapidly expanding into most of Europe, and the main suitable habitat in East Asia is gradually moving to southwestern and northeastern China. Our results indicate that a northerly climate would be suitable for *C. coggygria* in the future. The results under different scenarios indicated that *C. coggygria* tends to occupy a wider area of suitable habitat and higher latitudes under higher emission climate scenarios (Figure 3). This is in contrast to the results of previous analyses of many species, such as *Cunninghamia lanceolata* [60] and *Sapindus mukorossi* [61], which indicate that the area of the suitable habitat was positively related to the environmental friendliness of the climate scenario. Our results revealed that future climate conditions will be more suitable for *C. coggygria* than current climate conditions.

5. Conclusions

The results of this study revealed the adaptation of *C. coggyria* to environmental variables and differences in the suitable habitat under different climatic conditions between Europe and East Asia. Climatic variables had a significantly stronger effect on the suitable habitat of this species than soil variables and topographic variables. Bio11 and bio12 were particularly important variables. *C. coggygria* exhibits a disjunct East Asian–Tethyan distribution, and its core distribution region was concentrated in the Mediterranean and East Asia. During the Last Glacial Maximum, *C. coggygria* in Europe was much lower than that during the interglacial period and concentrated in glacial refugia in Southern Europe, whereas *C. coggygria* in East Asia was present in regions similar to interglacial habitats. Under future climate scenarios, the suitable habitat of *C. coggygria* gradually expand northward. As the climate scenarios became more extreme, the suitable habitat of *C. coggygria* shifted northward. The overall area of suitable habitat was larger in Europe than in East Asia. The area of highly suitable habitat was smaller in Europe than in East Asia, and the area of suitable habitat in Europe significantly differed under the four scenarios.

Supplementary Materials: The following supporting information can be downloaded at: https://www.mdpi.com/article/10.3390/plants14040547/s1. Figure S1: The 335 distribution records of *Cotinus coggygria*; Figure S2: Environmental variable screened. (a) Correlation and (b) contribution rate of 16 environmental variables involved in simulating; Figure S3: MaxEnt Simulations and Model Accuracy Evaluation. a Receiver operating characteristic (ROC) curves. b jackknife test of variable importance. c-r MaxEnt model response curves of 16 environmental variables; Table S1: The environmental variables used in this study; Table S2: The classification range of suitable habitat in each model; Table S3: Area of historical suitable habitat of each grade (10^6 km^2); Table S4: Area of future suitable habitat of each grade (10^6 km^2).

Author Contributions: Conceptualization: Z.Z., L.Z. and W.D.; Methodology: Z.Z.; Software: Z.Z., X.Y. and C.G.; Formal Analysis: Z.Z., D.L., X.Y. and C.G.; Data Curation: Z.Z., X.Y., D.L. and C.G.; Visualization: Z.Z; Writing—Original Draft: Z.Z., L.Z. and W.D.; Writing—Review and Editing: Z.Z., L.Z. and W.D.; Supervision: L.Z.; Project Administration: L.Z. and W.D. All authors have read and agreed to the published version of the manuscript.

Funding: This research was financially supported by the Science and Technology Basic Resources Investigation Program of China (Grant No. 2021FY100200), and the Survey of Herbaceous Germplasm Resources in Shandong Province (Grant No. Lu Financial [2021]1).

Data Availability Statement: The original contributions presented in the study are included in the article/Supplementary Materials, further inquiries can be directed to the corresponding author.

Conflicts of Interest: The authors declare no conflicts of interest.

References

1. Wu, D.; Milne, R.I.; Yang, H.; Zhang, Y.; Wang, Y.; Jia, S.; Li, J.; Mao, K. Phylogenomics shed light on the complex evolutionar history of a gymnosperm genus showing East Asian–Tethyan disjunction. *J. Syst. Evol.* **2025**. *early view*. [CrossRef]
2. Xie, L.; Yang, Z.-Y.; Wen, J.; Li, D.-Z.; Yi, T.-S. Biogeographic history of *Pistacia* (Anacardiaceae), emphasizing the evolution of th Madrean-Tethyan and the eastern Asian-Tethyan disjunctions. *Mol. Phylogenetics Evol.* **2014**, *77*, 136–146. [CrossRef] [PubMed]
3. Velitzelos, D.; Bouchal, J.M.; Denk, T. Review of the Cenozoic floras and vegetation of Greece. *Rev. Palaeobot. Palynol.* **2014**, *20* 56–117. [CrossRef]
4. Xia, M.; Cai, M.; Comes, H.P.; Zheng, L.; Ohi-Toma, T.; Lee, J.; Qi, Z.; Konowalik, K.; Li, P.; Cameron, K.M.; et al. An overlooke dispersal route of Cardueae (Asteraceae) from the Mediterranean to East Asia revealed by phylogenomic and biogeographica analyses of *Atractylodes*. *Ann. Bot.* **2022**, *130*, 53–64. [CrossRef] [PubMed]
5. Jiang, X.L.; Hipp, A.L.; Deng, M.; Su, T.; Zhou, Z.K.; Yan, M.X. East Asian origins of European holly oaks (*Quercus* section *Ile* Loudon) via the Tibet-Himalaya. *J. Biogeogr.* **2019**, *46*, 2203–2214. [CrossRef]
6. Sue, J.P. Origin and evolution of the Mediterranean vegetation and climate in Europe. *Nature* **1984**, *307*, 429–432.
7. Morley, R.J. Palynological evidence for Tertiary plant dispersals in the SE Asian region in relation to plate tectonics and climat *Biogeogr. Geol. Evol. SE Asia* **1998**, *1*, 211–234.
8. Pacifici, M.; Foden, W.B.; Visconti, P.; Watson, J.E.; Butchart, S.H.; Kovacs, K.M.; Scheffers, B.R.; Hole, D.G.; Martin, T.G.; Akçakaya H.R. Assessing species vulnerability to climate change. *Nat. Clim. Change* **2015**, *5*, 215–224. [CrossRef]
9. Walther, G.-R.; Post, E.; Convey, P.; Menzel, A.; Parmesan, C.; Beebee, T.J.; Fromentin, J.-M.; Hoegh-Guldberg, O.; Bairlein, I Ecological responses to recent climate change. *Nature* **2002**, *416*, 389–395. [CrossRef] [PubMed]
10. Miller, R.L. A Treatise on Limnology, Vol. 1: Geography, physics, and chemistry. *J. Geol.* **1958**, *66*, 710. [CrossRef]
11. Chauvier, Y.; Thuiller, W.; Brun, P.; Lavergne, S.; Descombes, P.; Karger, D.N.; Renaud, J.; Zimmermann, N.E. Influence of climate soil, and land cover on plant species distribution in the European Alps. *Ecol. Monogr.* **2021**, *91*, e01433. [CrossRef]
12. Qazi, A.W.; Saqib, Z.; Zaman-ul-Haq, M. Trends in species distribution modelling in context of rare and endemic plants: A systematic review. *Ecol. Process.* **2022**, *11*, 40. [CrossRef]
13. Wani, Z.A.; Negi, V.S.; Bhat, J.A.; Satish, K.; Kumar, A.; Khan, S.; Dhyani, R.; Siddiqui, S.; Al-Qthanin, R.N.; Pant, S. Elevation aspect, and habitat heterogeneity determine plant diversity and compositional patterns in the Kashmir Himalaya. *Front. For. Glob Change* **2023**, *6*, 1019277.
14. Karami, R.; Mehrabi, H.R.; Ariapoor, A. The effect of altitude and slope in the species diversity of herbaceous plants (case study Watershed Miandar Qarootag—Gilangharb). *J. Appl. Environ. Biol. Sci.* **2015**, *5*, 197–204.
15. Franklin, J. Species distribution modelling supports the study of past, present and future biogeographies. *J. Biogeogr.* **2023**, *50* 1533–1545. [CrossRef]
16. Araújo, M.B.; Guisan, A. Five (or so) challenges for species distribution modelling. *J. Biogeogr.* **2006**, *33*, 1677–1688. [CrossRef]
17. Franklin, J. Moving beyond static species distribution models in support of conservation biogeography. *Divers. Distrib.* **2010**, *16* 321–330. [CrossRef]
18. Sánchez-Mercado, A.; Ferrer-Paris, J.; Franklin, J. Mapping species distributions: Spatial inference and prediction. *Oryx* **2010**, *44*, 615 [CrossRef]
19. Phillips, S.J.; Anderson, R.P.; Schapire, R.E. Maximum entropy modeling of species geographic distributions. *Ecol. Model.* **2006** *190*, 231–259. [CrossRef]
20. Evans, J.S.; Murphy, M.A.; Holden, Z.A.; Cushman, S.A. Modeling species distribution and change using random forest. In *Predictive species and habitat modeling in landscape ecology: Concepts and applications*; Springer: New York, NY, USA, 2011; pp 139–159.
21. Elith, J.; Leathwick, J.R.; Hastie, T. A working guide to boosted regression trees. *J. Anim. Ecol.* **2008**, *77*, 802–813. [CrossRef] [PubMed]
22. Lai, J.; Wang, Y.; Huang, T.; Lyu, Y.; Zhao, Y.; Liu, J. Maximum entropy analysis of bird diversity and environmental variables in Nanjing Megapolis, China. *Sustainability* **2024**, *16*, 2139. [CrossRef]
23. Tripp, K.E. Considering *Cotinus*. 1994. Available online: https://www.cabidigitallibrary.org/doi/full/10.5555/19950309901 (accessed on 4 February 2025).

24. da Silva, J.A.T.; Pacholczak, A.; Ilczuk, A. Smoke tree (*Cotinus coggygria* Scop.) propagation and biotechnology: A mini-review. *South Afr. J. Bot.* **2018**, *114*, 232–240. [CrossRef]
25. Liu, Q.; Yang, N.; Dong, W.; Zhao, L. Molecular evolution and phylogenomic analysis of complete chloroplast genomes of *Cotinus* (Anacardiaceae). *Ecol. Evol.* **2023**, *13*, e10134. [CrossRef]
26. Emelyanova, O.; Firsov, A. Ecological and biological features and prospects of using *Cotinus coggygria* in breeding. *BIO Web Conf.* **2021**, *36*, 01016. [CrossRef]
27. Wang, W.; Tian, C.Y.; Li, Y.H.; Li, Y.; Peeters, T. Molecular data and ecological niche modelling reveal the phylogeographic pattern of *Cotinus coggygria*(Anacardiaceae) in China's warm-temperate zone. *Plant Biol.* **2014**, *16*, 1114–1120. [CrossRef] [PubMed]
28. Lei, Y.-K.; Wang, W.; Liu, Y.-P.; He, D.; Li, Y. Adaptive genetic variation in the smoke tree (*Cotinus coggygria* Scop.) is driven by precipitation. *Biochem. Syst. Ecol.* **2015**, *59*, 63–69. [CrossRef]
29. GBIF.org. GBIF Occurrence Download. Available online: https://doi.org/10.15468/dl.zmgrye (accessed on 24 October 2023).
30. Brown, J.L.; Anderson, B. SDMtoolbox: A python-based GIS toolkit for landscape genetic, biogeographic and species distribution model analyses. *Methods Ecol. Evol.* **2014**, *5*, 694–700. [CrossRef]
31. Gavinet, J.; Santonja, M.; Baldy, V.; Hashoum, H.; Peano, S.; Tchong, T.; Gros, R.; Greff, S.; Fernandez, C.; Bousquet-Mélou, A. Phenolics of the understory shrub *Cotinus coggygria* influence Mediterranean oak forests diversity and dynamics. *For. Ecol. Manag.* **2019**, *441*, 262–270. [CrossRef]
32. Zhao, Q.; Mi, Z.Y.; Lu, C.; Zhang, X.F.; Chen, L.J.; Wang, S.Q.; Niu, J.F.; Wang, Z.Z. Predicting potential distribution of *Ziziphus spinosa* (Bunge) HH Hu ex FH Chen in China under climate change scenarios. *Ecol. Evol.* **2022**, *12*, e8629. [CrossRef] [PubMed]
33. Warren, D.L.; Glor, R.E.; Turelli, M. ENMTools: A toolbox for comparative studies of environmental niche models. *Ecography* **2010**, *33*, 607–611. [CrossRef]
34. Zhang, K.; Yao, L.; Meng, J.; Tao, J. Maxent modeling for predicting the potential geographical distribution of two peony species under climate change. *Sci. Total Environ.* **2018**, *634*, 1326–1334. [CrossRef] [PubMed]
35. Phillips, S.J.; Dudík, M. Modeling of species distributions with Maxent: New extensions and a comprehensive evaluation. *Ecography* **2008**, *31*, 161–175. [CrossRef]
36. Swets, J.A. Measuring the accuracy of diagnostic systems. *Science* **1988**, *240*, 1285–1293. [CrossRef]
37. Canran, L.; Graeme, N.; Matt, W. On the selection of thresholds for predicting species occurrence with presence-only data. *Ecol. Evol.* **2016**, *6*, 337–348.
38. Eyring, V.; Bony, S.; Meehl, G.A.; Senior, C.A.; Stevens, B.; Stouffer, R.J.; Taylor, K.E. Overview of the Coupled Model Intercomparison Project Phase 6 (CMIP6) experimental design and organization. *Geosci. Model Dev.* **2016**, *9*, 1937–1958. [CrossRef]
39. Van Vuuren, D.P.; Stehfest, E.; Gernaat, D.E.; Doelman, J.C.; Van den Berg, M.; Harmsen, M.; de Boer, H.S.; Bouwman, L.F.; Daioglou, V.; Edelenbosch, O.Y. Energy, land-use and greenhouse gas emissions trajectories under a green growth paradigm. *Glob. Environ. Change* **2017**, *42*, 237–250. [CrossRef]
40. O'Neill, B.C.; Kriegler, E.; Riahi, K.; Ebi, K.L.; Hallegatte, S.; Carter, T.R.; Mathur, R.; Van Vuuren, D.P. A new scenario framework for climate change research: The concept of shared socioeconomic pathways. *Clim. Change* **2014**, *122*, 387–400. [CrossRef]
41. Van Vuuren, D.P.; Edmonds, J.; Kainuma, M.; Riahi, K.; Thomson, A.; Hibbard, K.; Hurtt, G.C.; Kram, T.; Krey, V.; Lamarque, J.-F. The representative concentration pathways: An overview. *Clim. Change* **2011**, *109*, 5–31. [CrossRef]
42. Scott, L.M.; Janikas, M.V. Spatial statistics in ArcGIS. In *Handbook of applied spatial analysis: Software tools, methods and applications*; Springer: Berlin/Heidelberg, Germany, 2010; pp. 27–41.
43. He, X.-H.; Si, J.-H.; Zhu, L.; Zhou, D.-M.; Zhao, C.-Y.; Jia, B.; Wang, C.-L.; Qin, J.; Zhu, X.-L. Modeling habitat suitability of *Hippophae rhamnoides* L. using MaxEnt under climate change in China: A case study of *H. r. sinensis* and *H. r. turkestanica*. *Front. For. Glob. Change* **2023**, *5*, 1095784. [CrossRef]
44. Fang, J.; Shi, J.; Zhang, P.; Shao, M.; Zhou, N.; Wang, Y.; Xu, X. Potential distribution projections for *Senegalia senegal* (L.) britton under climate change scenarios. *Forests* **2024**, *15*, 379. [CrossRef]
45. Bradie, J.; Leung, B. A quantitative synthesis of the importance of variables used in MaxEnt species distribution models. *J. Biogeogr.* **2016**, *44*, 1344–1361. [CrossRef]
46. Miao, C.-Y.; Li, Y.; Yang, J.; Mao, R.-L. Landscape genomics reveal that ecological character determines adaptation: A case study in smoke tree (*Cotinus coggygria* Scop.). *BMC Evol. Biol.* **2017**, *17*, 202. [CrossRef] [PubMed]
47. Beck, E.H.; Fettig, S.; Knake, C.; Hartig, K.; Bhattarai, T. Specific and unspecific responses of plants to cold and drought stress. *J. Biosci.* **2007**, *32*, 501–510. [CrossRef] [PubMed]
48. Pearce, R.S. Plant freezing and damage. *Ann. Bot.* **2001**, *87*, 417–424. [CrossRef]
49. Granda, E.; Scoffoni, C.; Rubio-Casal, A.E.; Sack, L.; Valladares, F. Leaf and stem physiological responses to summer and winter extremes of woody species across temperate ecosystems. *Oikos* **2014**, *123*, 1281–1290. [CrossRef]
50. Milne, R.I.; Abbott, R.J. The origin and evolution of tertiary relict floras. *Adv. Bot. Res.* **2002**, *38*, 281–314.
51. Tang, C.Q.; Matsui, T.; Ohashi, H.; Dong, Y.-F.; Momohara, A.; Herrando-Moraira, S.; Qian, S.; Yang, Y.; Ohsawa, M.; Luu, H.T.; et al. Identifying long-term stable refugia for relict plant species in East Asia. *Nat. Commun.* **2018**, *9*, 4488. [CrossRef]

52. Pierre Taberlet, R.C. Quaternary refugia and persistence of biodiversity. *Science* **2002**, *297*, 2009–2010. [CrossRef]
53. Bennett, K.D.; Tzedakis, P.C.; Willis, K.J. Quaternary refugia of north European trees. *J. Biogeogr.* **1991**, *18*, 103–115. [CrossRef]
54. Svenning, J.C.; Normand, S.; Kageyama, M. Glacial refugia of temperate trees in Europe: Insights from species distributio modelling. *J. Ecol.* **2008**, *96*, 1117–1127. [CrossRef]
55. Örücü, Ö.K.; Azadi, H.; Arslan, E.S.; Kamer Aksoy, Ö.; Choobchian, S.; Nooghabi, S.N.; Stefanie, H.I. Predicting the distributio of European Hop Hornbeam: Application of MaxEnt algorithm and climatic suitability models. *Eur. J. For. Res.* **2023**, *142*, 579–59 [CrossRef]
56. Rodríguez-Sánchez, F.; Arroyo, J. Reconstructing the demise of Tethyan plants: Climate-driven range dynamics of *Laurus* sinc the Pliocene. *Glob. Ecol. Biogeogr.* **2008**, *17*, 685–695. [CrossRef]
57. Li, Y.; Li, M.; Li, C.; Liu, Z. Optimized maxent model predictions of climate change impacts on the suitable distribution o *Cunninghamia lanceolata* in China. *Forests* **2020**, *11*, 302. [CrossRef]
58. Yan, H.; He, J.; Xu, X.; Yao, X.; Wang, G.; Tang, L.; Feng, L.; Zou, L.; Gu, X.; Qu, Y. Prediction of potentially suitable distribution of *Codonopsis pilosula* in China based on an optimized MaxEnt model. *Front. Ecol. Evol.* **2021**, *9*, 773396. [CrossRef]
59. Zhao, G.; Cui, X.; Sun, J.; Li, T.; Wang, Q.I.; Ye, X.; Fan, B. Analysis of the distribution pattern of Chinese *Ziziphus jujuba* unde climate change based on optimized biomod2 and MaxEnt models. *Ecol. Indic.* **2021**, *132*, 108256. [CrossRef]
60. Zhou, Y.; Zhang, Z.; Zhu, B.; Cheng, X.; Yang, L.; Gao, M.; Kong, R. MaxEnt modeling based on CMIP6 models to project potentia suitable zones for *Cunninghamia lanceolata* in China. *Forests* **2021**, *12*, 752. [CrossRef]
61. Li, Y.; Shao, W.; Jiang, J. Predicting the potential global distribution of *Sapindus mukorossi* under climate change based on MaxEn modelling. *Environ. Sci. Pollut. Res.* **2022**, *29*, 21751–21768. [CrossRef] [PubMed]

plants

MDPI

Article

MaxEnt-Based Predictions of Suitable Potential Distribution of *Leymus secalinus* Under Current and Future Climate Change

Shimeng Zhao †, Zongxian Zhang †, Changyu Gao, Yiding Dong, Zeyao Jing, Lixia Du and Xiangyang Hou *

Key Laboratory of Efficient Forage Production Mode, Ministry of Agriculture and Rural Affair, College of Grassland Science, Shanxi Agricultural University, Jinzhong 030801, China; zsmsxau@163.com (S.Z.); zongxian.zhang1@gmail.com (Z.Z.); 18434761726@163.com (C.G.); nice_1998@126.com (Y.D.); jing_zeyao@163.com (Z.J.); dulixia0328@126.com (L.D.)

* Correspondence: houxy16@vip.126.com

† These authors contributed equally to this work.

Abstract: Grassland degradation is a serious ecological issue in the farming–pastoral ecotone of northern China. Utilizing native grasses for the restoration of degraded grasslands is an effective technological approach. *Leymus secalinus* is a superior indigenous grass species for grassland ecological restoration in northern China. Therefore, the excavation of potential distribution areas of *L. secalinus* and important ecological factors affecting its distribution is crucial for grassland conservation and restoration of degraded grasslands. Based on 357 data points collected on the natural distribution of *L. secalinus*, this study employs the jackknife method and Pearson correlation analysis to screen out 23 variables affecting its spatial distribution. The MaxEnt model was used herein to predict the current suitable distribution area of *L. secalinus* and the suitable distribution of *L. secalinus* under different SSP scenarios (SSP1-26, SSP2-45, and SSP5-85) for future climate. The results showed the following: (1) Mean diurnal temperature range, annual mean temperature, precipitation of the wettest quarter, and elevation are the major factors impacting the distribution of *L. secalinus*. (2) Under the current climatic conditions, *L. secalinus* is mainly distributed in the farming–pastoral ecotone of northern China; in addition, certain suitable areas also exist in parts of Xinjiang, Tibet, Sichuan, Heilongjiang, and Jilin. (3) Under future climate change scenarios, the suitable areas for *L. secalinus* are generally the same as at present, with slight changes in area under different scenarios, with the largest expansion of 97,222 km^2 of suitable area in 2021–2040 under the SSP1-26 scenario and the largest shrinkage of potential suitable area in 2061–2080 under the SSP2-45 scenario, with 87,983 km^2. Notably, the northern boundary of the middle- and high-suitability areas is reduced, while the northeastern boundary and some areas of Heilongjiang and Jilin are expanded. The results of this study revealed the suitable climatic conditions and potential distribution range of *L. secalinus*, which can provide a reference for the conservation, introduction, and cultivation of *L. secalinus* in new ecological zones, avoiding the blind introduction of inappropriate habitats, and is also crucial for sustaining the economic benefits associated with *L. secalinus* ecological services.

Keywords: *Leymus secalinus*; species distribution modeling (SDM); shared socioeconomic pathway (SSP) scenarios; environmental variables; suitable area

1. Introduction

Grasslands, which cover 40% of the land area, are an essential global biodiversity repository, providing enormous material and non-material benefits to humans [1–3]. However, grassland degradation is very serious globally and is increasing in many areas, with about 49% of the world's grasslands being degraded to varying degrees to date, undermining their ability to maintain biodiversity, ecosystem services, and benefits to people [4]. China is one of the countries with the richest grassland resources in the world. Grassland resources not only provide an important material basis for the development of livestock husbandry, but also play an important ecological function in maintaining soil and water, preventing winds and stabilizing sands, nourishing water, protecting biodiversity, and maintaining the balance of ecosystems [5]. The farming–pastoral ecotone of northern China extends from the northeast to the southwest, following the line of the well-known Great Wall of China. It is a typical farming–pastoral zone in history, and nowadays, and it is also an important ecological security barrier in the middle-eastern region of China and an essential water conservation belt in the Beijing–Tianjin–Hebei region, which is of vital significance for the ecological security of the country [6]. Over the past half-century, ecological and environmental problems such as grassland degradation, land sanding, and salinization have arisen due to climate change and anthropogenic phenomena such as over-cultivation and overloaded grazing [7,8]. This has made the farming–pastoral ecotone one of the most serious ecological problems in China. In China, there are currently 264 million hectares of grassland, with 70% of the area being degraded [9,10]. Since the late 1990s, the Chinese Government has carried out a variety of ecological protection projects and policies to alleviate the degradation of grassland ecosystems in the farming–pastoral ecotone. These include the 'Returning Ploughland to Forests and Grassland Project', the 'Beijing-Tianjin Wind and Sand Source Control Project', the 'Three-North Protective Forests Protection Project', and the 'Grassland Ecological Protection Subsidy and Incentive Policies' [11–13]. The implementation of these measures has partially changed the way grasslands are utilized and curbed the deterioration of the ecological environment in the farming–pastoral ecotone. This has mitigated the trend of grassland degradation and improved the ecological environment. However, despite efforts to reverse the trend of grassland degradation, grassland protection and restoration remain challenging tasks [14,15].

The demand for high-quality grass species is rising due to grassland protection and construction projects, leading to stricter selection criteria [16]. While exotic breeds are often introduced, they struggle to adapt to local climates, increasing restoration costs and invasion risks [17]. In contrast, native plants, being well-adapted to local conditions, are resilient, cost-effective, and easier to maintain, making them the preferred choice for ecological construction [18]. Particularly in fragile farming–pastoral ecotones, selecting suitable native grass species can significantly enhance ecological functions and support grassland restoration [19].

Leymus secalinus, a perennial herb of the *Poaceae* family, is mainly distributed in the western part of Northeast China, Hebei, Shanxi, Ningxia, Sichuan, Qinghai, Xinjiang, and other provinces. It is an excellent pasture grass, with drought resistance, saline–alkaline tolerance, trampling resistance, and strong adaptability, making it ideal for ecological restoration in farming–pastoral ecotones [20]. Its underground rhizomes are effective for wind-breaking, sand-fixing, and soil and water conservation [21,22], playing an important role in combating desertification, improving saline–alkaline land, and restoring the ecological environment [23,24]. Additionally, *L. secalinus* holds significant economic and ecological value [25] and is a valuable genetic resource for wheat improvement due to its stress resistance and genetic diversity [26,27]. Therefore, it is crucial to study its distribution and exploitation for the protection and restoration of grassland in the farming–pastoral ecotone.

In recent years, climate change, ecological deterioration, human interference, and the "low spiking, low fruiting, low germination" issues of *L. secalinus* have severely limited its practical application. Collecting and conserving *L. secalinus* germplasm resources are a vital material basis for livestock production, breeding, and ecological restoration. Systematic surveys and scientific sampling across its distribution area are needed to protect and utilize these resources effectively. Studies like FRPS and others [28–30] have detailed its distribution, while classifying potential areas based on environmental factors can guide conservation and planting zoning. Although research has explored the genomic, chemical, and biological control of *L. secalinus* [31–37], its ecological suitability, spatial distribution, and response to climate change in China have not been previously reported. Addressing these gaps will enhance understanding of its distribution and adaptability under future climate change scenarios.

Species distribution models (SDMs) [38] are a method for estimating the distribution of species in geospatial space based on the actual distribution of species and are a mathematical model based on species distribution data and environmental variables, which have been broadly applied to the prediction of species' potential areas [39,40] and can still give good results when species distribution data are not complete [41,42]. With the advance of mathematical modeling methods and geographic information systems (GIS), numerous species distribution models have been established, and the popular species distribution models include the Maximum Entropy (MaxEnt) model, Bioclimatic Model (BIOCLM), Genetic Algorithm for Rule-set Production (GARP), and Generalized Linear Model (GLM) [43]. Among them, the MaxEnt model is the most widespread [44–47]. Compared to other species distribution models, the MaxEnt model exhibits excellent forecasting ability and accuracy. It is particularly effective in situations where species distribution data are scarce (e.g., when distribution points are few or sample sizes are small), as it can quantitatively describe the potential distribution area of species through screening of major ecological factors, so as to realize the simulation of species distribution. Therefore, it has a wider application in predicting the potential distribution of invasive species and assessing the impact of climate change on species and has become a mainstream prediction model [48–52].

In this work, we simulated the potential distribution of *L. secalinus* in China by means of the MaxEnt model. Specifically, this work aims to (1) predict the potential geographic distribution of *L. secalinus* in China under current climatic conditions, (2) identify the key environmental factors affecting the distribution of *L. secalinus* and quantitatively describe the environmental conditions suitable for its survival, and (3) predict the natural trends and changes in the suitable habitats of *L. secalinus* in future climate scenarios and provide recommendations for *L. secalinus* management and use.

2. Materials and Methods

2.1. Study Area

L. secalinus is a perennial herbaceous plant of the family *Poaceae*, primarily distributed in Xinjiang, Gansu, Qinghai, Shaanxi, Sichuan, Nei Mongol, Hebei, Shanxi, and the northeast in China, with a wide range of habitats and strong adaptability. The geographical coordinates of its main natural distribution areas are 75° E~128.9° E and 29.4° N~49.08° N.

Our study area includes the farming–pastoral ecotone, which is traditionally the main agricultural area to the east and south, and the main grassland animal husbandry to the west and north. In the vicinity of the farming–pastoral ecotone, however, farming and grazing co-exist and intermingle.

2.2. Collection of Occurrence Data

Data on the geographic distribution of *L. secalinus* are mainly from field investigation data in 2022–2023 and relevant distribution information recorded in the literature, the latter included records from the Chinese Virtual Herbarium (CVH), and a total of 493 distribution points of *L. secalinus* were collected. To avoid model overfitting, distribution points were filtered using a threshold of 100 m. When the distance between multiple points was less than 100 m, one point was randomly retained while the others were removed. Finally, 357 distribution points of *L. secalinus* in China were obtained (Figure 1).

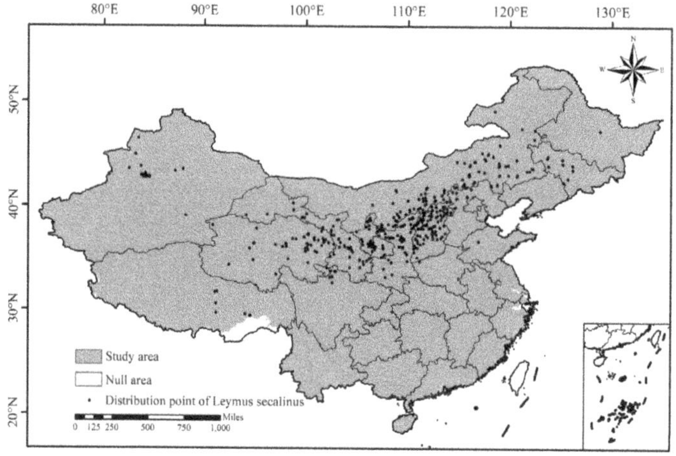

Figure 1. Distribution of *Leymus secalinus* in China.

Note: Taiwan Province and some other regions are not included in this study (the same as below).

2.3. Environmental Variables

This study used 52 environmental variables, including 19 bioclimatic factors, 30 soil factors, and 3 topographic factors (Table S1). Climate data were sourced from the World Climate Database (https://worldclim.org (accessed on 13 January 2024)). The environmental variables data include current (1970–2000) and future climate data (2021–2040, 2041–2060, 2061–2080, 2081–2100), with the future climate condition data containing data for three shared socioeconomic pathways (SSP1-26, SSP2-45, and SSP5-85) [53]. The 19 commonly used bioclimatic factors are denoted Bio1-Bio19. Soil factors were obtained from FAO SOILS PORTAL (https://www.fao.org/soils-portal/en/ (accessed on 13 January 2024)), and in this study, all the data above and below ground of the soil were separated to form a file in tiff format to ensure the spatial continuity of the data. The topographic factors include three factors: DEM (elevation), Slope, and Aspect. The Digital Elevation Model (DEM) data were obtained from the geospatial data cloud platform (https://www.gscloud.cn/ (accessed on 17 January 2024)), and Slope and Aspect data were obtained using the spatial analysis function of ArcGIS. The spatial resolution of all used environmental variables was transformed to 1 km^2.

Due to the varying degrees of correlation between environmental variables, it is essential to analyze and address multicollinearity to minimize its impact on predictive outcomes and the contributions of variables, thereby ensuring the model's precision and accuracy. We considered the importance of the variables obtained by the jackknife method, quantitatively evaluated the effect of environmental factors on the geographic distribution of *L. secalinus*, and tested the correlation of environmental factors using Pearson (Figure S1) [54], and environmental variables with high correlation ($r \geq 0.8$) and low contribution rates or 0 were removed [55–57]. Ultimately, 23 environmental factors were retained (Table 1).

Table 1. Contribution and permutation importance of environmental variables affecting the distribution of *L. secalinus*.

Variable	Percent Contribution/%	Permutation Importance/%
Bio2	27.1	16.8
Bio1	22.1	9.2
Bio16	18.8	22.6
DEM	7.3	12.6
Bio15	7.2	6.2
Bio8	3.5	3.9
Bio13	3.1	2.3
Slope	2.9	9
Bio3	2.9	9.4
Bio14	2.3	4.2
S_OC	0.4	0.4
S_CLAY	0.4	0.7
S_PH_H_2O	0.3	0.1
Aspect	0.3	0.8
S_TEB	0.3	0.4
S_GRAVEL	0.2	0.2
T_REF_BULK	0.2	0.3
T_GRAVEL	0.2	0.2
T_BS	0.1	0.2
T_SILT	0.1	0.2
S_CEC_CLAY	0.1	0
T_$CACO_3$	0	0.1
S_CEC_SOIL	0	0.2

2.4. Geographic Data Sources

The map data for this study are 1:1 million Chinese maps and administrative division maps downloaded from the National Basic Geographic Information Database (http://www.ngcc.cn/dlxxzy (accessed on 17 January 2024)) as the base map for analysis.

2.5. MaxEnt Model

The MaxEnt model is a sophisticated machine learning algorithm that emulates the probability of a species' existence from only the existing data and environmental variables, based on the principle of maximum entropy. This study employed MaxEnt version 3.4.1. Of the distribution point data, 75% was randomly selected as training data for model construction, while the other 25% was used as testing data for model validation. To reduce model uncertainty, the Bootstrapping method was used to perform 10 replicate runs for each subsample type, and the results were averaged, with the number of iterations set to 500 and background points set to 10,000. The jackknife test was employed to obtain the contribution percentage of environmental variables. Jackknife experiments were performed on all environmental variables to determine the main environmental factors influencing the distribution of *L. secalinus*. The MaxEnt model uses the ROC curve of the subject's work characteristics to measure model prediction accuracy, and the area enclosed by the ROC curve and the abscissa is known as the AUC value (ranging from 0.5 to 1.0), which is recognized as the optimal detection index of model accuracy [45,58]. The higher the AUC value, the better the simulation results. Based on the AUC value, model performance can be categorized as follows: 0.9–1, extremely accurate; 0.8–0.9, accurate; 0.7–0.8, moderately accurate; 0.6–0.7, general; and 0.5–0.6, failure [59].

2.6. Prediction of Suitable Area of *L. secalinus*

To more intuitively show the suitable habitat area and distribution characteristics, the model-predicted suitable habitat for *L. secalinus* was classified into four levels using Jenks natural breaks. The levels are as follows: In the high-suitability area (0.51–1), the distribution of *L. secalinus* is dominated by dominant species, and its habitat is especially suitable for *L. secalinus*. The middle-suitability area (0.26–0.51) has a large number of *L. secalinus* distribution, but *L. secalinus* is not a dominant species in the community. The low-suitability area (0.08–0.26) has a distribution of *L. secalinus*. The non-suitable area (0–0.08) is unsuitable for the growth of *L. secalinus*, and there may be no *L. secalinus* distribution.

3. Results and Analysis

3.1. Accuracy Evaluation of MaxEnt Model

The accuracy of the MaxEnt model's predictions for the potential suitable area of *L. secalinus* was evaluated using the ROC curve., and the ROC curve of the calculation results was obtained (Figure 2). It can be seen from Figure 3 that in this study, the average AUC of the data from the 10 trials' data was 0.931 with a standard deviation of 0.003. This shows that the prediction results of the model have high accuracy and reliability and can reasonably simulate the distribution of *L. secalinus* in China, which provides conditions for further research.

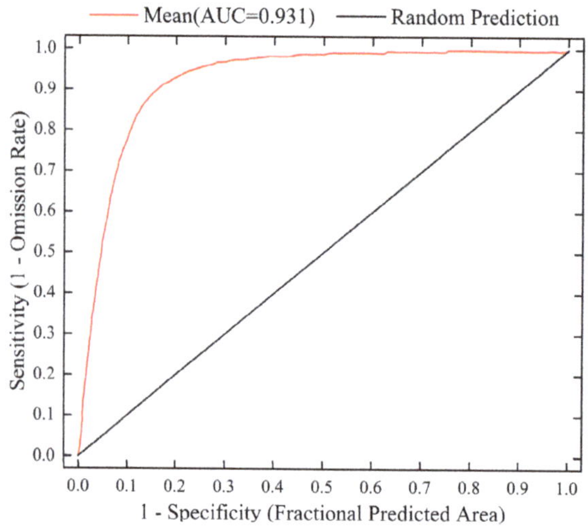

Figure 2. ROC curve of MaxEnt models for *L. secalinus*.

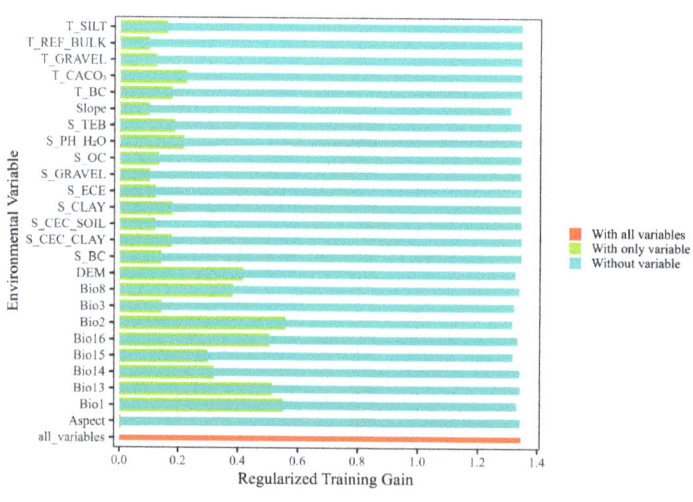

Figure 3. Regularized training gain of the MaxEnt model based on the jackknife test.

3.2. Analysis of Important Environmental Variables

Based on the contribution rate and permutation importance of environmental variables in the MaxEnt model (Table 1), the contribution rates of Bio2, Bio1, Bio16, DEM, and Bio15 for modeling were 27.1%, 22.1%, 18.8%, 7.3%, and 7.2%, respectively, and the cumulative contribution rate was 82.5%. This indicated that Bio2, Bio1, Bio16, DEM, and Bio15 were the major environmental factors influencing the potential suitable area of *L. secalinus.* Among the permutation importance of environmental variables, the permutation importance values of Bio16, Bio2, DEM, Bio3, and Bio1 were 22.6%, 16.8%, 12.6%, 9.4%, and 9.2%, respectively, with a cumulative importance of 70.6%, indicating that these five environmental variables play a key role in the modeling process. The results of the jackknife test (Figure 3) show that when only this factor is present, the five environmental factor variables that have the largest effect on the regularization training gain are Bio2, Bio1, Bio13, Bio16, and DEM. This shows that these environmental variables contain more useful information compared to others. The gain value of Aspect normalization training is the lowest, indicating that it has the least influence on the distribution of *L. secalinus.*

Looking at it from a holistic point of view, under the current climatic conditions, Bio2 (mean diurnal temperature range), Bio1 (annual mean temperature), Bio16 (precipitation of wettest quarter), and DEM (elevation) are the main environmental factors affecting the distribution of *L. secalinus.* The temperature factor had the greatest effect, followed by the precipitation factor, while topographic and soil factors have a relatively minor influence on the distribution of *L. secalinus.*

The species response curve reveals the relationship between environmental variables and the probability of species existence, and it shows biological tolerance to target species and habitat preferences [60]. It is generally believed that the occurrence probability threshold greater than 0.5 is regarded as the corresponding environment conducive to the growth of *L. secalinus* [61]. Based on response curves of four major climate variables (Figure 4), Bio1 of −3.3–11.6 °C, DEM of 160–3300 m, Bio16 of 90–400 mm, and Bio2 of 8.6–15 °C were suitable for the distribution of *L. secalinus.*

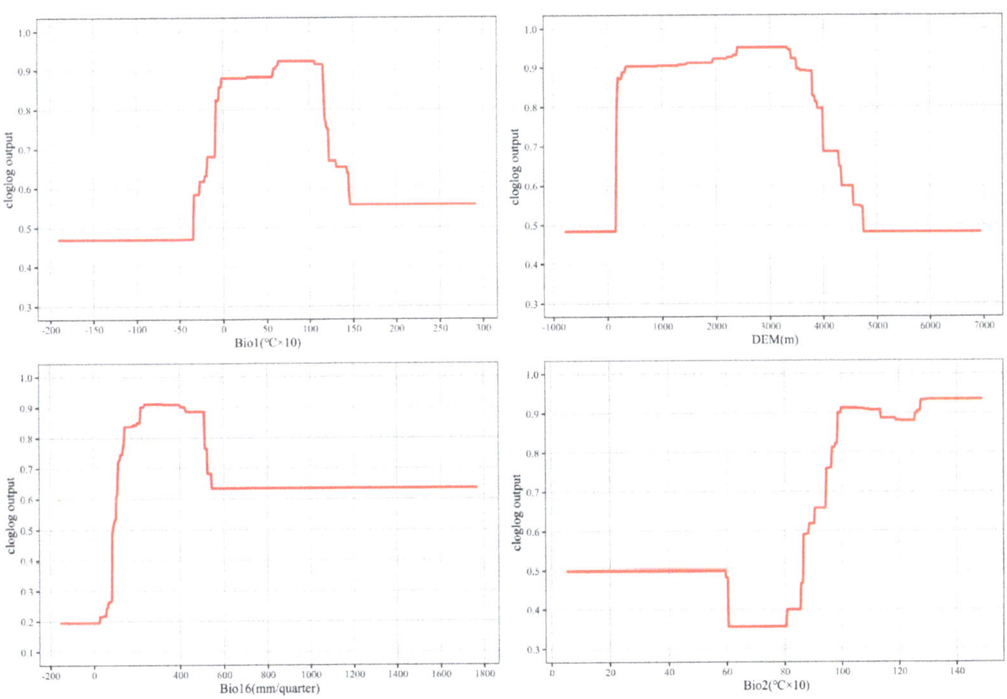

Figure 4. Response curve of main environmental variables in *L. secalinus* (Bio1, Bio2 units are °C × 10).

3.3. *The Suitable Distribution Area of L. secalinus in China Under the Current Climate Scenario*

Using the MaxEnt model for simulation, classification of "presence and absence" according to MTTS thresholds, and using ArcGIS to map a visual analysis chart. Under the current climatic conditions, the distribution of suitable areas of *L. secalinus* in China is shown in Figure 5. The red areas represent the high-suitability area of *L. secalinus*, the orange areas represent the medium-suitability area, the blue areas represent the low-suitability area, and the gray areas represent the non-suitable area. As shown in Figure 5, the potential suitable distribution areas of *L. secalinus* are located in 18 provinces and cities in Northwest China and North China, mainly concentrated in the range of the farming–pastoral ecotone.

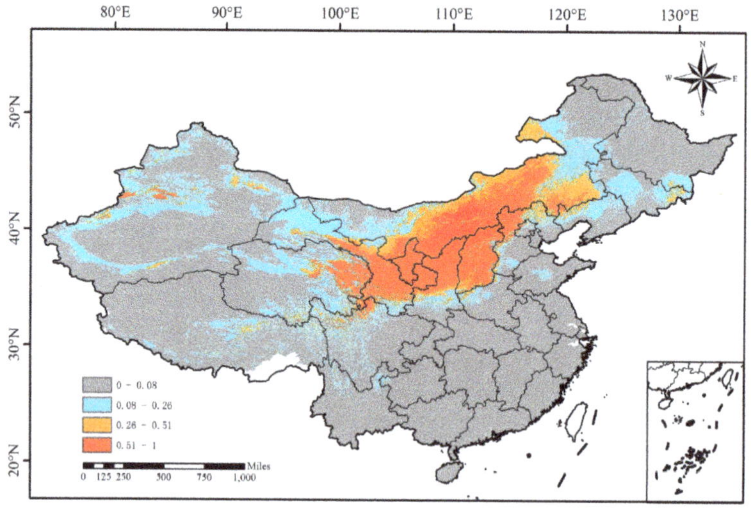

Figure 5. Predictions of the potentially suitable area of *L. secalinus* under current climate conditions based on the MaxEnt model.

The highly suitable areas of *L. secalinus* are primarily distributed in Shanxi Province, Inner Mongolia, Shaanxi, Ningxia, Gansu, eastern Qinghai, western and northeastern Xinjiang, northwestern Sichuan, northwestern Hebei, western Liaoning, and sporadically in central Tibet, western Henan, northwestern Beijing, and eastern Jilin, accounting for 8.89% of China's total area (Table 2).

Table 2. Potential suitable area of *L. secalinus* in China under future scenarios.

Scenario	Period	Non-Suitable Area	Lowly Suitable Area	Moderately Suitable Area	Highly Suitable Area	Suitable Area Change Ratio
	current	68.28	15.75	7.08	8.89	
SSP1-26	2021–2040	66.92	17.05	7.00	9.03	+4.29%
	2041–2060	68.58	16.57	6.36	8.49	−0.95%
	2061–2080	68.56	15.62	6.80	9.02	−0.88%
	2081–2100	68.22	15.76	7.60	8.42	+0.19%
SSP2-45	2021–2040	69.20	15.46	7.27	8.07	−2.90%
	2041–2060	68.70	15.31	6.87	9.12	−1.32%
	2061–2080	69.61	15.17	6.96	8.26	−4.19%
	2081–2100	68.34	16.49	7.09	8.08	−0.19%
SSP5-85	2021–2040	67.74	16.66	7.20	8.40	+1.70%
	2041–2060	68.47	17.04	5.89	8.60	−0.60%
	2061–2080	69.24	15.49	6.99	8.28	−3.03%
	2081–2100	67.18	17.37	6.65	8.80	+3.47%

The distribution range of the suitable area of *L. secalinus* is usually located around the high-suitability area, accounting for 7.08% of China's total area.

The low-suitability areas of *L. secalinus* are primarily located in parts of north-central and southern Xinjiang, southeastern Tibet, central and southern Qinghai, northwest and southern Gansu, western and eastern Inner Mongolia, the middle of Shaanxi, northwestern Henan, southwestern and Northeastern Hebei, the middle of Shandong, northern Liaoning, central and eastern Jilin, and western and southern Heilongjiang, with small portions located in northern Yunnan and western Guizhou, accounting for 15.75% of China's total area.

The other provinces and cities are non-suitable for planting *L. secalinus*, accounting for 68.28% of China's total area.

3.4. The Suitable Distribution Area of *L. secalinus* in China Under Future Climate Scenarios

This study is based on the three socioeconomic pathways proposed by the IPCC (SSP1-26, SSP2-45, SSP5-85); the MaxEnt model was used to predict the possible geographical distribution of *L. secalinus* in 2021–2040, 2041–2060, 2061–2080, and 2081–2100 (Figure 6). The results of this study found that the range of potential suitable areas of *L. secalinus* under current climatic conditions and different climatic scenarios in the future is basically the same as the range of the farming–pastoral ecotone. In addition, there are also suitable areas in Xinjiang, Tibet, Sichuan, Heilongjiang, and parts of Jilin. The overall behavior of the potentially suitable areas for *L. secalinus* in the three simulated scenarios was largely similar, with slight differences in the details of the predictions (Table 2). We observed that among all scenarios and years, SSP1-26 had the highest rate of expansion of suitable area in 2021–2040 and SSP5-85 in 2081–2100, with 4.29% and 3.47%, respectively (Figure 6(A1,C4)), and in all years of SSP2-45, the suitable areas were reduced compared to the current period, with 2061–2080 having the greatest shrinkage of potential suitable areas at 4.19% (Figure 6(B3)). In predicting potential changes in high- and medium-suitability areas, we also observed a significant decrease in potential suitability areas along the northern limit and an increase in eastern Inner Mongolia (including Xing'anmeng and southern Hulunbeier), southern Heilongjiang (including Mudanjiang and the western part of Jixi City), and eastern Jilin (Figure 7). A trend towards the northeast was present.

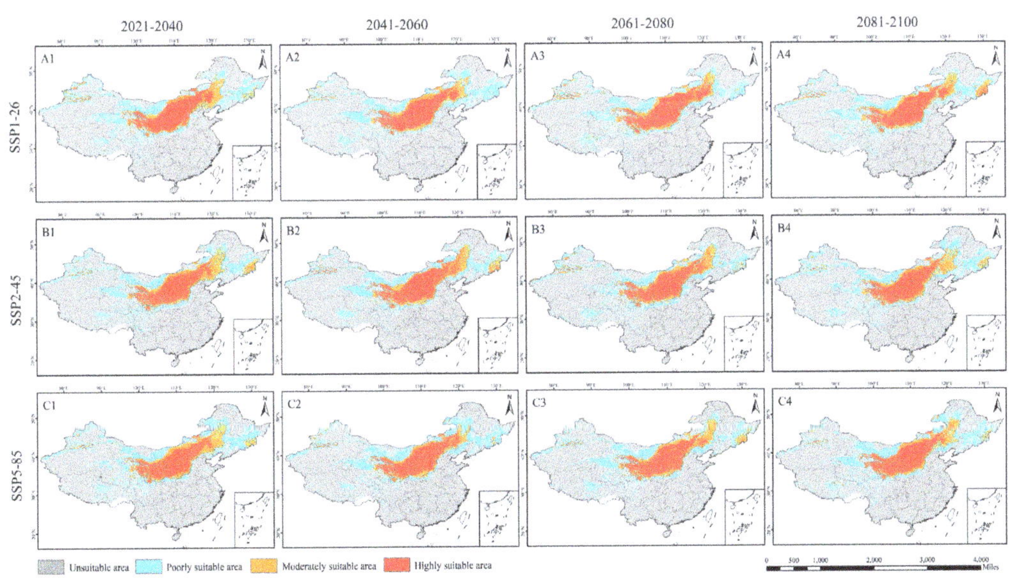

Figure 6. Potential distribution of *L. secalinus* under the SSP1-26 (**A1–A4**), SSP2-45 (**B1–B4**) and SSP5-85 (**C1–C4**).

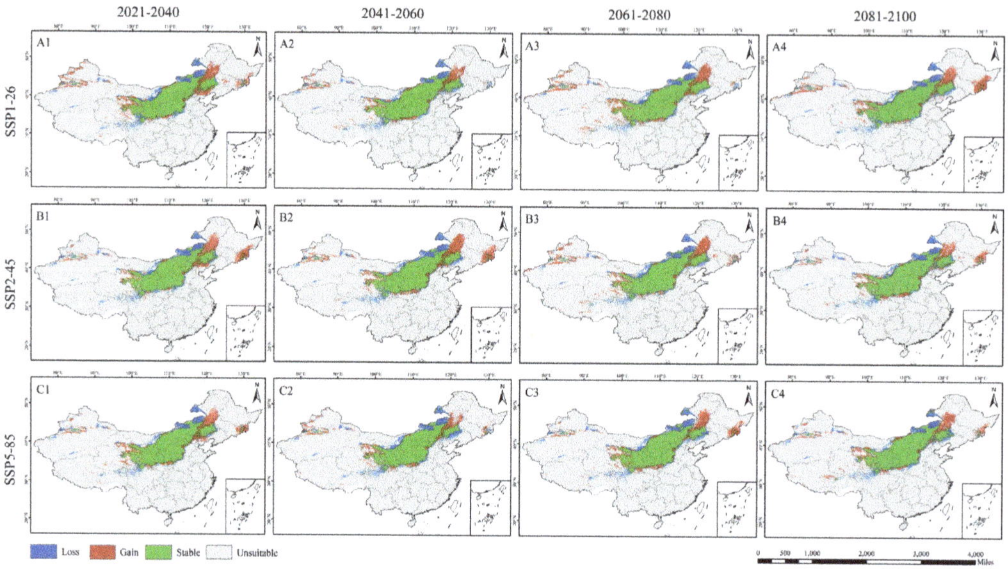

Figure 7. Potential changes in high- and medium-suitability areas for *L. secalinus* considering different SSPs. Changes in distribution highlighted in different colors; for gain (range expansion), red; loss (range contraction), blue; unsuitable, grey; and stable, green. (**A1–A4**) SSP1-26, (**B1–B4**) SSP2-45 and (**C1–C4**) SSP5-85.

4. Discussion

4.1. Discussion on MaxEnt Model to Predict the Current Distribution of L. secalinus

Despite differences in simulation results from different climate models, the MaxEnt model remains an essential research tool for evaluating and forecasting past and future changes in species' geographic distributions [62]. Based on the MaxEnt model combined with ArcGIS, this study predicted the potential distribution area of *L. secalinus*, providing an intuitive understanding of the current spatial distribution structure and the distribution of suitable areas in China. This research lays a basis for future studies on the geographic distribution of *L. secalinus*, as well as provides scientific evidence for future management and exploitation of *L. secalinus*. This study analyzed the percentage contribution of each ecological factor to the model and employed Pearson correlation coefficient analysis to address the collinear effect.

This study used the ROC curve to evaluate the species distribution model [63,64], running the MaxEnt model ten times with all data and taking the average AUC value from these runs to assess the model's accuracy [65,66]. It is generally believed that a higher AUC value suggests a closer alignment between the model's predicted species distribution and the actual distribution. The findings show that the average AUC value of the model is 0.931 with a standard deviation of 0.003, which proves that the prediction results are accurate and the model fitting is excellent, which can reflect the potential geographic distribution of *L. secalinus* in China extremely accurately.

The prediction results show that *L. secalinus* is primarily distributed in Shanxi, Inner Mongolia, Shaanxi, Ningxia, Gansu, Qinghai, Xinjiang, Sichuan, Hebei, and Northeast China, which is in line with the actual situation of the major production areas of *L. secalinus* in China. According to the relevant literature and the investigation report of *L. secalinus* resources in recent years [28,29], the comparison between the prediction results and the distribution status of *L. secalinus* in China shows that the existing distribution conforms to the prediction results and is mostly in the high-suitability area, which confirms the reliability of the MaxEnt model predictions.

The MaxEnt model, with its high predictive accuracy, flexibility, and reliance solely on species presence data, has been widely applied in the fields of ecological conservation and resource management. Additionally, MaxEnt can quantify the importance of environmental variables, assisting managers in identifying key driving factors and providing a scientific basis for optimizing management strategies. However, the application of MaxEnt also has certain limitations, such as potential uncertainties in predictions caused by data biases, uneven sampling distributions, or the omission of critical variables. Moreover, the selection of environmental variables in model construction requires comprehensive consideration of multiple factors, including climatic, ecological, biological, and anthropogenic influences. Therefore, as an efficient predictive tool, the outputs of MaxEnt should be integrated with actual ecological and social factors to enhance the scientific rigor and feasibility of management decisions.

4.2. Discussion on the Main Ecological Factors

Climate is a determining factor in the distribution of species, with changes in species distribution area being the most direct and clear reflection of climate change [67]. Climate, historical distribution, topography, soil, and other factors have significant effects on species distribution at different spatial scales. Among these factors, temperature and precipitation emerge as the predominant drivers affecting the growth of most plants [68]. Temperature factors affect species distribution by affecting plant photosynthesis and respiration, while precipitation conditions also constrain plant growth and development. Among the 23 environmental variables utilized for modeling in this study, the 4 key factors restricting the current potential suitable distribution of *L. secalinus* are mean diurnal temperature range (8.6–15 °C), annual mean temperature (−3.3–11.6 °C), precipitation of wettest quarter (90–400 mm), and elevation (160–3300 m), accounting for 75.3% of the total contribution rate. These factors reflect the temperature conditions, precipitation conditions, and topographic conditions required for the growth of *L. secalinus*. This is consistent with the habits of *L. secalinus*: *L. secalinus* is not only slightly fond of humidity, but also tolerant of cold and drought, and has a wide range of adaptation [69,70]. The increased frequency of extreme weather events in the future is expected to result in significant temperature fluctuations [71], whereas *L. secalinus* is suitable for living in habitats with cooler temperatures and is more widely distributed on the Qinghai–Tibet Plateau [72]. The climatic conditions of alpine grassland make *L. secalinus* seriously stressed by alpine hypoxia, intense solar radiation, and rapid climate change. The ability of *L. secalinus* to adapt to the alpine growth environment indicates that it has its own cold tolerance mechanisms. Large temperature fluctuations may lead to a reduction in or even loss of suitable areas, so proper precautions need to be taken.

In addition, precipitation is also a crucial variable affecting the establishment and survival of species, with higher precipitation generally being favorable for seedling growth; however, the connection between moisture availability and species distribution is contingent upon local conditions and the ecological demand of related species. Notably, excessive water in the soil reduces the root respiration rate, thus limiting plant growth. The response curves showed that the suitable area for *L. secalinus* increases up to a precipitation threshold of 400 mm in the wettest quarter. However, an increase in precipitation beyond this threshold results in a sharp decline in the suitable area. For instance, the main climatic factors affecting the distribution pattern of the recently reported *L. chinensis*, which is in the same genus as *L. chinensis* [73], are the precipitation of the coldest quarter, precipitation of the driest quarter, and precipitation of the driest month. The distribution value increases as cold-season precipitation rises. When precipitation during the cold season reaches 44.53 mm, the distribution value decreases with the increase in precipitation; precipitation during the driest quarter and driest month show the same pattern with their distribution values, which is the same as the results of the present study and both indicate the impact of precipitation on the establishment and survival of species and that the expansion or reduction of natural populations depends on future precipitation.

4.3. Discussion on the Suitable Area of Species Under SSP Model

A growing body of research has found that under the global climate change scenario, some species (*Acorus calamus* [74], *Pistacia chinensis* [75], *Phoebe sheareri* [76], *Leersia hexandra* [66]) will likely expand their distribution area, and for most species, their distribution area will decrease [47,60,77,78]. This study shows that *L. secalinus* suitable areas performed relatively consistently in different climate change scenarios, with generally similar distributional ranges. Projections of potential suitable areas showed a loss in all years under the SSP2-45 scenario, but only up to a maximum of 4.19% from the current (2061–2080), with the smallest range loss in 2081–2100; in contrast, the SSP1-26 (2021–2040) and SSP5-85 (2081–2100) scenarios had the largest range expansions (4.29% and 3.47%, respectively). Therefore, these projections suggest minimal overall changes in distribution patterns, indicating that *L. secalinus* has strong stress resistance to future climate change, has certain adaptability, and can survive and maintain its distribution to a greater extent. In addition, most of the high- and medium-suitability areas will not change considerably in the future and will be identical to the current period, with the exception of a partial shrinkage of the northern boundary. Previous studies have shown that under future climate scenarios, the extent of some species' suitable areas will not change considerably, and the overall pattern will be consistent with the current distribution [79,80]. These findings are consistent with the conclusions of this study that the ranges of some suitable areas will remain unchanged with climate change, indicating that these species are expected to retain the largest potential distribution area in the current and future climate change scenarios.

This study explains changes in the geographic distribution of suitable *L. secalinus* habitat by considering the expansion, contraction, and preservation of suitable areas. Under future climate change scenarios, the general trend of changes in the distribution of suitable *L. secalinus* is consistent. The areas that have been stably distributed include Shanxi, Shaanxi, Ningxia, Gansu, Inner Mongolia, Qinghai, and some areas of Hebei, indicating that these areas are conducive to growth and maintain the distribution of *L. secalinus*. The expanding areas are located in parts of Jilin and Heilongjiang in the northeast, as well as in parts of eastern Inner Mongolia. The shrinking area is located at the northern boundary of the current suitable area, and the geographic distribution changes are more significant, suggesting that *L. secalinus* is not adapted to this climatic environment and is sensitive to climate change, which may be attributed to the increase in precipitation and the rise in temperature in North China and Northwest China [81]. Measures should be taken to protect the existing *L. secalinus* resources in these areas to prevent the loss of valuable germplasm resources.

In conclusion, this study offers valuable information on the potential distribution of *L. secalinus* and provides a basis for conservation and management initiatives. It successfully predicted the potential distribution area of *L. secalinus* and revealed the main geographical distribution area of *L. secalinus* in northern China. A comprehensive analysis of various environmental factors identified the main climatic variables that affect species distributions. By considering the impact of different climate change scenarios on species distributions, it will be possible to develop more favorable and adaptive management strategies.

The shortcomings of this study are as follows: (i) the distribution modeling did not take into account non-environmental factors, such as light, wind direction, wind speed, human disturbance, biological interaction, etc.; (ii) *L. secalinus* is distributed in Russia, North Korea, and Japan, in addition to China, and the distribution points of *L. secalinus* obtained in this study were limited; and (iii) the projections under different shared socioeconomic pathways (SSPs) may not fully capture the complexity of future climate scenarios, and this study primarily focuses on short-term climate projections (up to 2100). Regrettably, we could not collect complete data on these factors, which should be taken into account in future studies to more fully consider the effects of non-environmental factors, socioeconomic changes, and longer time scales, integrating them into the distribution model, with more long-term and continuous scientific investigation and statistics, to further enhance the accuracy of the model and reduce the variability from the actual distribution area, to make better-informed decisions on managing and utilizing L. secalinus.

5. Conclusions

In this study, the MaxEnt model was applied to successfully predict the suitable distribution area of *L. secalinus* based on different environmental factors. Our results show that under the current climate scenario, *L. secalinus* is mainly distributed in the farming–pastoral ecotone of northern China, and the adaptation range is quite wide. The distribution of *L. secalinus* is mainly affected by the mean diurnal temperature range (Bio2), annual mean temperature (Bio1), precipitation of the wettest quarter (Bio16), and elevation (dem). Although the distribution area of *L. secalinus* will change slightly under different scenarios in the future, the overall distribution range is not expected to change significantly, and *L. secalinus* will migrate to the northeast in the future. The predicted results of this study can provide considerable reference value for the exploration and utilization of *L. secalinus* resources, help to understand the distribution law of *L. secalinus*, promote the protection and resource utilization of *L. secalinus*, promote the management work, and improve the ecological and economic value of *L. secalinus*.

Supplementary Materials: The following supporting information can be downloaded at https://www.mdpi.com/article/10.3390/plants14020293/s1: Table S1. A list of environmental variables. Figure S1. Spatial autocorrelation heat maps for (a) 19 contemporary climate data, (b) 19 future climate data, (c) 30 soil data.

Author Contributions: S.Z.: investigation, data curation, writing—original draft, writing—review and editing. Z.Z.: software, formal analysis, visualization. C.G.: visualization, writing—review and editing. Y.D.: investigation, data curation. Z.J.: investigation, data curation. L.D.: supervision, writing—review and editing. X.H.: conceptualization, supervision, funding acquisition, writing—review and editing. All authors have read and agreed to the published version of the manuscript.

Funding: This study was financially supported by the National Key R&D Plan Program (Grant No. 2022YFF1302803), the Shanxi Province Key R&D Plan Program (Grant No. 202102140601006), and the Shanxi Forage and Grass-industry Technology Research System (Grant No. 2023CYJSTX11).

Data Availability Statement: The original contributions presented in this study are included in the article/Supplementary Materials; further inquiries can be directed to the corresponding authors.

Conflicts of Interest: The authors declare no conflicts of interest.

References

1. Suttie, J.M.; Reynolds, S.G.; Batello, C. *Grasslands of the World*; Food and Agriculture Organization of the United Nations: Rome, Italy, 2005. [CrossRef]
2. O'Mara, F.P. The role of grasslands in food security and climate change. *Ann. Bot.* **2012**, *110*, 1263–1270. [CrossRef] [PubMed]
3. Wilsey, B.J. *The Biology of Grasslands*; Oxford University Press: Oxford, UK, 2018. [CrossRef]

4. Bardgett, R.D.; Bullock, J.M.; Lavorel, S.; Manning, P.; Schaffner, U.; Ostle, N.; Chomel, M.; Durigan, G.; LFry, E.; Johnson, I Combatting global grassland degradation. *Nat. Rev. Earth Environ.* **2021**, *2*, 720–735. [CrossRef]
5. Liu, H.L.; Lu, W.H.; Chen, C. Research Progress of Grassland Degraded Succession and Diagnosis. *Acta Agrestia Sin.* **201** *19*, 865–871. [CrossRef]
6. Huang, D.; Wang, K.; Wu, W.L. Problems and strategies for sustainable development of farming and animal husbandry in th Agro-Pastoral Transition Zone in Northern China (APTZNC). *Int. J. Sustain. Dev. World Ecol.* **2007**, *14*, 391–399. [CrossRef]
7. Xu, D.Y.; Li, C.L.; Song, X.; Ren, H.Y. The dynamics of desertification in the farming-pastoral region of North China over the pas 10years and their relationship to climate change and human activity. *Catena* **2014**, *123*, 11–22. [CrossRef]
8. Li, S.J.; Sun, Z.G.; Tan, M.H.; Guo, L.L.; Zhang, X.B. Changing patterns in farming–pastoral ecotones in China between 1990 an 2010. *Ecol. Indic.* **2018**, *89*, 110–117. [CrossRef]
9. Wang, S.Y. China Natural Resources Statistics Bulletin 2022. In *Resources Guide*; Ministry of Natural Resources of the People' Republic of China: Beijing, China, 2023; pp. 6–7.
10. Pan, Q.M.; Yang, Y.H.; Huang, J.H. Limiting Factors of Degraded Grassland Restoration in China and Related Basic Scientifi Issues. *Bull. Natl. Nat. Sci. Found. China* **2023**, *37*, 571–579. [CrossRef]
11. Wang, F.; Pan, X.B.; Wang, D.F.; Shen, C.Y.; Lu, Q. Combating desertification in China: Past, present and future. *Land Use Polic* **2013**, *31*, 311–313. [CrossRef]
12. Zhou, X.; Yamaguchi, Y.; Arjasakusuma, S. Distinguishing the vegetation dynamics induced by anthropogenic factors usin vegetation optical depth and AVHRR NDVI: A cross-border study on the Mongolian Plateau. *Sci. Total Environ.* **2018**, *616–61* 730–743. [CrossRef]
13. Peng, D.L.; Wu, C.Y.; Zhang, B.; Huete, A.; Zhang, X.Y.; Sun, R.; Lei, L.P.; Huang, W.J.; Liu, L.Y.; Liu, X.J.; et al. The Influences o Drought and Land-Cover Conversion on Inter-Annual Variation of NPP in the Three-North Shelterbelt Program Zone of Chin Based on MODIS Data. *PLoS ONE* **2017**, *11*, e0158173. [CrossRef] [PubMed]
14. He, J.S.; Liu, Z.P.; Yao, T.; Sun, S.C.; Lu, Z.; Hu, X.W.; Cao, G.M.; Wu, X.W.; Li, L.; Bu, H.Y.; et al. Analysis of the main constraint and restoration techniques of degraded grassland on the Tibetan Plateau. *Sci. Technol. Rev.* **2020**, *38*, 66–80. [CrossRef]
15. Wang, D.L.; Wang, L.; Xin, X.P.; Li, L.H.; Tang, H.J. Systematic Restoration for Degraded Grasslands: Concept, Mechanisms an Approaches. *Sci. Agric. Sin.* **2020**, *53*, 2532–2540. [CrossRef]
16. Mao, P.S.; Hou, L.Y.; Wang, M.Y. Limited factors and key technologies of forage seed production in the northern of China. *Chin Sci. Bull.* **2016**, *61*, 250–260. [CrossRef]
17. Wang, H.B.; Sun, J.; Yu, Y.X. The impact of biological invasion on biodiversity and grassland agroecosystem. *Pratacultural Sc* **2007**, *24*, 68–72. [CrossRef]
18. Nan, Z.B.; Yan, W.; Fu, H.; Guo, Z. *Stress Tolerance Biology of Native Plants*; Science Publications: New York, NY, USA, 2021.
19. Wang, L.X.; Shi, Y.L.; Zhang, H.W.; Bi, X.L.; Shen, W.M.; Ma, W.D. Analysis of Vegetation Ecological Function Changes and Driving Factors in Farming-pastoral Ecotone in Northern China from 2000 to 2020. *Ecol. Environ. Sci.* **2021**, *30*, 1990–1998 [CrossRef]
20. Ye, X.H.; Yu, F.H.; Dong, M. A trade-off between guerrilla and phalanx growth forms in Leymus secalinus under different nutrien supplies. *Ann. Bot.* **2006**, *98*, 187–191. [CrossRef] [PubMed]
21. Wang, B.; Zhang, G.H.; Yang, Y.F.; Li, P.P.; Liu, J.X. Response of soil detachment capacity to plant root and soil properties in typical grasslands on the Loess Plateau, Agriculture. *Ecosyst. Environ.* **2018**, *266*, 68–75. [CrossRef]
22. Wang, B.; Li, P.P.; Huang, C.H.; Liu, G.B.; Yang, Y.F. Effects of root morphological traits on soil detachment for ten herbaceous species in the Loess Plateau. *Sci. Total Environ.* **2020**, *754*, 142304. [CrossRef]
23. Kang, J.J.; Zhao, W.Z.; Zhao, M. Remediation of blowouts by clonal plants in Maqu degraded alpine grasslands of northwest China. *J. Plant Res.* **2017**, *130*, 291–299. [CrossRef] [PubMed]
24. Jian, T.K.; Xia, Y.; He, R.P.; Zhang, J. The influence of planting Carex praeclara and Leymus secalinus on soil properties and microbial community in a Zoige desertified alpine grassland. *Glob. Ecol. Conserv.* **2022**, *34*, e02002. [CrossRef]
25. Shen, H.; Dong, S.K.; Xiao, J.N.; Zhi, Y.L. Short-term warming and N deposition alter the photosynthetic pigments trade-off in leaves of Leymus secalinus growing in different alpine grassland habitats on Qinghai-Tibetan plateau. *Environ. Sci. Pollut. Res* **2023**, *30*, 15282–15292. [CrossRef] [PubMed]
26. Liu, Z.P.; Chen, Z.Y.; Pan, J.; Li, X.F.; Su, M.; Wang, L.J.; Li, H.J.; Liu, G.S. Phylogenetic relationships in Leymus (Poaceae: Triticeae) revealed by the nuclear ribosomal internal transcribed spacer and chloroplast trnL-F sequences. *Mol. Phylogenetics Evol.* **2008**, *46*, 278–289. [CrossRef] [PubMed]
27. Pradheep, K.; Singh, M.; Sultan, S.M.; Singh, K.; Parimalan, R.; Ahlawat, S.P. Diversity in wild relatives of wheat: An expedition collection from cold-arid Indian Himalayas. *Genet. Resour. Crop Evol.* **2019**, *66*, 275–285. [CrossRef]
28. Liu, Y.P.; Liu, T.; Lü, T.; Zhou, Y.H.; Zhang, X.Y.; Su, X. Geographical distribution, origin, and dispersal of Leymus Hochst (Poaceae). *Plant Sci. J.* **2017**, *35*, 305–317. [CrossRef]
29. Li, Z.M. *Studies on the Systematics of Domestic Leymusand Plant Geography*; Qinghai Normal University: Qinghai, China, 2016.

30. Zhi, L.; Teng, Z.H. Classification and geographical distribution of Leymus in China. *Bull. Bot. Res.* **2005**, *25*, 22. [CrossRef]
31. Zhu, Y.J.; Dong, M.; Huang, Z.Y. Caryopsis germination and seedling emergence in an inland dune dominant grass Leymus secalinus. *Flora—Morphol. Distrib. Funct. Ecol. Plants* **2007**, *202*, 249–257. [CrossRef]
32. Li, J.L.; Li, H.M.; Yin, N.; Quan, X.Y.; Wang, W.B.; Shan, Q.L.; Wang, S.Q.; Bermudez, R.S.; He, W.X. Identification of LsPIN1 gene and its potential functions in rhizome turning of Leymus secalinus. *BMC Genom.* **2022**, *23*, 753. [CrossRef] [PubMed]
33. Zhu, Y.J.; Yang, X.J.; Baskin, C.C.; Baskin, J.M.; Dong, M.; Huang, Z.Y. Effects of amount and frequency of precipitation and sand burial on seed germination, seedling emergence and survival of the dune grass Leymus secalinus in semiarid China. *Plant Soil* **2014**, *374*, 399–409. [CrossRef]
34. Zhu, Y.J.; Dong, M.; Huang, Z.Y. Response of Seed Germination and Seedling Growth to Sand Burial of Two Dominant Perennial Grasses in Mu-Us Sandy Grassland, Semiarid China. *Rangel. Ecol. Manag.* **2009**, *62*, 337–344. [CrossRef]
35. Wang, H.; Xie, Y.S.; She, X.Y.; Jiang, Q.L.; Zhang, Y.L.; Yang, Y.L.; Chuai, J.F. Autotoxocity and Allelopathic effects of Artemisia sacrorum L. on dominant spp (Stipa bungeana, Stipa grandis, Thymus mongolicus and Leymus secalinus) of steppe from Loess Plateau, China. *Allelopath. J.* **2016**, *37*, 41–54.
36. Li, J.; Hou, F.J.; Ren, J.Z. Grazing Intensity Alters Leaf and Spike Photosynthesis, Transpiration, and Related Parameters of Three Grass Species on an Alpine Steppe in the Qilian Mountains. *Plants* **2021**, *10*, 294. [CrossRef]
37. Xia, Y.; He, R.P.; Xu, W.R.; Zhang, J. The Zoige pioneer plant Leymus secalinus has different endophytic bacterial community structures to adapt to environmental conditions. *Peerj* **2023**, *11*, e15363. [CrossRef] [PubMed]
38. Guo, Y.L.; Wei, H.Y.; Lu, C.Y.; Gao, B.; Gu, W. Predictions of potential geographical distribution and quality of Schisandra sphenanthera under climate change. *PeerJ* **2016**, *4*, e2554. [CrossRef]
39. Luciano, B.; Danilo, R.; Mirko, D.F.; Gennaro, C.; Astolfo, Z. Potential distribution of Xylella fastidiosa in Italy: A maximum entropy model. *Phytopathol. Mediterr.* **2016**, *55*, 62–72. [CrossRef]
40. Zhou, G.S. Adaptation and Vulnerability of Vegetation/Terrestrial Ecosystems to Climate Change in China. *Meteorol. Mon.* **2015**, *41*, 1048.
41. Chen, X.M.; Lei, Y.C.; Zhang, X.Q.; Jia, H.Y. Effects of Sample Sizes on Accuracy and Stability of Maximum Entropy Model in Predicting Species Distribution. *Sci. Silvae Sin.* **2012**, *48*, 53–59. [CrossRef]
42. Phillips, S.J.; Anderson, R.P.; Schapire, R.E. Maximum entropy modeling of species geographic distributions. *Ecol. Model.* **2006**, *190*, 231–259. [CrossRef]
43. Zhao, J.C.; Wang, C.; Shi, X.Y.; Bo, X.Z.; Li, S.; Shang, M.F.; Chen, F.; Chu, Q.Q. Modeling climatically suitable areas for soybean and their shifts across China. *Agric. Syst.* **2021**, *192*, 103205. [CrossRef]
44. Xu, L.; Fan, Y.; Zheng, J.H.; Guan, J.Y.; Lin, J.; Wu, J.G.; Liu, L.; Wu, R.; Liu, Y.J. Impacts of climate change and human activity on the potential distribution of Aconitum leucostomum in China. *Sci. Total Environ.* **2023**, *912*, 168829. [CrossRef]
45. Zhang, K.L.; Yao, L.J.; Meng, J.S.; Tao, J. Maxent modeling for predicting the potential geographical distribution of two peony species under climate change. *Sci. Total Environ.* **2018**, *634*, 1326–1334. [CrossRef] [PubMed]
46. Qiu, L.; Jacquemyn, H.; Burgess, K.S.; Zhang, L.G.; Zhou, Y.D.; Yang, B.Y.; Tan, S.L. Contrasting range changes of terrestrial orchids under future climate change in China. *Sci. Total Environ.* **2023**, *895*, 165128. [CrossRef] [PubMed]
47. YWei, Q.; Zhang, L.; Wang, J.N.; Wang, W.W.; Niyati, N.; Guo, Y.L.; Wang, X.F. Chinese caterpillar fungus (Ophiocordyceps sinensis) in China: Current distribution, trading, and futures under climate change and overexploitation. *Sci. Total Environ.* **2021**, *755*, 142548. [CrossRef]
48. Shabani, F.; Tehrany, M.S.; Solhjouy-fard, S.; Kumar, L. A comparative modeling study on non-climatic and climatic risk assessment on Asian Tiger Mosquito (*Aedes albopictus*). *PeerJ* **2018**, *6*, e4474. [CrossRef] [PubMed]
49. Zafirah, A.L.N.; Zulkifli, Y.; Mazlan, H.; Jamilah, M.S.; Shinya, N. Predicting the Habitat Suitability of Melaleuca cajuputi Based on the MaxEnt Species Distribution Model. *Forests* **2021**, *12*, 1449. [CrossRef]
50. Kong, F.; Tang, L.; He, H.; Yang, F.X.; Tao, J.; Wang, W.C. Assessing the impact of climate change on the distribution of Osmanthus fragrans using Maxent. *Environ. Sci. Pollut. Res. Int.* **2021**, *28*, 34655–34663. [CrossRef] [PubMed]
51. Yan, H.Y.; Feng, L.; Zhao, Y.F.; Feng, L.; Wu, D.; Zhu, C.P. Prediction of the spatial distribution of Alternanthera philoxeroides in China based on ArcGIS and MaxEnt. *Glob. Ecol. Conserv.* **2020**, *21*, e00856. [CrossRef]
52. Anibaba, Q.A.; Dyderski, M.K.; Jagodziński, A.M. Predicted range shifts of invasive giant hogweed (Heracleum mantegazzianum) in Europe. *Sci. Total Environ.* **2022**, *825*, 154053. [CrossRef]
53. Thrasher, B.; Wang, W.; Michaelis, A.; Melton, F.; Lee, T.; Nemani, R. NASA Global Daily Downscaled Projections, CMIP6. *Sci. Data* **2022**, *9*, 262. [CrossRef] [PubMed]
54. Cui, X.Y.; Wang, W.J.; Yang, X.Q.; Li, S.; Qin, S.Y.; Rong, J. Potential distribution of wild Camellia oleifera based on ecological niche modeling. *Biodiv. Sci.* **2016**, *24*, 1117–1128. [CrossRef]
55. Che, L.; Cao, B.; Bai, C.K.; Wang, J.J.; Zhang, L.L. Predictive distribution and habitat suitability assessment of Notholirion bulbuliferum based on MaxEnt and ArcGIS. *Chin. J. Ecol.* **2014**, *33*, 1623–1628. [CrossRef]

56. Shi, F.N.; Liu, S.L.; An, Y.; Sun, Y.X.; Zhao, S.; Liu, Y.X.; Li, M.Q. Climatic factors and human disturbance influence ungulat species distribution on the Qinghai-Tibet Plateau. *Sci. Total Environ.* **2023**, *869*, 161681. [CrossRef] [PubMed]
57. Lemke, D.; Hulme, P.E.; Brown, J.A.; Tadesse, W. Distribution modelling of Japanese honeysuckle (Lonicera japonica) invasion in the Cumberland Plateau and Mountain Region, USA. *For. Ecol. Manag.* **2011**, *262*, 139–149. [CrossRef]
58. Lobo, J.M.; Jiménez-Valverde, A.; Real, R. AUC: A misleading measure of the performance of predictive distribution model. *Glob. Ecol. Biogeogr.* **2008**, *17*, 145–151. [CrossRef]
59. Swets, J.A. Measuring the accuracy of diagnostic systems. *Science* **1988**, *240*, 1285–1293. [CrossRef]
60. Soilhi, Z.; Sayari, N.; Benalouache, N.; Mekki, M. Predicting current and future distributions of Mentha pulegium L. in Tunisia under climate change conditions, using the MaxEnt model. *Ecol. Inform.* **2022**, *68*, 101533. [CrossRef]
61. Hu, W.; Wang, Y.; Zhang, D.; Yu, W.; Chen, G.; Xie, T.; Liu, Z.; Ma, Z.; Du, J.; Chao, B.; et al. Mapping the potential of mangrove forest restoration based on species distribution models: A case study in China. *Sci. Total Environ.* **2020**, *748*, 142321. [CrossRef] [PubMed]
62. Liu, L.; Guan, L.L.; Zhao, H.X.; Huang, Y.; Mou, Q.Y.; Liu, K.; Chen, T.T.; Wang, X.Y.; Zhang, Y.; Wei, B.; et al. Modeling habitat suitability of Houttuynia cordata Thunb (Ceercao) using MaxEnt under climate change in China. *Ecol. Inform.* **2021**, *63*, 101324. [CrossRef]
63. Han, Y.Y.; Wang, Y.; Xiang, Y.; Ye, J.R. Prediction of potential distribution of Bursaphelenchus xylophilus in China based on Maxent ecological niche model. *J. Nanjing For. Univ. (Nat. Sci. Ed.)* **2015**, *39*, 6–10. [CrossRef]
64. Wang, C.; Hawthorne, D.; Qin, Y.J.; Pan, X.B.; Li, Z.H.; Zhu, S.F. Impact of climate and host availability on future distribution of Colorado potato beetle. *Sci. Rep.* **2017**, *7*, 4489. [CrossRef] [PubMed]
65. Zhao, Y.; Deng, X.W.; Xiang, W.H.; Chen, L.; Ouyang, S. Predicting potential suitable habitats of Chinese fir under current and future climatic scenarios based on Maxent model. *Ecol. Inform.* **2021**, *64*, 101393. [CrossRef]
66. Cao, B.; Bai, C.K.; Xue, Y.; Yang, J.J.; Gao, P.F.; Liang, H.; Zhang, L.L.; Che, L.; Wang, J.J.; Xu, J.; et al. Wetlands rise and fall: Six endangered wetland species showed different patterns of habitat shift under future climate change. *Sci. Total Environ.* **2020**, *731*, 138518. [CrossRef] [PubMed]
67. Du, Z.Y.; He, Y.M.; Wang, H.T.; Wang, C.; Duan, Y.Z. Potential geographical distribution and habitat shift of the genus Ammopiptanthus in China under current and future climate change based on the MaxEnt model. *J. Arid Environ.* **2021**, *184* 104328. [CrossRef]
68. Jiang, H.L.; Xu, X.; Guan, M.X.; Wang, L.F.; Huang, Y.M.; Jiang, Y. Determining the contributions of climate change and human activities to vegetation dynamics in agro-pastural transitional zone of northern China from 2000 to 2015. *Sci. Total Environ.* **2020**, *718*, 134871. [CrossRef] [PubMed]
69. Yi, J.; Gu, A.L.; Jia, G.H.; Wu, X. Studies on the Drought Hardiness in Seedling of Leymus Hochst. *Stud. Drought Hardiness Seedl Leymus Hochst* **2001**, *15*, 47–50.
70. Zhao, D.M. *Effect of Precipitation Variability on Soil Respiration and Net Primary Productivity of Vegetation on a Semiarid Grassland in the Loess Plateau of China*; Lanzhou University: Lanzhou, China, 2016.
71. Patricola, C.M.; Ping, C.; Saravanan, R. Impact of Atlantic SST and high frequency atmospheric variability on the 1993 and 2008 Midwest floods: Regional climate model simulations of extreme climate events. *Clim. Chang.* **2015**, *129*, 397–411. [CrossRef]
72. He, W.X.; Yang, Z.R.; Cao, Y.; Chen, F. Effects of severied rhizome on clonal growth of Leymus secalinus and Carex praeclara of alpine desertification grassland in Northwestern Sichuan. *Chin. J. Ecol.* **2005**, *6*, 607–612.
73. Chen, J.S.; Liu, J.L.; Zhu, R.F.; Li, J.K.; Di, G.L.; Zhang, Q.; Mao, D.J.; Kong, X.L. Analysis of Suitable Bioclimatic Characteristics of Leymus chinensis Using MaxEnt Model. *Acta Agrestia Sin.* **2019**, *27*, 35–42. [CrossRef]
74. Yue, C.; Li, H.; Shi, X. Geographical Distribution Dynamics of Acorus calamus in China Under Climate Change. *Plants* **2024**, *13*, 3352. [CrossRef] [PubMed]
75. Guo, Y.L.; Zhao, Z.F.; Zhu, F.X.; Gao, B. The impact of global warming on the potential suitable planting area of Pistacia chinensis is limited. *Sci. Total Environ.* **2023**, *864*, 161007. [CrossRef] [PubMed]
76. Wang, Y.; Ma, X.H.; Lu, Y.F.; Hu, X.G.; Lou, L.H.; Tong, Z.K.; Zhang, J.H. Assessing the current genetic structure of 21 remnant populations and predicting the impacts of climate change on the geographic distribution of Phoebe sheareri in southern China. *Sci. Total Environ.* **2022**, *846*, 157391. [CrossRef] [PubMed]
77. Shi, J.; Xia, M.; He, G.; Gonzalez, N.C.T.; Zhou, S.; Lan, K.; Ouyang, L.; Shen, X.; Jiang, X.; Cao, F.; et al. Predicting Quercus gilva distribution dynamics and its response to climate change induced by GHGs emission through MaxEnt modeling. *J. Environ. Manag.* **2024**, *357*, 120841. [CrossRef]
78. Puchałka, R.; Paź-Dyderska, S.; Woziwoda, B.; Dyderski, M.K. Climate change will cause climatic niche contraction of Vaccinium myrtillus L. and V. *vitis-idaea L. Europe. Sci. Total Environ.* **2023**, *892*, 164483. [CrossRef] [PubMed]
79. Ngarega, B.K.; Masocha, V.F.; Schneider, H. Forecasting the effects of bioclimatic characteristics and climate change on the potential distribution of Colophospermum mopane in southern Africa using Maximum Entropy (Maxent). *Ecol. Inform.* **2021**, *65*, 101419. [CrossRef]

▪. Cao, Z.; Zhang, L.; Zhang, X.X.; Guo, Z.J. Predicting the Potential Distribution of Hylomecon japonica in China under Current and Future Climate Change Based on Maxent Model. *Sustainability* **2021**, *13*, 11253. [CrossRef]

.. Fu, Y.H.; Jiang, D.B. Climate Change over China with a 2 °C Global Warming. *Chin. J. Atmos. Sci.* **2011**, *10*, 212. [CrossRef]

plants

MDPI

Article

Central Asia Cold Case: Siberian Pine Fingers New Suspects in Growth Decline CA 1700 CE

David M. Meko [1], Dina F. Zhirnova [2,3], Liliana V. Belokopytova [2,3,*], Yulia A. Kholdaenko [2,3], Elena A. Babushkina [2,3], Nariman B. Mapitov [4] and Eugene A. Vaganov [3,5]

[1] Laboratory of Tree-Ring Research, University of Arizona, 1215 E. Lowell Street, Tucson, AZ 85721-0045, USA; dmeko@arizona.edu
[2] Khakass Technical Institute, Siberian Federal University, 27 Schetinkina Street, Abakan 655017, Russia; dina-zhirnova@mail.ru (D.F.Z.); kropacheva_yulechka@mail.ru (Y.A.K.); babushkina70@mail.ru (E.A.B.)
[3] Institute of Ecology and Geography, Siberian Federal University, 79 Svobodny Pr., Krasnoyarsk 660041, Russia; eavaganov@hotmail.com
[4] Department of Biology and Ecology, Toraighyrov University, 64 Lomov Street, Pavlodar 140008, Kazakhstan; mapitov@mail.ru
[5] Department of Dendroecology, V.N. Sukachev Institute of Forest, Siberian Branch of the Russian Academy of Science, 50/28 Akademgorodok, Krasnoyarsk 660036, Russia
* Correspondence: white_lili@mail.ru

Abstract: Tree-ring width chronologies of *Pinus sibirica* Du Tour from near the upper treeline in the Western Sayan, Southern Siberia are found to have an exceptional (below mean–3SD) multi-year drop near 1700 CE, highlighted by the seven narrowest-ring years in a 1524–2022 regional chronology occurring in the short span of one decade. Tree rings are sometimes applied to reconstruct seasonal air temperatures; therefore, it is important to identify other factors that may have contributed to the growth suppression. The spatiotemporal scope of the "nosedive" in tree growth is investigated with a large network of *P. sibirica* (14 sites) and *Larix sibirica* Ledeb. (61 sites) chronologies, as well as with existing climatic reconstructions, natural archives, documentary evidence (e.g., earthquake records), and climate maps based on 20th-century reanalysis data. We conclude that stress from low summer temperatures in the Little Ice Age was likely exacerbated by tree damage associated with weather extremes, including infamous Mongolian "dzuds", over 1695–1704. A tropical volcanic eruption in 1695 is proposed as the root cause of these disturbances through atmospheric circulation changes, possibly an amplified Scandinavia Northern Hemisphere teleconnection pattern. Conifer tree rings and forest productivity recorded this event across all of Altai–Sayan region.

Keywords: conifers; tree-ring width; severe growth suppression; abiotic factors; stress event

1. Introduction

Tree rings are a well-known and widespread proxy for seasonal climate reconstructions [1–3]. However, growth suppression or release in trees can also be caused by abrupt events—pest infestations, earthquakes, volcanic eruptions, landslides, floods, fires, etc. [4–10]. Such events may or may not be causally linked to anomalies of seasonal climate. For reliable climate reconstructions, we must distinguish as much as possible growth departures driven by the reconstructed variable from those due to other factors. Case studies of exceptional fluctuations in climate proxies through comparison with documentary evidence and other natural archives recording environmental conditions can pave the way in flagging possible inaccuracies in the climatic reconstructions.

Recently, we documented an exceptional decadal decline, or 'nosedive', in tree-ring width (TRW) and wood anatomical parameters of Siberian pine (*Pinus sibirica* Du Tour) throughout the Western Sayan highlands, beginning ca 1700 CE [11,12]. The finding spurred us to investigate whether the phenomenon was limited to *P. sibirica* and certain habitats, and whether a seasonal climate (temperature) anomaly or some other natural extreme or event was responsible. Tree-ring chronologies from continental temperate Asia have been widely used to reconstruct temperatures [1,13–16]. Some chronologies in these studies have an abrupt drop in growth around 1700, but the feature went unnoticed, not standing out against the background of other fluctuations, such as a long-term growth decrease during the 19th century (end of Little Ice Age) and the subsequent faster growth due to contemporary warming [1,13,15]. The near-1700 nosedive in *P. sibirica* growth is too sharp, however, to assume that it was driven by seasonal temperature fluctuation alone, so we hypothesized that there was some other stressful event that caused or enhanced this growth suppression.

This study explores possible alternative explanations for the near-1700 nosedive in *P. sibirica* growth. Growth fluctuations in *P. sibirica and* Siberian larch (*Larix sibirica* Ledeb.), another climate-sensitive and co-existing conifer frequently used for reconstructions [17], are studied in the context of natural factors and events that could plausibly have negatively impacted tree growth in the western Sayan Mountains near 1700 across the eastern part of Greater Central Asia.

2. Results

Since this phenomenon occurred several centuries ago, the lack of direct instrumental data made us investigate its possible causes through circumstantial evidence. This evidence is in the form of natural proxies recording environmental fluctuations (tree rings, lake sediments, glaciers' ice cores etc.), climatic reconstructions based on those proxies, and historical chronicles. Here we assess the near-1700 CE anomaly in our chronologies of *P. sibirica* tree-ring width in the context of this available evidence.

2.1. Fingerprint of Event

The primary tree-ring data for our analysis consist of 14 pine (*P. sibirica*) and 61 larch (*L. sibirica*) chronologies with time coverage back to at least 1690 CE available in a study area (40–60° N 60–120° E), centered on the Western Sayan Mountains (Figure 1a; Table A1, Figure A1). We defined the "nosedive" as the sharp multi-year suppression of growth starting ca 1700 CE. The nosedive was evident in 10 of the pine and 19 of the larch chronologies, and most seriously expressed (lowest values of Z-scores) in four near-timberline Western Sayan *P. sibirica* chronologies, some of which were previously shown as sensitive to temperature [11], A regional chronology PISIreg, developed from these four sites and containing more than 25 individual tree series covering 1690, is therefore used here to characterize the ca-1700 nosedive. The seven lowest values of PISIreg in the 499 years of its coverage (1524–2022 CE) are the consecutive years 1699–1705, and in six of these seven years, the Z-score of PISIreg is below -3 (Figure 1b). A less pronounced growth suppression is also present in several *P. sibirica* chronologies from the Altai Mountains and the Mongolian Plateau.

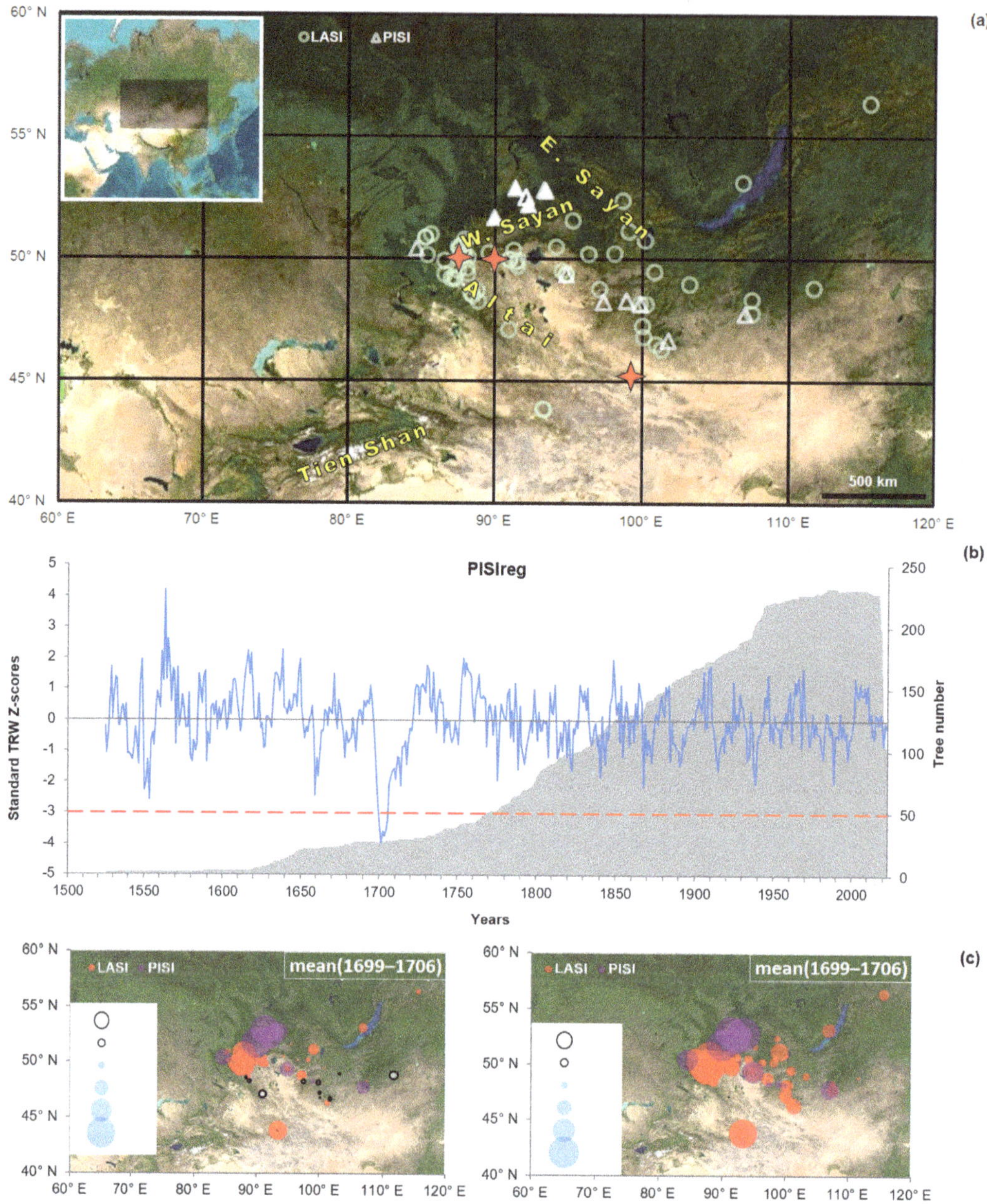

Figure 1. Growth suppression in tree-ring width of *Pinus sibirica* Du Tour (PISI) and *Larix sibirica* Ledeb. (LASI). (**a**) Map of the study area with marked locations of PISI (triangles) and LASI (circles) sampling sites, and epicenters of the strongest earthquakes (asterisks). Four filled triangles represent PISI chronologies with the most severe growth suppression used to develop regional chronology PISIreg. Insert map shows the location of the study area in Asia; (**b**) Regional chronology PISIreg, 1524–2022, as Z-scores. Red line represents the Z = −3.0 threshold of growth suppression; grey shading represents sample depth; (**c**) Geographical distribution of mean (left) and minimum (right) Z-score values for all 14 PISI and 61 LASI chronologies. Colored circles represent negative Z-scores for PISI (purple) and LASI (red); white circles represent positive Z-scores; circle diameters (see legend) represent absolute value.

The nosedive is characterized in pine and larch tree rings of individual trees (selected from chronologies where it occurred) as an abrupt initial drop in growth in 1698–1699 CE, followed by several years of very low growth and a gradual recovery toward normal growth (Figure 2a,b). The entire episode covered roughly 15 years in pine and 8 years in larch. Years with the largest departures were similar for pine (1699–1706) and larch (1699, 1700, 1702). For both species, growth suppression was accentuated at higher elevations. It is also the most expressed growth departure common for those trees since 1500, especially for pine; the only other deviation of similar magnitude and duration is the recent positive trend in larch growth (Figure 2c,d).

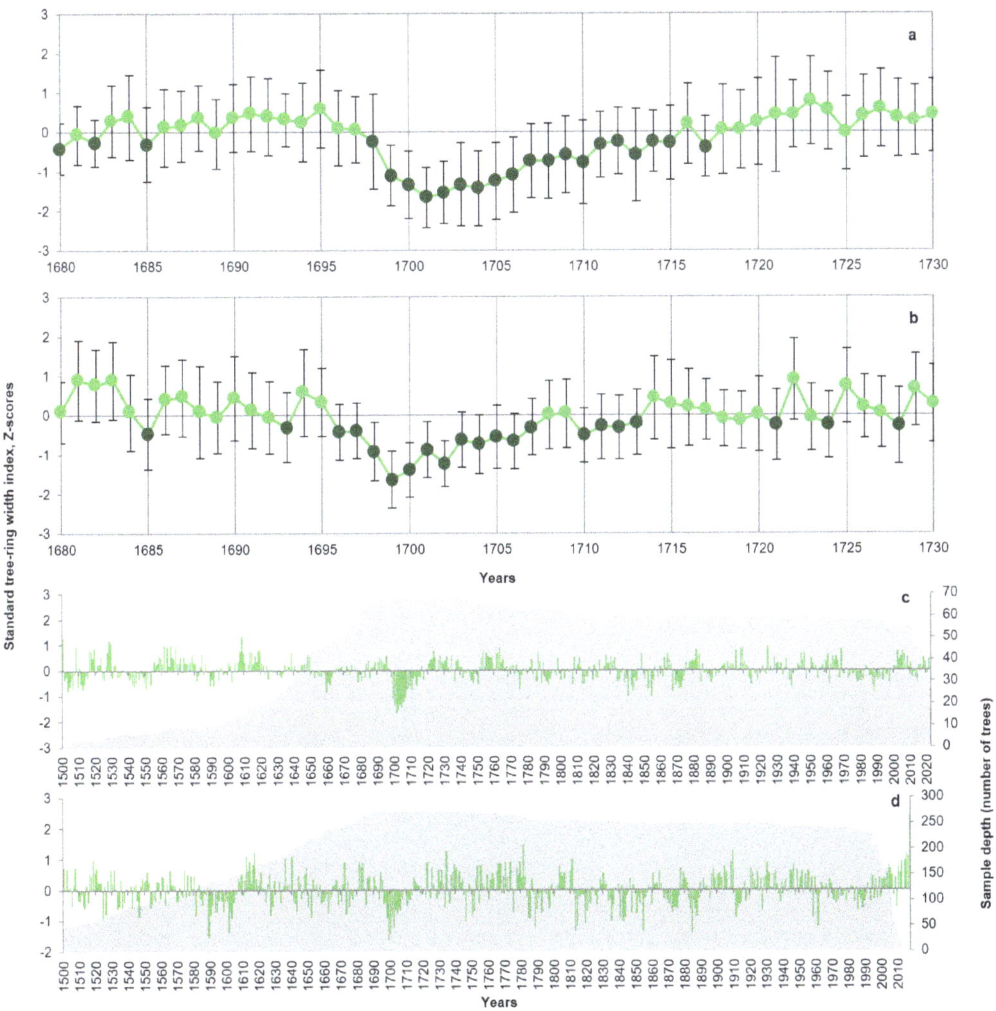

Figure 2. Growth suppression in tree-ring width of *Pinus sibirica* (**a**,**c**) and *Larix sibirica* (**b**,**d**) individual trees covering the period of the near-1700 CE nosedive. Trees were selected from growth-suppressed chronologies, and indices were transformed into Z-scores (standard series for 68 trees from 10 pine chronologies, 272 trees from 19 larch chronologies). (**a**,**b**) Mean values (markers) and SD (standard deviation; error whiskers) for 1680–1730; years with mean Z-scores below zero (mean + SE < 0; SE, standard error) are presented by darker shade of markers. (**c**,**d**) Time series of mean tree-ring index (Z-scores; bars) and sample depth (grey shading) for those trees since 1500.

Suppression of growth in larch and pine chronologies in the study area during the nosedive window 1699–1706 is most pronounced in the Western Sayan Mountains (pine) and the eastern part of the Russian Altai Mountains (larch) (Figure 1c), as supported by other studies [7,11,17,18]. Mean of Z-scores below −2 and minimum values below −3 (extreme suppression) were observed only in one chronology outside that territory, but moderate suppression extended to China and Mongolia (Table A1).

2.2. Paleoclimatic and Tectonic Context

Summer temperature reconstructions in the study area, based on TRW [19,20], blue intensity [16] and maximum latewood density [15] of tree species other than *P. sibirica*, are significantly ($p < 0.05$) positively correlated with PISIreg both at high and low frequencies (Table 1). Correlations are higher for 1650–1750 than for the full overlap of records and generally increase with the smoothing of the time series. These reconstructions have negative z-scores for the nosedive window, 1699–1706, indicating cool summers at the time. Correlation with PISIreg is consistently higher for reconstructions from larch than from spruce (*Picea schrenkiana* Fisch. & C.A.Mey.). The temperature reconstructions reach maximum correlation with PISIreg at lag = 0 before smoothing and at high frequencies, indicating synchrony in annual departures, but at lags of several years at low frequencies (Figure 2).

Table 1. Statistics summarizing the relationships between PISIreg and temperature reconstructions based on other tree-ring data.

Temperature Reconstruction				Z-Score Mean/min, 1699–1706	Correlations with PISIreg, 1650–1750/Total Overlap Period [b]				
Publication	Used Proxy [a]	Season	Location		Annual	Smoothed [c]	Smoothed (Lagged) [d]	Lag, Years	High-Pass [e]
[19]	LASI TRW	June–August	Altai	−1.56/−2.68	**0.52/0.30**	**0.73/0.20**	**0.78/0.22**	3/2	**0.41/0.41**
[20]	LASI TRW	June–July	Mongolia	−1.49/−2.45	**0.49/0.27**	**0.70/0.20**	**0.71/0.21**	1/2	**0.35/0.33**
[15]	PCSH MXD	July–August	Tien-Shan	−1.34/−1.97	0.35/0.04	0.40/0.06	0.82/0.23	7/8	−0.13/−0.03
[16]	LASI BI	June–August	Mongolia	−0.82/−2.01	**0.25/0.27**	**0.23/0.21**	**0.30/0.23**	4/3	**0.25/0.28**

[a] Tree species: LASI—*Larix sibirica*, PCSH—*Picea schrenkiana* Fisch. & C.A.Mey. Tree-ring parameters: TRW—tree-ring width; MXD—maximum wood density; BI—blue intensity (proxy for wood density). [b] Bold correlation coefficients are significant at $p < 0.05$. [c] Time series were smoothed by an 11-year moving average. [d] Maximum cross-correlation between smoothed time series, with PISIreg delayed by "lag" years after temperature reconstruction. Plots for 1650–1750 are presented in Figure A2. [e] High-pass filter applied to both time series by subtracting smoothed series from original.

Documentary records list neither volcanic eruptions nor any earthquakes in the study area at the time of the nosedive. Strong earthquakes (intensity of 10–11, Mercalli scale) were reported in recent history in the Altai Mountains in 1761 CE, 1957 CE and 2003 CE. Epicenters are marked in Figure 1a. Eight-year-growth departures of pine and larch, as Z-scores, after these earthquakes are much less severe than the PISIreg departure during the 1699–1706 nosedive (Table 2). Suppression of tree growth after these earthquakes was moreover observed in only a few chronologies.

2.3. Climate Teleconnections

Only two 20th-century years, 1938 and 1988, had PISIreg z-scores lower than −2.0; these are the 14th and 15th lowest values of PDSIreg over the full 1524–2022 CE record, although not nearly as extreme as in the nosedive. The composite 500 mb height and 2 m air temperature anomaly patterns for those years highlight cyclonic flow and cold conditions in the study area in both seasons as signatures for extremely narrow *P. sibirica* rings in the Western Sayan (Figure 3). In the warm season, PISIreg is under anomalous northerly flow and is near the core of a well-defined central Asia cold anomaly. In the cool season, PISIreg is near the center of a broad zonally oriented core of anomalous cold stretching across southern Siberia and is directly under very strong anomalous northerly flow at 500 mb.

Table 2. Comparison of growth departures in tree-ring width chronologies of pine and larch for 8-year periods after the investigated event and after the three strongest known earthquakes in the Altai mountains.

Type of Event [b]	Calendar Date	Epicenter Coordinates		Z-Scores [a]		
		Lat. N	Lon. E	Range of Means		Period of Calculation [c]
				Larch	Pine	
Unknown	ca 1698–1699	~47–53°	~85–95°	−2.61. . .0.95	−3.39. . .0.53	1699–1706
Earthquake	9 December 1761	~50°	~90°	−0.69. . .2.03	−0.32. . .1.48	1762–1769
Earthquake	4 December 1957	45.5°	99.5°	−1.16. . .1.00	−1.72. . .1.22	1958–1965
Earthquake	27 September 2003	50.038°	87.813°	−0.15. . .0.77	−1.01. . .1.54	2004–2011

[a] Z-scores represent a complete set of 14 pine and 61 larch chronologies in the study area. [b] Event, factor, or their combination that led to the nosedive in tree growth investigated in this study. [c] Since all three earthquakes occurred after growth season, tree growth could react to them only beginning in the next year.

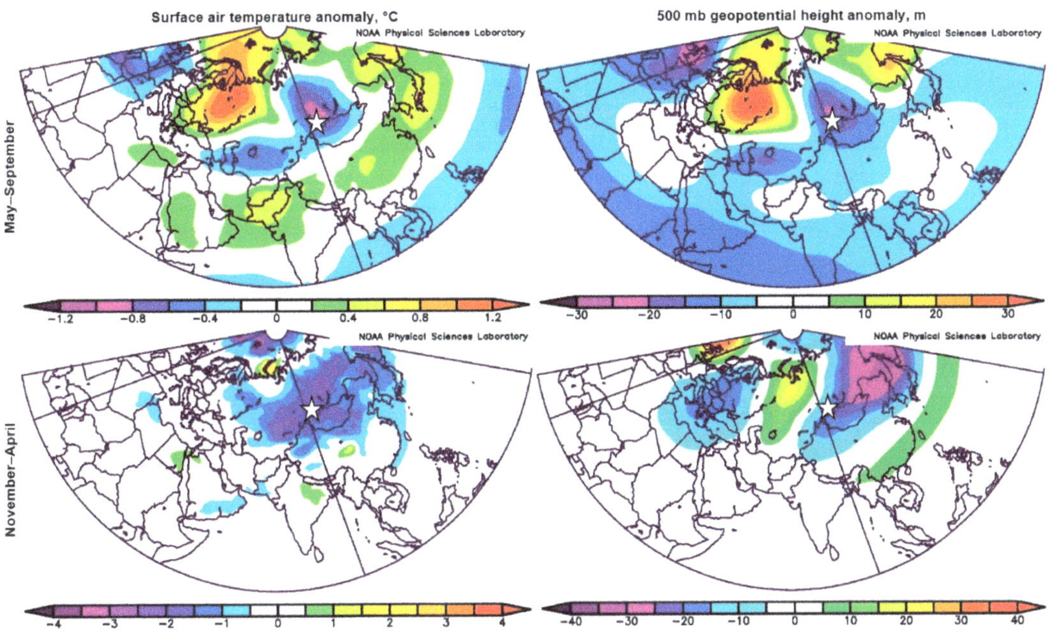

Figure 3. Composite anomaly maps of surface air temperature (**left**) and 500 mb geopotential height (**right**) in warm (May–September, **top**) and preceding cold (November–April, **bottom**) seasons for the years 1938 and 1988. These are the years with the lowest *P. sibirica* growth in the 20th century according to regional chronology PISIreg (white asterisk). Maps were developed by the web-based 20th Century Reanalysis V2 tool (https://psl.noaa.gov/cgi-bin/data/composites/plot20thc.v2.pl, accessed on 14 October 2024).

The anomalous 500 mb flow in the 1938/1988 composite is strongly meridional, with centers of action (highs or lows of anomalous heights) in the warm season resembling the positive mode of the "Eurasia 1", or "Scandinavia" teleconnection circulation pattern [21]. This is one of 10 preferred modes of Northern Hemisphere circulation that can explain the simultaneous occurrence of abnormal weather patterns across vast distances (http://www.cpc.ncep.noaa.gov/data/teledoc/telecontents.shtml; accessed on 15 November 2024) and is known to significantly affect the study area [22]. The cool season 500 mb composite map has no clear resemblance to the Scandinavia pattern.

3. Discussion

3.1. Prime Suspect: Cold Spell

A decadal-scale exceptionally cold period is the natural suspect in the PISIreg nosedive. Energy limitation is after all the generally accepted foundation for temperature reconstruction from the ring width of boreal and subalpine trees [23,24]. Moreover, a long cold period could impact forest vitality and resistance to other stress factors [4,25]. Growth reduction in conifer forests in the study area in cold periods increases with elevation [26,27], which is consistent with amplified reduction in the subalpine *P. sibirica* chronologies of PISIreg. The nosedive occurred during the Little Ice Age, and the several decades before and after the nosedive are estimated to have been cold, possibly exacerbated by the Maunder solar activity minimum of 1645–1715 [19,28,29]. Summer temperature reconstructions and temperature-sensitive tree-ring chronologies from different species and tree-ring variables (TRW, wood density, and blue intensity) support 1699–1706 as a cold period in the study area [11,13–20]. Anomalies consistent with the nosedive were also observed in temperature-sensitive wood cell parameters of *P. sibirica* in Western Sayan [12]. It is likely that all growth processes were inhibited at the time—both cell division in the cambium and carbon accumulation during tracheid differentiation, since wood density is determined by cell structure [30,31].

The concentration of $NH4^{+}$ and $HCOO^{-}$ ions, originating from forest emissions, in the Belukha ice cores reached a historical minimum in the decade 1710–1720 [29]. A comprehensive temperature reconstruction based on Lake Teletskoe sediments (Altai), including bromine reflecting the state of the vegetation, also had a minimum in ~1710 [32]. These data indicate a regional decline in the productivity of the forest ecosystems soon after the investigated event on a decadal scale.

However, not all proxy data support an extreme cold event ca 1700 CE in the study area (Figure 4). Pollen in the Hulun Lake sediments [33] recorded cold July temperatures in Mongolia for a century before that, but no sharp drop matching the nosedive. The 10-year-smoothed March–November temperatures in the Altai, reconstructed from the d18O in the Belukha glacier ice cores [28], had only a moderate negative deviation in 1691–1720 and a stronger one in 1741–1750. Of more than 90 reconstructions assembled by Christiansen and Ljungqvist [1] to describe variation in Northern Hemisphere temperature, six temperature reconstructions are in our study area [13,14,28,32,34,35], and only one of them [35] indicates moderate cooling ca 1700 CE.

3.2. Extenuating Circumstances

From the evidence just given, we conclude that warm season cold spells, as in 1937 and 1988, probably contributed to the low in *P. sibirica* growth ca 1700 CE. The remarkable occurrence of six consecutive years of PISIreg is more than three standard deviations below the mean; however, this likely requires additional environmental impact, such as damage to *P. sibirica* trees from weather or another factor. We dismiss earthquakes as a likely suspect, since even the strongest known earthquakes in Altai did not exhibit comparable consequences. Drought and related causes, including wildfires [4,36–38] can also be dismissed because the habitat of the most affected trees is moist and cool [27,39,40]. Defoliation of trees by insects could severely suppress growth [4,6], and the study area is home to the gypsy moth and Siberian silk moth [41]. The most growth-suppressed trees in our study, however, are outside the distribution range of those insects due to short cool summers and frosty winters and would be more so during a cold period [4,42].

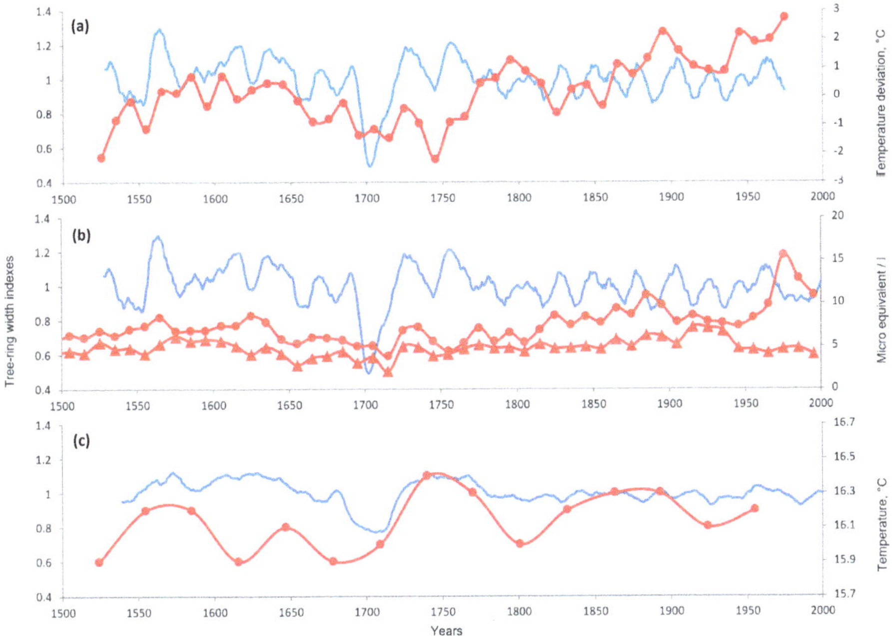

Figure 4. Growth suppression ca 1700 CE in PISIreg (blue line without markers) in the context of climate reconstructions and natural archives of decadal resolution (red lines with markers). (**a**) July temperature at the Hulun Lake (Mongolia) reconstructed from pollen in lake sediments (average resolution ~31 years) [33]. (**b**) March–November temperature, smoothed by a 10-year average, reconstructed from dO18 in ice cores of the Belukha glacier, Altai [28]. (**c**) Concentrations of $HCOO^-$ (triangles) and NH_4^+ ions (circles) in the ice core at the same glacier [29]. Lines with markers represent temperature reconstruction or natural archive; line without markers represents PISIreg smoothed by 10-year (**a**,**b**) or 31-year (**c**) moving average.

Stress from extreme weather events is on the other hand a likely factor in the nosedive. Mongolian historical records mention cold-related events at the time over vast areas [43]: multiple winter dust storms and snowstorms were registered in 1695, 1697 and 1705, and "dzuds" were reported in 1696 and 1704. The term "dzud" embraces several winter weather extremes—including heavy snowfalls, severe frosts, and thaws followed by the formation of thick ice crust—leading to a lack of forage and die-off of livestock and thus important for Mongolian nomadic herders [44]. Some of these phenomena (heavy snowfall, thaw–freeze cycles), as well as snowstorms, indicate the establishment of winter cyclones or atmospheric fronts during 1695–1705, in contrast to the typical Siberian High anticyclone [13,45].

Snowstorms and strong winds can adversely affect forests through crown damage and defoliation [6,9,23,46]. Extreme freeze–thaw cycles could also lead to stress in evergreen conifers through physiological drought and desiccation of needles [47]. The strengthening of the nosedive with elevation is explained by the greater amount of snow and stronger winds. The observed more severe and prolonged nosedive in pine than in larch (Figure 2) could be explained by ecophysiological factors. The average 5–8 years retention of needles by *P. sibirica* [48] would slow down recovery after defoliation. The large crown and dense bunches of needles in *P. sibirica* could also present a better target for wind and for accumulation of ice and snow in the crown, increasing the risk of mechanical damage.

3.3. *Accomplice from the Tropics*

Although our literature search revealed no local volcanic eruptions in the mountains of south-central Asia at the time of the nosedive, a large tropical eruption ca 1695 CE has been implicated in catastrophic summer cooling ca 1700 CE in Scotland and most of Northern Europe [5,8,49,50]. The resemblance of the warm season composite 1938/1988 500 mb anomaly pattern (Figure 3) to the positive mode of the Scandinavia pattern suggests possible influence from the 1695 volcano on the *P. sibirica* growth in the nosedive through amplification of the positive-phase Scandinavia pattern. The cool-season linkage of the Scandinavia pattern to the 1938/1988 composite anomaly maps of 500 mb height and 2 m air temperature is less clear. We can speculate that winter circulation may have been disrupted after the 1695 volcano to favor high winds, heavy snow, and tree damage through pathways already mentioned. Such damage would exacerbate growth suppression in *P. sibirica* beyond that due to energy shortage in the unusually cool, wet summers.

4. Materials and Methods

4.1. *Data Sources*

To analyze the geographical distribution of the ca 1700 CE nosedive in conifer tree-ring width (TRW), a study area (40–60° N, 60–120° E) was defined that extends roughly 10° to the south and north and 30° to the west and east of the central Western Sayan Mountains, where the phenomenon was initially discovered (Figure 1a).

The authors' own dendrochronological data, starting no later than 1690, include TRW measurements of Siberian stone pine (*Pinus sibirica*) at 7 sites in the Western Sayan, and Siberian larch (*Larix sibirica*) at one site within the above-mentioned region. Collection and processing of samples and dendrochronological measurements, chronology development, and dendroclimatic analysis for most of the chronologies have been published elsewhere [11,51].

This study also uses a TRW dataset of *Pinus sibirica* and *Larix sibirica* species publicly available in The International Tree-Ring Data Bank (ITRDB; https://www.ncei.noaa.gov/products/paleoclimatology/tree-ring, accessed on 11 September 2024), selected by geographic screening for location within 40–60° N 60–120° E and start year no later than 1690. Screening yielded 60 sampled sites of *Larix sibirica* and 7 sites of *Pinus sibirica*. (One of the *Larix* chronologies is listed in the ITRDB as *Larix gmelinii* Rupr., but the site is in the distribution area of *Larix sibirica*. The complete TRW dataset broadly covers the Altai–Sayan Mountain complex and the Mongolian Plateau. All TRW chronologies used in the study are "standard" (autocorrelation not modeled out) chronologies with ring-width trends removed by conservative approaches to avoid the removal of low-frequency climate signals [24].

The ITRDB, with the same geographic and time-coverage screening as above and selection of the "climate reconstructions" category, was also the source of time series representing temperature variation in the study area. Reconstructions with varying temporal resolution and intra-seasonal frames available from the ITRDB were used and analyzed. This selection also provided us with ion concentration data in ice cores from the Belukha glacier [29]. In addition, other published temperature reconstructions were used qualitatively, where publications (references in the main text) include temporal plots.

The list of Holocene volcanoes, their coordinates and information on eruptions were taken from the database 'Volcanoes of the World' (https://volcano.si.edu, accessed on 7 October 2024). Documentary historical evidence for the study area was taken from generalized sources for the territories of Russia and Mongolia on social and natural phenomena [43,52] and on earthquakes (NGDC/WDS Global Significant Earthquake Database https://www.ngdc.noaa.gov/hazel/view/hazards/earthquake/search, accessed on 10 October 2024).

Maps of anomaly fields of geopotential height (500 mb) and temperature (2 m) for 20th to 21st century years with lowest PISIreg were generated using a web-based NOAA tool for mapping "20th Century Reanalysis Monthly Composites" (https://psl.noaa.gov/cgi-bin/data/composites/plot20thc.v2.pl, accessed on 14 October 2024). From the online menu, we specified "20CRV3" as the data source and used the menu options to tailor the map domain. Information on Northern Hemisphere Teleconnections, including maps with centers of action and associated patterns of climate anomalies for the "Scandinavia" pattern, were accessed at the NOAA Climate Prediction website (https://www.cpc.ncep.noaa.gov/data/teledoc/teleintro.shtml, accessed on 7 October 2024).

4.2. Data Processing and Analysis

For all raw TRW measurements, standardization was performed in the ARSTAN program [53] by smoothing with a 67% cubic spline, then averaging all the series within the site by bi-weighted mean. To identify the presence and intensity of the growth decline, tree-ring chronologies were transformed into Z-scores (mean = 0; SD = 1), since chronologies have different ranges of variability.

For the chronologies of *Pinus sibirica* with a visible decline in growth around 1700 (Figure A1), a standard series of individual trees covering the period from 1690 were analyzed using the superimposed epoch method, revealing the period of the most pronounced suppression as 1699–1706 (Z-scores less than −1, including the range ± SE, standard error). The presence of a long-term (or any) suppression in growth in the chronologies under consideration was assessed by mean (minimum) values of Z-scores of local standard chronologies for 1699–1706, also comparing them with the threshold of −1 (i.e., a drop in growth by SD or more relative to the long-term average). For *Larix sibirica* chronologies with suppressed growth, a series of individual trees were also investigated using the superimposed epoch method to estimate years of suppression for this species.

The similarity between tree-ring chronologies among themselves and with other time series was estimated using paired Pearson correlation coefficients, either for the entire period of overlapping series or for the century-long period 1650–1750 including the phenomenon of interest. In addition, an 11-year smoothing filter (moving average) was applied to a time series with an annual time resolution, dividing the series into low-frequency and (by subtracting the smoothed series from the original) high-frequency components, for which a correlation analysis was also performed. Cross-correlation analysis was performed to find if there is a lag of tree-ring chronologies relative to other time series, separately for low- and high-frequency components.

For the pollen-based temperature reconstruction, having an uneven temporal distribution of data points, or nonuniform time step, the average temporal resolution was estimated by dividing the coverage period by a number of intervals between data points and rounding to an integer number of years. For comparison of the TRW chronology PISIreg (annual resolution) with lower-resolution time series, we smoothed chronology with a moving average of appropriate window width.

The selection of 1938 and 1988 as years for mapping 500 mb height anomalies and 2 m temperature anomalies was made by (1) converting the 1524–2022 time series of PISIreg to Z scores, using the mean and standard deviation for that full length, (2) sorting the Z-scores from smallest to largest, and (3) selecting the 20th century years with the lowest Z-scores. Years 1938 and 1988 are the only years with Z < −2.0 and are low outliers in that they are much lower than the next-lowest (Z = −1.6 in 1914).

Outside of specialized tools, data processing and statistical analysis were performed in Microsoft Excel 2007 (v12.0).

5. Conclusions

Case studies of extreme events in climate reconstruction are important because those events often draw the most interest and are highlighted in placing current and projected climate variation in a long-term context. Lacking instrumental observations, we present evidence from available natural and historical archives as proxies for environmental change and events. Such evidence is not irrefutable, but points to: (1) tree damage by winter weather extremes as the most likely cause of the extreme growth suppression of conifers near the upper treeline in the Sayan–Altai region at the start of the 18th century; (2) a tropical volcanic eruption contributed to the severity of the event; and (3) atmospheric teleconnections, primarily through an amplified Scandinavia pattern, is a global-scale transfer mechanism that helps explain the exacerbation of tree-growth suppression in our study area. We invite the international scientific community to prove or disprove our opinion.

Author Contributions: Conceptualization, E.A.V.; methodology, D.M.M., L.V.B. and E.A.B.; validation, N.B.M. and E.A.V.; formal analysis, D.M.M., L.V.B. and Y.A.K.; investigation, D.F.Z., L.V.B., Y.A.K. and N.B.M.; resources, D.F.Z.; data curation, L.V.B. and Y.A.K.; writing—original draft preparation, D.M.M., D.F.Z. and L.V.B.; writing—review and editing, D.M.M. and L.V.B.; visualization, L.V.B. and Y.A.K.; supervision, E.A.B. and E.A.V.; project administration, D.F.Z. and E.A.B.; funding acquisition, E.A.B. and E.A.V. All authors have read and agreed to the published version of the manuscript.

Funding: This research was funded by the Russian Science Foundation, grant no. 23-44-00067, and the National Natural Science Foundation of China, grant no. 42261134537, in the framework of a joint Russian–Chinese project (original data collection), by the Russian Ministry of Science and Higher Education, grant no. FSRZ-2023-0007 (data analysis).

Data Availability Statement: Analyzed data from publicly archived datasets include tree-ring width measurements, climatic reconstructions and glacier ice composition data from the ITRDB (https://www.ncei.noaa.gov/products/paleoclimatology, accessed on 11 September 2024), data on volcanoes and their eruptions from the database 'Volcanoes of the World' (https://volcano.si.edu). data on earthquakes from the NGDC/WDS Global Significant Earthquake Database https://www.ngdc.noaa.gov/hazel/view/hazards/earthquake/search, accessed on 10 October 2024), and climatic data from the NOAA websites (https://psl.noaa.gov/cgi-bin/data/composites/plot20thc.v2.pl, accessed on 14 October 2024, and https://www.cpc.ncep.noaa.gov/data/teledoc/teleintro.shtml, accessed on 7 October 2024). Original data included in this study (tree-ring width measurements) will be available on request from the corresponding author.

Acknowledgments: The authors are grateful to the Joint Directorate of the State Natural Biosphere Reserve "Sayano-Shushensky" and the National Park "Shushensky Bor", to the administration of the Natural Park "Ergaki" for permission and assistance in sampling on their territories.

Conflicts of Interest: The authors declare no conflicts of interest. The funders had no role in the design of the study; in the collection, analyses, or interpretation of data; in the writing of the manuscript; or in the decision to publish the results.

Appendix A

Table A1. List of used tree-ring width data sources.

No	Tree Species [a]	Country	ITRDB Code or Chronology Code [b]	Coordinates Lat. N	Long. E	Elev. [c], m a.s.l.	Mean/min in 1699–1706 [d]	R [e]
1	LASI	Russia	RUSS127	50.15	85.37	1750	−0.47/**−1.64**	−0.03
2	LASI	Russia	RUSS129	51	85.63	1450	−0.31/**−1.43**	0.18
3	LASI	Russia	RUSS130	50.87	85.23	1450	−0.65/**−1.39**	**0.21**
4	LASI	Russia	RUSS135	50.42	87.58	2000	**−1.23/−2.66**	**0.48**
5	LASI	Russia	RUSS137	50.48	87.65	2150	**−1.20/−2.24**	**0.31**

Table A1. *Cont.*

No	Tree Species [a]	Country	ITRDB Code or Chronology Code [b]	Coordinates			Mean/min in 1699–1706 [d]	R [e]
				Lat. N	Long. E	Elev. [c], m a.s.l.		
6	LASI	Russia	RUSS226	50.803	87.933		**−2.61/−3.42**	**0.65**
7	LASI	Russia	RUSS227	49.617	88.1		**−2.09/−3.26**	**0.59**
8	LASI	Russia	RUSS228	50.867	85.233		−0.18/−0.93	0.14
9	LASI	Russia	RUSS229	50.3	87.833		−0.83/**−2.50**	**0.26**
10	LASI	Russia	RUSS230	50.267	87.833		−0.26/**−1.17**	0.08
11	LASI	Russia	RUSS231	49.167	87.283		−0.50/**−1.32**	0.04
12	LASI	Russia	RUSS232	50.167	87.833		**−1.86/−3.28**	**0.62**
13	LASI	Russia	RUSS233	50.117	87.917		−0.57/**−1.75**	**0.22**
14	LASI	Russia	RUSS234	50.483	87.483		**−1.27/−2.31**	**0.37**
15	LASI	Russia	RUSS235	50.683	87.967		**−1.12/−1.88**	0.19
16	LASI	Russia	RUSS241	52.399	98.685	2020	−0.15/−0.78	0.06
17	LASI	Russia	RUSS247	49.23	87.23	2200	**−1.19/−2.52**	**0.40**
18	LASI	Russia	RUSS248	49.2	87.02	2250	−0.70/**−1.70**	**0.31**
19	LASI	Russia	RUSS249	51.58	95.31	2060	−0.23/−0.64	0.18
20	LASI	Russia	RUSS250	49.51	87.5	2250	**−1.49/−3.45**	**0.50**
21	LASI	Russia	RUSS251	49.36	86.57	2200	−0.92/**−2.13**	**0.30**
22	LASI	Russia	RUSS252	50.36	91.28	2170	−0.66/**−2.68**	0.06
23	LASI	Russia	RUSS253	50.2	96.39	2254	−0.23/−1.00	**0.25**
24	LASI	Russia	RUSS254	50.24	89.59	2280	**−1.49/−2.94**	**0.52**
25	LASI	Russia	RUSS255	50.07	88.17	2100	−0.61/**−1.98**	0.17
26	LASI	Russia	RUSS256	50.49	94.18	2130	−0.31/**−1.13**	0.11
27	LASI	Russia	RUSS257	49.39	88.14	2250	**−1.16/−2.78**	**0.37**
28	LASI	Russia	RUSS258	50.22	98.14	2200	−0.62/**−1.12**	**0.35**
29	LASI	Russia	RUSS259	50.04	87.54	2250	**−1.37/−2.22**	**0.30**
30	LASI	Russia	RUSS288	56.412	115.589	465	−0.62/**−1.29**	**0.30**
31	LASI	Russia	RUSS292	53.144	106.832	810	**−1.03/−1.68**	0.18
32	LASI	Russia	AKK [b]	49.917	86.55	2050	**−1.75/−3.09**	**0.55**
33	LASI	China	CHIN029	43.85	93.3	2810	0.11/**−1.54**	−0.04
34	LASI	China	CHIN030	43.833	93.383	2840	**−2.10/−3.41**	**0.49**
35	LASI	Mongolia	MONG001	50.77	100.2	2300	−0.09/−0.56	0.15
36	LASI	Mongolia	MONG006	47.78	107.5	1415	0.07/**−1.45**	−0.04
37	LASI	Mongolia	MONG009	49.92	91.57	2500	−0.26/−0.88	0.10
38	LASI	Mongolia	MONG010	47.27	100.03	2500	0.36/**−1.65**	−0.06
39	LASI	Mongolia	MONG011	48.15	100.28	1900	−0.30/**−1.38**	0.17
40	LASI	Mongolia	MONG012	48.98	103.23	1400	0.20/−0.94	0.00
41	LASI	Mongolia	MONG013	48.77	97.12	1840	**−1.14/−1.53**	**0.25**
42	LASI	Mongolia	MONG014	49.48	100.83	1800	−0.20/**−1.20**	0.10
43	LASI	Mongolia	MONG015	48.17	99.87	2060	0.14/**−1.13**	−0.02
44	LASI	Mongolia	MONG016	48.6	88.367		0.36/**−1.24**	**−0.24**
45	LASI	Mongolia	MONG017	49.967	91		**−1.06/−2.00**	**0.20**
46	LASI	Mongolia	MONG018	49.967	90.983		**−1.03/−1.92**	**0.33**
47	LASI	Mongolia	MONG019	47.1	90.967		0.85/0.03	−0.11
48	LASI	Mongolia	MONG020	48.267	88.867		0.45/−0.45	−0.13
49	LASI	Mongolia	MONG021	48.35	107.467		−0.28/−0.71	0.11
50	LASI	Mongolia	MONG023	49.5	94.583		−0.38/−0.92	0.02
51	LASI	Mongolia	MONG024	48.5	88.5		−0.07/**−1.22**	0.10
52	LASI	Mongolia	MONG025	48.7	88.8		−0.49/**−1.13**	0.06
53	LASI	Mongolia	MONG026	46.817	100.117		0.14/−0.87	0.01
54	LASI	Mongolia	MONG027	46.317	101.317		−0.67/**−1.81**	**0.30**
55	LASI	Mongolia	MONG028	48.833	111.683		0.95/−0.38	**−0.26**
56	LASI	Mongolia	MONG029	49.867	91.433		−0.76/**−1.72**	−0.03
57	LASI	Mongolia	MONG030	49.383	94.883		−0.52/**−2.27**	0.17
58	LASI	Mongolia	MONG032	46.517	100.95		−0.27/**−1.41**	**0.21**
59	LASI	Mongolia	MONG033	49.37	94.88	2229	−0.74/−0.99	**0.51**
60	LASI	Mongolia	MONG040	51.15	99.083	2400	**−1.29/−2.20**	**0.49**
61	LAGM	Mongolia	MONG007	49.7	91.55	2000	−0.44/**−1.90**	−0.10
62	PISI	Russia	RUSS222	50.417	84.617	1898	**−1.83/−2.54**	**0.64**
63	PISI	Russia	SPass20 [b]	51.71	89.96	2000	**−3.39/−3.82**	**0.91**
64	PISI	Russia	GladSW [b]	52.91	91.36	1620	**−3.17/−3.80**	**0.86**
65	PISI	Russia	SAR [b]	52.23	92.25	1630	**−2.46/−3.27**	**0.85**
66	PISI	Russia	MGol8 [b]	52.533	92.05	920	**−1.56/−2.20**	**0.49**
67	PISI	Russia	MGol5 [b]	52.55	92.117	570	**−1.02/−2.01**	0.19
68	PISI	Russia	ErgV [b]	52.8	93.433	1650	**−2.07/−3.99**	**0.75**
69	PISI	Russia	ErgR [b]	52.833	93.333	1450	**−1.17/−1.87**	**0.57**
70	PISI	Mongolia	MONG041	48.17	99.87	2060	0.53/−0.67	0.00
71	PISI	Mongolia	MONG042	46.68	101.77	2125	0.42/−0.32	−0.12
72	PISI	Mongolia	MONG002	47.77	107	1755	**−1.45/−2.24**	0.13
73	PISI	Mongolia	MONG003	48.3	98.93	2420	−0.58/−0.93	**0.25**
74	PISI	Mongolia	MONG008	49.37	94.88	2229	**−1.66/−2.58**	**0.46**
75	PISI	Mongolia	MONG031	48.25	97.4		0.51/−0.72	−0.20

[a] Species: LASI—*Larix sibirica* Ledeb.; LAGM—*Larix gmelinii* Rupr. (although, this is probably mislabeled LASI, considering species distribution areas); PISI—*Pinus sibirica* Du Tour. [b] Marked are chronologies collected by authors. Not marked are chronologies from The International Tree-Ring Data Bank (ITRDB; https://www.ncei.noaa.gov/products/paleoclimatology/tree-ring, accessed on 11 September 2024). [c] Elevation data are listed if they are mentioned in the ITRDB record or measured by authors. [d] Mean/minimal values are presented after transformation of chronologies into Z-scores. Bold values are those below −1 (i.e., below mean–SD in original standard chronologies), indicating long-term/any growth suppression. [e] R is correlation with PISIreg (generalized TRW chronology of *Pinus sibirica* from sampling sites SPass20, GladSW, SAR, ErgV) over 1650–1750. Bold values are significant at $p < 0.05$.

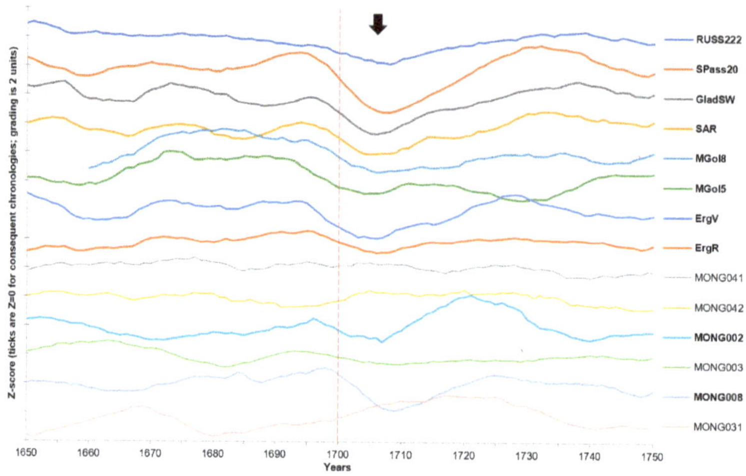

Figure A1. Standard chronologies of *Pinus sibirica* (1650–1750) smoothed by 11-year moving average. Chronologies are plotted in Z-score units and shifted sequentially by 2 units along vertical axis for clarity. The same relative y-scale applies to all plots, tick marks represent Z = 0 for each chronology. Vertical dashed line marks calendar year 1700; vertical arrow marks growth suppression; chronologies with synchronous growth suppression after 1698 are represented by thick lines and marked with bold labels.

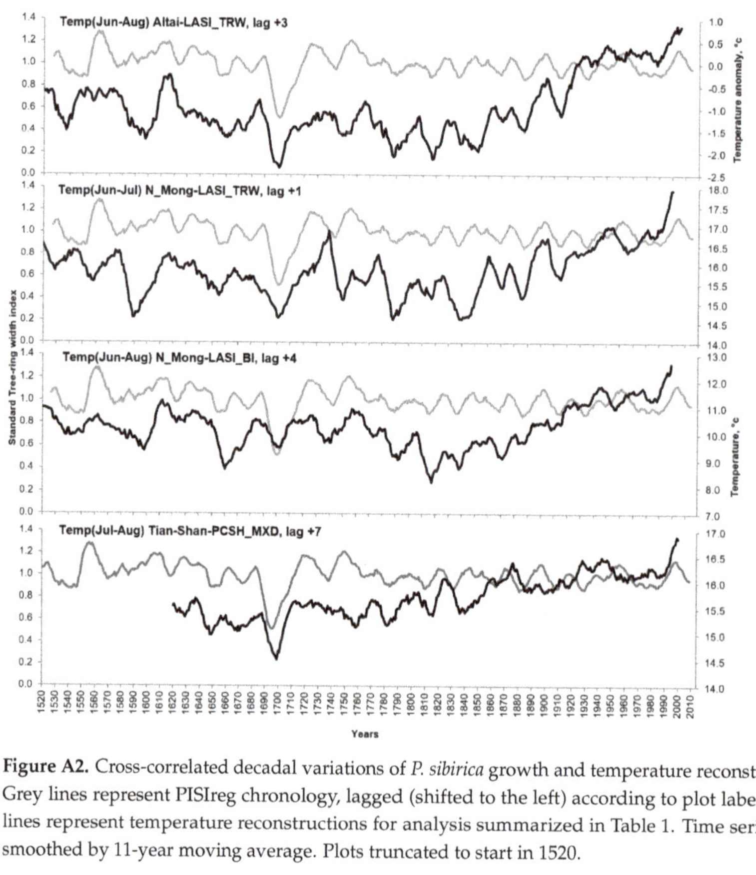

Figure A2. Cross-correlated decadal variations of *P. sibirica* growth and temperature reconstructions. Grey lines represent PISIreg chronology, lagged (shifted to the left) according to plot labels; black lines represent temperature reconstructions for analysis summarized in Table 1. Time series were smoothed by 11-year moving average. Plots truncated to start in 1520.

References

1. Christiansen, B.; Ljungqvist, F.C. The extra-tropical Northern Hemisphere temperature in the last two millennia: Reconstructions of low-frequency variability. *Clim. Past* **2012**, *8*, 765–786. [CrossRef]
2. Esper, J.; Krusic, P.J.; Ljungqvist, F.C.; Luterbacher, J.; Carrer, M.; Cook, E.; Davi, N.K.; Harti-Meier, C.; Kirdyanov, A.; Konter, O.; et al. Ranking of tree-ring based temperature reconstructions of the past millennium. *Quat. Sci. Rev.* **2016**, *145*, 134–151. [CrossRef]
3. Ljungqvist, F.C.; Piermattei, A.; Seim, A.; Krusic, P.J.; Büntgen, U.; He, M.; Kirdyanov, A.V.; Luterbacher, J.; Schneider, L.; Seftingen, K.; et al. Ranking of tree-ring based hydroclimate reconstructions of the past millennium. *Quat. Sci. Rev.* **2020**, *230*, 106074. [CrossRef]
4. Dale, V.H.; Joyce, L.A.; McNulty, S.; Neilson, R.P.; Ayres, M.P.; Flannigan, M.D.; Hanson, P.J.; Irland, L.C.; Lugo, A.E.; Peterson, C.J.; et al. Climate change and forest disturbances: Climate change can affect forests by altering the frequency, intensity, duration, and timing of fire, drought, introduced species, insect and pathogen outbreaks, hurricanes, windstorms, ice storms, or landslides. *BioScience* **2001**, *51*, 723–734. [CrossRef]
5. Crowley, T.J.; Zielinski, G.; Vinther, B.; Udisti, R.; Kreutz, K.; Cole-Dai, J.; Castellano, E. Volcanism and the little ice age. *PAGES News* **2008**, *16*, 22–23. [CrossRef]
6. Stoffel, M.; Bollschweiler, M.; Butler, D.R.; Luckman, B.H. *Tree Rings and Natural Hazards: A State-of-Art*; Springer: Dordrecht, The Netherlands, 2010; 505p. [CrossRef]
7. Nikolaeva, S.A.; Savchuk, D.A.; Bocharov, A.Y. Influence of different factors on tree growth of *Pinus sibirica* in the highlands of the central Altai Mountains. *J. Sib. Fed. Univ. Biol.* **2015**, *8*, 299. [CrossRef]
8. Sigl, M.; Winstrup, M.; McConnell, J.R.; Welten, K.C.; Plunkett, G.; Ludlow, F.; Büntgen, U.; Caffee, M.; Chellman, N.; Dahl-Jensen, D.; et al. Timing and climate forcing of volcanic eruptions for the past 2500 years. *Nature* **2015**, *523*, 543–549. [CrossRef]
9. Klesse, S.; Etzold, S.; Frank, D. Integrating tree-ring and inventory-based measurements of aboveground biomass growth: Research opportunities and carbon cycle consequences from a large snow breakage event in the Swiss Alps. *Eur. J. For. Res.* **2016**, *135*, 297–311. [CrossRef]
10. Gao, S.; Liang, E.; Liu, R.; Lu, X.; Rossi, S.; Zhu, H.; Piao, S.; Penuelas, J.; Camarero, J.J. Shifts of forest resilience after seismic disturbances in tectonically active regions. *Nat. Geosci.* **2024**, *17*, 189–196. [CrossRef]
11. Zhirnova, D.F.; Belokopytova, L.V.; Krutovsky, K.V.; Kholdaenko, Y.A.; Babushkina, E.A.; Vaganov, E.A. Spatial-coherent dynamics and climatic signals in the radial growth of Siberian stone pine (*Pinus sibirica* Du Tour) in subalpine stands along the Western Sayan Mountains. *Forests* **2022**, *13*, 1994. [CrossRef]
12. Zhirnova, D.F.; Belokopytova, L.V.; Upadhyay, K.K.; Tripathi, S.K.; Babushkina, E.A.; Vaganov, E.A. 495-year wood anatomical record of Siberian stone pine (*Pinus sibirica* Du Tour) as climatic proxy on the timberline. *Forests* **2022**, *13*, 247. [CrossRef]
13. Jacoby, G.C.; D'Arrigo, R.D.; Davaajamts, T. Mongolian tree rings and 20th-century warming. *Science* **1996**, *273*, 771–773. [CrossRef] [PubMed]
14. D'Arrigo, R.; Jacoby, G.; Frank, D.; Pederson, N.; Cook, E.; Buckley, B.; Nachin, B.; Mijiddorj, R.; Dugarjav, C. 1738 years of Mongolian temperature variability inferred from a tree-ring width chronology of Siberian pine. *Geophys. Res. Lett.* **2001**, *28*, 543–546. [CrossRef]
15. Chen, F.; Yuan, Y.; Yu, S.; Chen, F. A 391-year summer temperature reconstruction of the Tien Shan, reveals far-reaching summer temperature signals over the midlatitude Eurasian continent. *J. Geophys. Res. Atmos.* **2019**, *124*, 11850–11862. [CrossRef]
16. Davi, N.K.; Rao, M.P.; Wilson, R.; Andreu-Hayles, L.; Oelkers, R.; D'Arrigo, R.; Nachin, B.; Buckley, B.; Pederson, N.; Leland, C.; et al. Accelerated recent warming and temperature variability over the past eight centuries in the Central Asian Altai from blue intensity in tree rings. *Geophys. Res. Lett.* **2021**, *48*, e2021GL092933. [CrossRef]
17. Kirdyanov, A.V.; Arzac, A.; Kirdyanova, A.A.; Arosio, T.; Ovchinnikov, D.V.; Ganyushkin, D.A.; Katjutin, P.N.; Myglan, V.S.; Nazarov, A.N.; Slyusarenko, I.Y.; et al. Tree-Ring Chronologies from the Upper Treeline in the Russian Altai Mountains Reveal Strong and Stable Summer Temperature Signals. *Forests* **2024**, *15*, 1402. [CrossRef]
18. Nazarov, A.B.; Myglan, V.S. The possibility of construction of the 6000-year chronology for Siberian pine in the Central Altai. *J. Sib. Fed. Univ. Biol.* **2012**, *5*, 70–88. (In Russian)
19. Büntgen, U.; Myglan, V.S.; Ljungqvist, F.C.; McCormick, M.; Di Cosmo, N.; Sigl, M.; Jungclaus, J.; Wagner, S.; Krusic, P.J.; Esper, J.; et al. Cooling and societal change during the Late Antique Little Ice Age from 536 to around 660 AD. *Nat. Geosci.* **2016**, *9*, 231–236. [CrossRef]
20. Davi, N.K.; D'Arrigo, R.; Jacoby, G.C.; Cook, E.R.; Anchukaitis, K.J.; Nachin, B.; Rao, M.P.; Leland, C. A long-term context (931–2005 CE) for rapid warming over Central Asia. *Quat. Sci. Rev.* **2015**, *121*, 89–97. [CrossRef]
21. Barnston, A.G.; Livezey, R.E. Classification, seasonality and persistence of low-frequency atmospheric circulation patterns. *Mon. Weather Rev.* **1987**, *115*, 1083–1126. [CrossRef]
22. Bueh, C.; Nakamura, H. Scandinavian pattern and its climatic impact. *J. Atmos. Sci. Appl. Meteorol. Phys. Oceanogr.* **2007**, *133*, 2117–2131. [CrossRef]

23. Fritts, H.C. *Tree Rings and Climate*; Academic Press: London, UK, 1976; 567p.
24. Cook, E.R.; Kairiukstis, L.A. *Methods of Dendrochronology: Applications in the Environmental Sciences*; Springer: Dordrecht, Th Netherlands, 1990; 394p. [CrossRef]
25. Niinemets, Ü. Responses of forest trees to single and multiple environmental stresses from seedlings to mature plants: Past stres history, stress interactions, tolerance and acclimation. *For. Ecol. Manag.* **2010**, *260*, 1623–1639. [CrossRef]
26. Gerasimova, O.V.; Zharnikov, Z.Y.; Knorre, A.A.; Myglan, V.S. Climatically induced dynamic of radial growth of Siberian ston pine and Siberian fir in the mountain-taiga belt in "Ergaki" National Park. *J. Sib. Fed. Univ. Biol.* **2010**, *3*, 18–29. (In Russian)
27. Fang, K.; Davi, N.; Gou, X.; Chen, F.; Cook, E.; Li, J.; D'Arrigo, R. Spatial drought reconstructions for central High Asia based o tree rings. *Clim. Dyn.* **2010**, *35*, 941–951. [CrossRef]
28. Eichler, A.; Olivier, S.; Henderson, K.; Laube, A.; Beer, J.; Papina, T.; Gäggeler, H.W.; Schwikowski, M. Temperature response i the Altai region lags solar forcing. *Geophys. Res. Lett.* **2009**, *36*, L01808. [CrossRef]
29. Eichler, A.; Brütsch, S.; Olivier, S.; Papina, T.; Schwikowski, M. A 750 year ice core record of past biogenic emissions from Siberia boreal forests. *Geophys. Res. Lett.* **2009**, *36*, L18813. [CrossRef]
30. Björklund, J.; Seftigen, K.; Schweingruber, F.; Fonti, P.; von Arx, G.; Bryukhanova, M.V.; Cuny, H.E.; Carrer, M.; Castagneri, D Frank, D.C. Cell size and wall dimensions drive distinct variability of earlywood and latewood density in Northern Hemispher conifers. *New Phytol.* **2017**, *216*, 728–740. [CrossRef] [PubMed]
31. Rathgeber, C.B. Conifer tree-ring density interannual variability—Anatomical, physiological and environmental determinants *New Phytol.* **2017**, *216*, 621–625. [CrossRef] [PubMed]
32. Kalugin, I.A.; Daryin, A.V.; Babich, V.V. Reconstruction of annual air temperatures for three thousand years in Altai region by lithological and geochemical indicators in Teletskoe Lake sediments. *Dokl. Earth Sci.* **2009**, *426*, 681–684. [CrossRef]
33. Wen, R.; Xiao, J.; Chang, Z.; Zhai, D.; Xu, Q.; Li, Y.; Itoh, S. Holocene precipitation and temperature variations in the East Asiar monsoonal margin from pollen data from Hulun Lake in northeastern Inner Mongolia, China. *Boreas* **2010**, *39*, 262–272. [CrossRef]
34. Wang, S.; Gong, D.; Zhu, J. Twentieth-century climatic warming in China in the context of the Holocene. *Holocene* **2001**, *11*, 313–321 [CrossRef]
35. Esper, J.; Shiyatov, S.G.; Mazepa, V.S.; Wilson, R.J.S.; Graybill, D.A.; Funkhouser, G. Temperature-sensitive Tien Shan tree ring chronologies show multi-centennial growth trends. *Clim. Dyn.* **2003**, *21*, 699–706. [CrossRef]
36. Allen, C.D.; Macalady, A.K.; Chenchouni, H.; Bachelet, D.; McDowell, N.; Vennetier, M.; Kitzberger, T.; Rigling, A.; Breshears D.D.; Hogg, E.H.; et al. A global overview of drought and heat-induced tree mortality reveals emerging climate change risks for forests. *For. Ecol. Manag.* **2010**, *259*, 660–684. [CrossRef]
37. Ponomarev, E.I.; Shvetsov, E.G.; Kharuk, V.I. Fires in the Altai-Sayan region: Landscape and ecological confinement. *Izv. Atmos Ocean. Phys.* **2016**, *52*, 725–736. [CrossRef]
38. Hammond, W.M.; Williams, A.P.; Abatzoglou, J.T.; Adams, H.D.; Klein, T.; López, R.; Saenz-Romero, C.; Hartmann, H.; Dreshears D.D.; Allen, C.D. Global field observations of tree die-off reveal hotter-drought fingerprint for Earth's forests. *Nat. Commun.* **2022** *13*, 1761. [CrossRef] [PubMed]
39. Bocharnikov, M.V. Climate-related gradients on vegetation diversity of the Altai-Sayan orobiome (Southern Siberia). *Geogr. Environ. Sustain.* **2023**, *15*, 17–31. [CrossRef]
40. Babushkina, E.; Belokopytova, L.; Zhirnova, D.; Barabantsova, A.; Vaganov, E. Divergent growth trends and climatic response of *Picea obovata* along elevational gradient in Western Sayan mountains, Siberia. *J. Mt. Sci.* **2018**, *15*, 2378–2397. [CrossRef]
41. Kostyakova, T.V.; Belokopytova, L.V.; Zhirnova, D.F.; Babushkina, E.A.; Vaganov, E.A. Dendrochronological indication of phyllophages' outbreaks by larch radial growth in the forest-steppe zone of the Republic of Tyva. *Contemp. Probl. Ecol.* **2021**, *14*, 37–48. [CrossRef]
42. Olatinwo, R.; Guo, Q.; Fei, S.; Otrosina, W.; Klepzig, K.D.; Streett, D. Climate-induced changes in vulnerability to biological threats in the Southern United States. In *Climate Change Adaptation and Mitigation Management Options: A Guide for Natural Resource Managers in Southern Forest Ecosystems*; Vose, J.M., Klepzig, K.D., Eds.; CRC Press—Taylor & Francis Group: Boca Raton, FL, USA, 2013; pp. 127–172. [CrossRef]
43. Levi, K.G.; Zadonina, N.V.; Berdnikova, N.E.; Voronin, V.I.; Glyzin, A.V.; Yazev, S.A.; Baasandzhav, B.; Ninzhbadgar, S.; Balzhinnyam, B.; Buddo, V.Y. *Modern Geodynamics and Heliogeodynamics: A 500-Year History of Anomalous Phenomena in the Natural and Social Life of Siberia and Mongolia*; Irkutsk State Technical University: Irkutsk, Russia, 2003; 382p. (In Russian)
44. Tachiiri, K.; Shinoda, M.; Klinkenberg, B.; Morinaga, Y. Assessing Mongolian snow disaster risk using livestock and satellite data. *J. Arid Environ.* **2008**, *72*, 2251–2263. [CrossRef]
45. Cohen, J.; Saito, K.; Entekhabi, D. The role of the Siberian high in Northern Hemisphere climate variability. *Geophys. Res. Lett.* **2001**, *28*, 299–302. [CrossRef]
46. Lafon, C.W.; Speer, J.H. Using dendrochronology to identify major ice storm events in oak forests of southwestern Virginia. *Clim. Res.* **2002**, *20*, 41–54. [CrossRef]

7. Mayr, S. Limits in water relations. In *Trees at Their Upper Limit: Treelife Limitation at the Alpine Timberline*; Wieser, G., Tausz, M., Eds.; Springer: Dordrecht, The Netherlands, 2007; pp. 145–162. [CrossRef]

8. CABI. *Pines of Silvicultural Importance: Compiled from the Forestry Compendium*; CABI: Wallingford, UK, 2001; 531p.

9. Jones, P.D.; Briffa, K.R.; Barnett, T.P.; Tett, S.F.B. High-resolution palaeoclimatic records for the last millennium: Interpretation, integration and comparison with General Circulation Model control-run temperatures. *Holocene* **1998**, *8*, 455–471. [CrossRef]

0. D'Arrigo, R.; Klinger, P.; Newfield, T.; Rydval, M.; Wilson, R. Complexity in crisis: The volcanic cold pulse of the 1690s and the consequences of Scotland's failure to cope. *J. Volcanol. Geotherm. Res.* **2020**, *389*, 106746. [CrossRef]

1. Zhirnova, D.F.; Belokopytova, L.V.; Upadhyay, K.K.; Koshurnikova, N.N.; Mapitov, N.B.; Kholdaenko, Y.A.; Vaganov, E.A.; Babushkina, E.A. Climatic reactions in the radial growth of *Pinus sibirica* Du Tour from the lower to the upper limit along the Western Sayan Mountains, Siberia. *For. Ecol. Manag.* **2024**, *563*, 121995. [CrossRef]

2. Myglan, V.S. *Climate and Society of Siberia During the Little Ice Age*; Siberian Federal University: Krasnoyarsk, Russia, 2010; 230p. (In Russian)

3. Cook, E.R.; Krusic, P.J. *Program ARSTAN: A Tree-Ring Standardization Program Based on Detrending and Autoregressive Time Series Modeling, with Interactive Graphics*; Lamont-Doherty Earth Observatory, Columbia University: Palisades, NY, USA, 2005; 14p.

plants

MDPI

Article

Physiological and Biochemical Responses of Pseudocereals with C3 and C4 Photosynthetic Metabolism in an Environment with Elevated CO_2

Bruna Evelyn Paschoal Silva [1], Stefânia Nunes Pires [1], Sheila Bigolin Teixeira [1], Simone Ribeiro Lucho [1], Natan da Silva Fagundes [1], Larissa Herter Centeno [2], Filipe Selau Carlos [2], Fernanda Reolon de Souza [3], Luis Antonio de Avila [3] and Sidnei Deuner [1,*]

1 Department of Botany, Biology Institute, Federal University of Pelotas, Pelotas 96010-610, RS, Brazil; brunabiologia89@hotmail.com (B.E.P.S.); stefanianunespires@gmail.com (S.N.P.); sheilabigolin@gmail.com (S.B.T.); simonibelmonte@gmail.com (S.R.L.); natanfagundes@gmail.com (N.d.S.F.)
2 Department of Soils, Faculty of Agronomy Eliseu Maciel, Federal University of Pelotas, Pelotas 96010-610, RS, Brazil; larissa.centeno@ufpel.edu.br (L.H.C.); filipeselaucarlos@hotmail.com (F.S.C.)
3 Department of Plant and Soil Sciences, Mississippi State University, Mississippi State, MS 39762, USA; fr278@msstate.edu (F.R.d.S.); luis.avila@pss.msstate.edu (L.A.d.A.)
* Correspondence: sdeuner@yahoo.com.br

Abstract: The present work aimed to investigate the effect of increasing CO_2 concentration on the growth, productivity, grain quality, and biochemical changes in quinoa and amaranth plants. An experiment was conducted in open chambers (OTCs) to evaluate the responses of these species to different levels of CO_2 {$a[CO_2]$ = 400 ± 50 μmol mol^{-1} CO_2 for ambient CO_2 concentration, $e[CO_2]$ = 700 ± 50 μmol mol^{-1} CO_2 for the elevated CO_2 concentration}. Growth parameters and photosynthetic pigments reflected changes in gas exchange, saccharolytic enzymes, and carbohydrate metabolism when plants were grown under $e[CO_2]$. Furthermore, both species maintained most of the parameters related to gas exchange, demonstrating that the antioxidant system was efficient in supporting the primary metabolism of plants under $e[CO_2]$ conditions. Both species were taller and had longer roots and a greater dry weight of roots and shoots when under $e[CO_2]$. On the other hand, the panicle was shorter under the same situation, indicating that the plants invested energy, nutrients, and all mechanisms in their growth to mitigate stress in expense of yield. This led to a reduction on panicle size and, ultimately, reducing quinoa grain yield. Although $e[CO_2]$ altered the plant's metabolic parameters for amaranth, the plants managed to maintain their development without affecting grain yield. Protein levels in grains were reduced in both species under $e[CO_2]$ in the average of two harvests. Therefore, for amaranth, the increase in CO_2 mainly contributes to lowering the protein content of the grains. As for quinoa, its yield performance is also affected, in addition to its protein content. These findings provide new insights into how plants C3 (amaranth) and C4 (quinoa) respond to $e[CO_2]$, significantly increasing photosynthesis and its growth but ultimately reducing yield for quinoa and protein content in both species. This result ultimately underscore the critical need to breed plants that can adapt to $e[CO_2]$ as means to mitigate its negative effects and to ensure sustainable and nutritious crop production in future environmental conditions.

Keywords: *Amaranthus* spp.; *Chenopodium quinoa* (Willd); climate change; photosynthetic parameters; carbohydrate metabolism

1. Introduction

With the advent of innovative techniques, it has become possible to detect even minor climate changes; however, these alterations have occurred over the past 2 million years [1]. In the past, gas releases may have occurred naturally from massive seabed deposits of methane hydrates, emissions from volcanic eruptions, or decay of vegetation associated with asteroid impacts. Nowadays, anthropogenic emissions of greenhouse gases (GHGs) may have similar effects. These emissions have massive effects on the global carbon cycle and are causing significant climate changes [2].

The concentration of GHGs in the Earth's atmosphere is directly linked to the average global temperature of the Earth, which, at the same time, has been rising steadily since the time of the Industrial Revolution, and the most abundant gas, accounting for two-thirds of GHGs is CO_2, the most common byproduct of fossil fuel burning [3].

Results of large-scale experiments presented significant variations and showed that an increase in CO_2 does not necessarily promote plant growth, varying from species to species [4,5]. These changes, considered positive in a short time, mainly promote the improvement of water use efficiency (amount of organic matter produced by the amount of water used) in both C4 and C3 plants.

The impacts of climate change on plant physiology and the increase in global population have raised serious concerns about food security [6]. Moreover, Brazil is a special case in this context because agriculture is a key economic activity, and most people working in agriculture in the country are family farmers. Smallholders help to maintain the diversity of agricultural products and nutrients worldwide, collaborating with food security in this scenario [7].

The most direct and simplest adaptive measure would be to encourage these farmers to conduct research that supports the implementation of different cultures [8]. In addition, smallholders should preferentially cultivate products with higher added value, such as Pseudocereals. The most likely replacement products are derived from protein-rich crops, including pseudocereals. Although these crops have declined in production and consumption over several decades, they have achieved significant recognition in recent years due to their effectiveness and sustainability [9].

According to the American Heritage Dictionary of the English language, the term pseudocereals can be defined as any plant that does not belong to the grass family. On the other hand, these plants produce fruits and seeds that are usually used as flour for bread and other staple foods [10]. Besides that, like cereals, they have starchy, dry seeds and are much more protein-rich. Plants usually included in the non-systematic grouping of pseudocereals are dicotyledonous and belong to different families. Consequently, plants such as quinoa and amaranth are placed in the Chenopodiaceae and Amaranthaceae family, which belongs to the Caryophyllales order, a subclass of Caryophyllidae [11,12].

Given the growing evidence that the research about changes and impacts in climate is leading to a better understanding of the potential pressures on the ability to ensure an adequate food supply for the human population, comprehending how plants respond is essential to precede studies related to these effects on the plant physiology [13].

Evidence that elevated carbon dioxide concentration {$e[CO_2]$} can cause rapid growth has been proved over the years since the first research in the 1970s and 1980s [14,15]. These alterations happen because photosynthetic carbon gain results in an enhanced carbohydrate metabolism. Consequently, there are alterations in plants' assimilate partitioning and sink strength [16,17].

As an important component of source–sink, the model of Munch postulates that unloading and loading of the conducting tissue are mainly driven by concentration and/or osmotic gradients [18,19]. From another point of view, the source–sink can be considered as the competitive ability of an organ to import photoassimilates [20]. However, studies about how the carbohydrate metabolism of pseudocereals responds to $e[CO_2]$ and what may be the impacts of the alteration of the source–sink function resulting from the effect of the increase in the concentration of atmospheric carbon on the nutrition, metabolism, and productivity of quinoa and amaranth plants are incipient.

Therefore, this study aimed to verify the effects of $e[CO_2]$ in amaranth (*Amaranthus* spp.) and quinoa (*Chenopodium quinoa*, Willd) plants, considering that these species have C3 and C4 metabolism, respectively. Here, it was demonstrated in a 2-year experiment that the relationship of physiological parameters such as growth and photosynthesis are intimately linked with source–sink, sucrose metabolism, and nutrient content influencing the grains' productivity.

Considered the main greenhouse gas (GHG), atmospheric carbon dioxide (CO_2) concentration has increased progressively since the Industrial Revolution. The last IPCC report indicated that the National Oceanic and Atmospheric Administration (NOAA) estimated a global [CO_2] was 421 ppm [2]. In addition to being a "greenhouse" gas, CO_2 is also the source of carbon for plant photosynthesis and growth, and the ongoing increase in its concentration has been shown to stimulate a wide range of plant species. However, the extent of any stimulation can be related to photosynthetic biochemistry, with C3 plants generally showing a more robust response than C4 plants. Indirect effects, primarily through *e*[CO_2] closing stomata, can also improve water use efficiency for both C3 and C4 species [4,21].

Understanding the behavior and adaptation of alternative crops to climate change is essential because crop diversity is key to adapting to climate change [7,13]. This is especially true for small family farms [8]. One important option for diversity in small family farms is to grow protein-rich crops, including pseudocereals. These pseudocereals have been recognized in recent years due to their nutritional values [9]. Quinoa and amaranth are important pseudocereals belonging to the Caryophyllales order, a subclass of Caryophyllidae from the Chenopodiaceae and Amaranthaceae families, respectively [12]. They differ in photosynthetic biochemistry and potential response to *e*[CO_2], with quinoa being a C3 species and amaranth a C4.

Quinoa and amaranth are considered inexpensive and abundant sources of digestive fiber, protein containing methionine and lysine, vitamin C, carotenoids, and minerals. It is also a great source of antioxidant pigments, such as betacyanin, betaxanthin, betalain, amaranthine, and bioactive phytochemicals, including flavonoids and phenolic acids. These bioactive components of natural origin can quench ROS [22]. Humans and animals have extensively consumed these crops in the Andes for millions of years [23] and have been studied recently for their ability to develop in adverse conditions such as high temperature, drought, and salinity [24,25]. Although, at present, there are data regarding how quinoa and amaranth can respond to rising CO_2 and climate change, there are still fundamental unknowns regarding primary metabolism, including changes in nutritional quality, antioxidant capacity, and associations with photosynthetic and growth parameters. Therefore, this study aimed to assess the effects of future high CO_2 *e*[CO_2] on quinoa and amaranth.

2. Materials and Methods

2.1. Growth Conditions

Seeds of the cultivar BRS Alegria (amaranth) and BRS Piabiru (quinoa) were sown in polystyrene trays on commercial substrate (Plantmax®, Cascavel, PR, Brazil). After the second pair of true leaves appeared, the seedlings were transplanted into 8 L polyethylene pots filled with soil, which was previously analyzed for its physical and chemical attributes, amended, and fertilized according to technical recommendations. In the transition from the vegetative to the reproductive stage, leaves were collected for subsequent growth, physiological, and biochemical analyses.

2.1.1. Description of the Experimental Site (OTCs)

The experiments were conducted using open-top chambers (OTC) from the Weed Science Center of the Federal University of Pelotas (Capão do Leão—RS). These chambers are equipped with sensors, an automated CO_2 concentration control center, coolers responsible for homogenizing the air inside them, and a gas injection and distribution valve system in each chamber. The OTC has a useful area of 4 m^2 and 2.15 m in height, is coated with a 150-micron thick, transparent polyethylene plastic film, and is equipped with a top-opening reducer to deflect the air and prevent the dilution of the desired concentration of CO_2 inside the chamber. Carbon dioxide (Messer®, Canoas, RS, Brazil) used was 99.9% pure and was supplied through a storage cylinder (capacity of 25 kg CO_2) coupled to the injection and distribution system of the chambers.

2.1.2. Elevated CO_2 Treatment

Two experiments were conducted separately in 2019/2020 and 2020/2021. Four OTCs with two different pseudocereals (amaranth and quinoa) were employed in each experimental run. The plants were grown at two levels, 400 ± 50 μmol mol^{-1} of CO_2 {*ambient* CO_2 concentration = *a*[CO_2]} and 700 ± 50 μmol mol^{-1} of CO_2 {*elevated* CO_2 concentration, = *e*[CO_2]} until the experiment was finished (Supplementary Figure S1). In short, two studies have been conducted with pseudocereal species at different levels of CO_2 (400 and 700 μmol mol^{-1}). The internal temperature of the OTCs within each experimental year was monitored daily using a data logger (HOBO Pro v2), as shown in Supplementary Figure S1A,B. The mean daily relative humidity (RH) from December 1 to 31 was 62.92% in 2019/2020 and 71.75% in 2020/2021. Concerning RH (minimum/maximum') was 36.70/80.00%' in 2019/2020 and 53.50/87.70% in 2020/2021.

2.2. *Growth Parameters*

Shoot (SL) and root length (RL): Measured using a graduated ruler (cm). Shoot (SDM) and root (RDM) dry matter: Obtained by drying the samples in an oven at 65 °C until constant weight (mg plant^{-1}). Stem diameter: Measured using a digital caliper (mm). Panicle length (PL): Evaluated using a graduated ruler (cm). Branches per panicle (BPP): Obtained by counting the number of branches per panicle per plant. Leaf area (cm^2) was estimated using the following methodologies:

Amaranth leaf area (LA): Measured using the equation 2HC/3, as described by Monteiro et al. [26], where 2/3 is the form factor determined for amaranth leaves, while H and C indicate the largest leaf dimensions in the longitudinal and transversal directions. The last pair of expanded leaves were used.

Quinoa leaf area (LA): Measured using the equation F (LxC), where F corresponds to the correction factor 0.6079 described by Benincasa [27] and L and C indicate the largest leaf dimensions in the longitudinal and transversal directions. The last pair of expanded leaves were used.

2.3. *Physiological and Biochemical Parameters*

Photosynthetic pigments: The photosynthetic pigments were quantified according to the methodology proposed by Welburn [28]. For this purpose, one leaf disc was obtained from two young-expanded leaves per experimental unity, sampling ten repetitions per treatment. The leaves were cut into small segments, using 0.01 g of fresh sample inserted into test tubes containing 3.5 mL of dimethyl sulfoxide (DMSO) neutralized with 5% calcium carbonate. Then, the tubes were incubated in a water bath at a temperature of 65 °C for 1 h, protected from light, and then cooled in the dark until reaching room temperature. After, absorbance readings at 480 nm, 649 nm, and 665 nm were taken in a spectrophotometer. Chlorophyll *a*, *b*, total, and carotenoid contents were calculated based on the equations: chlorophyll *a* = (12.47 × A665) − (3.62 × A649); chlorophyll *b* = (25.06 × A649) − (6.5 × A665); carotenoids = (1000 × A480) − (1.29 × chlorophyll a) − (53.78 × chlorophyll b)/220; and the results were expressed in mg g^{-1} FW.

Leaf gas exchange: The leaf gas exchange was determined using an Infrared Gas Analyzer LI 6400 XT (LI-COR Environmental, Lincoln, NE, USA). The evaluation was carried out between 8:30 and 10:00 a.m. The concentration of CO_2 in the chamber was matched for each treatment (400 and 700 μmol mol^{-1} CO_2), and the photon flux density was regulated to 1500 μmol of photons m^{-2} s^{-1} with a light source attached to the measuring chamber. Moreover, the leaf temperature ranged from 26 to 27 °C in 2019/2020 and 26 to 29 °C in 2020/2021. Regarding the VPD leaf, the mean was 1.76 and 1.89 in 2019/2020 and 2020/2021, respectively. Net CO_2 assimilation (A), stomatal conductance (g_s), internal concentration of CO_2 (Ci), and transpiration rate (E) were measured using the middle third of the youngest expanded leaf. Water use efficiency (WUE) was obtained through the A/E ratio. Leaf gas exchange measurements were performed only once for each experiment. The analysis was performed in the transition period, i.e., vegetative to flowering.

Total soluble sugars (TSS), starch, sucrose (SUC), and total soluble amino acids (SAA) in leaves: The collected material was standardized, using approximately 250 mg of the middle third of two fully expanded leaves, with four repetitions per treatment. After being weighed, the material was macerated in 8 mL of extracting solution M:C:W (methanol: chloroform: ultra-pure water in the proportion of 12:5:3) and stored in amber flasks for 24 h in the dark. After this period, 2 mL of M:C:W solution was added, and the extract was centrifuged at 2500× *g* for 30 min. After centrifugation, 8 mL of the supernatant was transferred to Falcon tubes, and 2 mL of chloroform and 3 mL of milli-Q water were added. The falcons were centrifuged again for 30 min at 2500× *g* for phase separation. The upper phase was collected and concentrated by evaporation to approximately 50% of the volume at 30 °C to eliminate the excess methanol and chloroform residues present. The extract obtained at the end was later used for quantification of TSS [29], SUC [30]and SAA [31].

After drying at room temperature, the precipitate obtained from the first centrifugation was resuspended in 8 mL of 10% (*w:v*) trichloroacetic acid (TCA). In the above precipitate, 10 mL of 30% perchloric acid was added. After stirring for 30 min, the tubes containing the reaction medium were centrifuged at 2500 rpm for 30 min. Starch was quantified from the collected supernatant [29].

Quantification of TSS was completed using test tubes with screw caps bathed in ice. After adding the extracts diluted in pure water, 1.5 mL of anthrone solution (0.15% in concentrated sulfuric acid) was added to each tube. After 15 min, the tubes were shaken and incubated at 90 °C for 20 min. After that, the tubes were kept in the dark until reaching room temperature. Starch determination was performed in the same way as TSS. At the end of the process, the values obtained were multiplied by the correction factor of 0.9 for conversion into starch contents. The determination of PSA was carried out using the same methodology as AST. Readings were performed in a spectrophotometer at wavelengths of 620 nm for total soluble sugars, starch, water-soluble polysaccharides, and sucrose and 570 nm for total soluble amino acids. For the quantification of sucrose, test tubes with screw caps bathed in ice, extracts were used 100 µL of 30% KOH was transferred to tubes. The tubes were incubated in a water bath for 10 min at 100 °C. After reaching room temperature, 3 mL of anthrone (0.15% in 70% sulfuric acid) were incubated again in a water bath at 40 °C for 15 min.

SAA contents were determined from extracts plus 0.5 mL of 0.2 M citrate buffer pH 5.0, 0.2 mL of 5% ninhydrin reactive in ethylene glycol monomethyl ether, and 1 mL of 2% (*v/v*) KCN in methyl cellosolve (prepared from the 0.01 M KCN solution in pure water). The capped test tubes were incubated in a water bath at 100 °C for 20 min. After 20 min at room temperature, 1.3 mL of 60% ethanol was added.

Determination of sucrose metabolism-related enzyme activity: Leaf samples from the apical portions of the plants (about 0.4 g) were ground to a fine powder in the presence of liquid N_2. The extraction of the neutral/alkaline invertase (CINV) and the acid invertase enzymes (CWINV and VINV) followed the methodology described by Zeng et al. [32], with minor modifications. In each sample, 1.5 mL of extractor medium containing potassium phosphate buffer (200 mM, pH 7.5), PMSF (1 mM), $MgCl_2$ (5 mM), DTT (1 mM), and ascorbic acid (50 mM) was added and then centrifuged at 18,000× *g* for 20 min at 4 °C. The supernatant solution was collected to measure soluble invertase activity (VINV and CINV), and the precipitate was collected to measure insoluble invertase (CWINV). In addition to the reagents used for the soluble invertases, NaCl (1 M) and Tritonne-X-100 (1%) were also added for CWINV. The enzyme extract (500 µL) was added to a 1000 µL assay medium containing 500 µL sodium acetate buffer (pH 4.5 for VINV and CWIN activity and pH 7.5 for CINV activity), 200 mM sucrose, and 5 mM $MgCl_2$. The incubation temperature was 37 °C, and 200 µL aliquots were collected after 10 and 40 min to determine enzymatic activity. Enzymatic activity was evaluated by quantifying reducing sugars produced according to the dinitrosalicylic acid (DNS) method described by Miller [33]. All enzyme activities were determined in triplicate and expressed in micromoles of glucose per gram of fresh weight per min (µmol glucose g^{-1} FW min^{-1}).

Susy's activity was determined according to Lowell et al. [34], with some modifications. The enzymatic extract of Susy was prepared using 0.5 g of homogenized samples, 0.05 M HEPES (pH 7.0), 1 mM EDTA (Ethylenediamine tetraacetic acid), 2 mM $MgCl_2$ and DTT (dithiothreitol), 0.1 M ascorbic acid and water. The homogenate was centrifuged at 13,000× *g* for 20 min at 4 °C. Then, 100 µL of extract (supernatant) was added to 1900 µL of the medium containing 0.1 M morpholino ethanesulfonic acid (MES) buffer (pH 6.0), 0.005 M $MgCl_2$, 0.3 M sucrose, 0.005 M uridine 5′diphosphoglucose disodium (UDP) and water. The determination of Susy enzyme activity was the same as that for the invertases.

Antioxidant activity: The ability of the extracts to scavenge the 1,1-diphenyl-2-picrylhydrazyl radical (DPPH) was determined according to Pérez-Tortosa et al. [35]. Briefly, 50 µL of a series of diluted thyme extracts were added to 1 mL of a 100 µM methanol solution of DPPH. An absorbance at 517 nm was measured after a 30-minute incubation period at room temperature in the dark, and the readings were compared. The absorbance readings were compared to a calibration curve constructed using caffeic acid (0–1500 µM). The results were expressed as micromoles of reduced DPPH per gram fresh weight using an extinction coefficient of 12,500 M^{-1} cm^{-1} at 517 nm.

Nutrient contents in leaves: The mixture of leaves from different parts of the plants was collected and placed in a forced air oven at 65 °C until constant weight and then double ground in a mill, according to Tedesco et al. [36]. Approximately 200 mg were weighed on an analytical balance for subsequent sulfuric digestion of macronutrients and 500 mg for nitrous-perchloric digestion of micronutrients. From the digested material, the reading of nitrogen (N)—Kjeldhal method [37]; phosphorus (P)—spectrophotometer at 660 nm; potassium (K)—flame photometer B 462 (Micronal, São Paulo, Brazil); calcium (Ca); magnesium (Mg); zinc (Zn); copper (Cu); manganese (Mn) and iron (Fe)—Flame atomic absorption spectrophotometer Model AA 990F (PG Instruments Limited, Woodway lane, Alma park, Leicestershire, UK).

2.4. Yield Components

Yield components per pot: Ten replicates per treatment were used for the grain yield components, where each pot with a plant was considered a replicate. The weight of one thousand grains pot^{-1}, number of panicle $grains^{-1}$, and weight of grains pot^{-1} were determined.

Crude protein: This was determined using the Kjeldahl method [37] based on three steps: digestion, distillation, and titration. A 200 mg amount of grain flour was used in duplicate.

The calculation for the determination of total nitrogen was as follows:

$$NT = (Va - Vb) \times F \times 0.1 \times 0.014 \times 100/P1$$

Being:
NT—Total nitrogen content in the sample, in percentage;
Va—Volume of hydrochloric acid solution used in sample titration;
Vb—Volume of hydrochloric acid solution used in blank titration;
F—Correction factor for hydrochloric acid;
P1—Sample mass (in grams).

To determine the total seed protein content, the value of total nitrogen verified by the Kjeldahl method [37] was multiplied by the factor conversion of nitrogen into protein; in this case, the value used was 6.75 (amaranth) and 6.25 (quinoa). The formula below was used to determine seed protein content: PT = NT x Fc where PT—total protein; NT—total nitrogen; Fc—conversion factor.

2.5. Experimental Design and Data Analyses

Two independent experiments (2019/2020 and 2020/2021) were carried out in a completely randomized design. Each pseudocereal species (amaranth or quinoa) was exposed to two treatments: (1) $a[CO_2]$—400 ppm, and (2) $e[CO_2]$—700 ppm, with ten individual plants to each treatment, totalizing 20 quinoa or 20 amaranth plants. In the current study, the ten pots planted with amaranth or quinoa in each OTC were treated as biological replicates (thus, $n = 10$ for $a[CO_2]$ and $n = 10$ for $e[CO_2]$. The data obtained were tested for homoscedasticity using the Bartlett test and for normality using the Shapiro–Wilk test. The analysis of variance (ANOVA) was carried out using the statistical software R (www.r-project.org/). All data were analyzed for statistical differences between the $a[CO_2]$ and the $e[CO_2]$ conditions. Afterward, if F was significant, the means were compared to the control by the *t*-test ($p \leq 0.05$). The data were expressed as the mean ± standard error (SE) of 10 replicates.

3. Results

3.1. Growth Parameters

Results presented in Figure 1 demonstrate a significant difference among treatments ($p \leq 0.05$) for all the growth parameters. When cultivated under $e[CO_2]$, plants grew more, with longer shoots (Figure 1A), longer roots (Figure 1C), and higher shoot (Figure 1B) and root (1D) dry weight (Figure 1) when compared to plants grown under $a[CO_2]$. When comparing the two species, the quinoa plants showed higher values, especially for the SDW and RDW parameters.

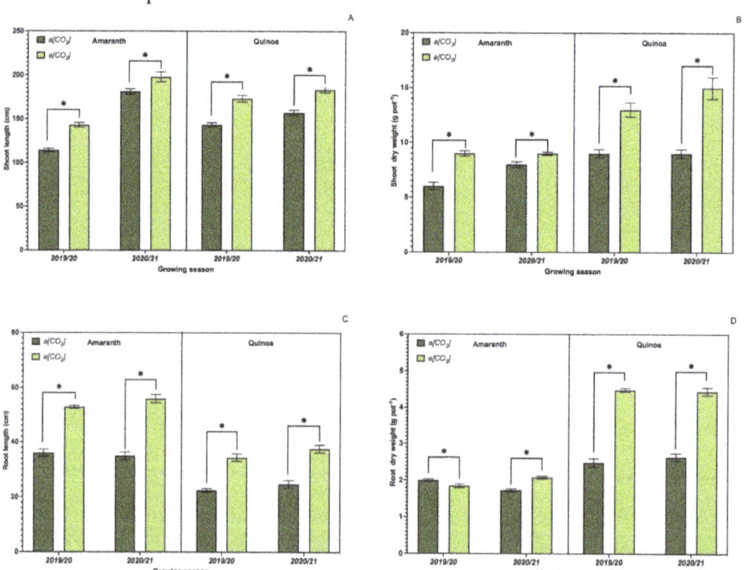

Figure 1. Effect of CO_2 on plant growth parameters. Shoot length (**A**); shoot dry matter (**B**); root length (**C**); root system dry matter (**D**) of amaranth and quinoa plants. $a[CO_2]$ = plants grown in OTC with 400 ± 50 μmol mol^{-1} CO_2; $e[CO_2]$ = plants grown in OTC with 700 ± 50 μmol mol^{-1} CO_2. The experiment was conducted during the 2019/2020 and 2020/2021 growing seasons. Error bars correspond to the 95% confidence interval of the mean. * Indicates significant difference between CO_2 concentration (*t* test $p \leq 0.05$, n = 10).

Data analysis showed significant differences between treatments when comparing the values obtained for the growth parameters (Figure 2A–D). Amaranth plants showed higher values in all parameters when subjected to $e[CO_2]$; the same occurred for quinoa plants, except for panicle length (Figure 2A) and number of branches per panicle (Figure 2B), showing higher values in $a[CO_2]$.

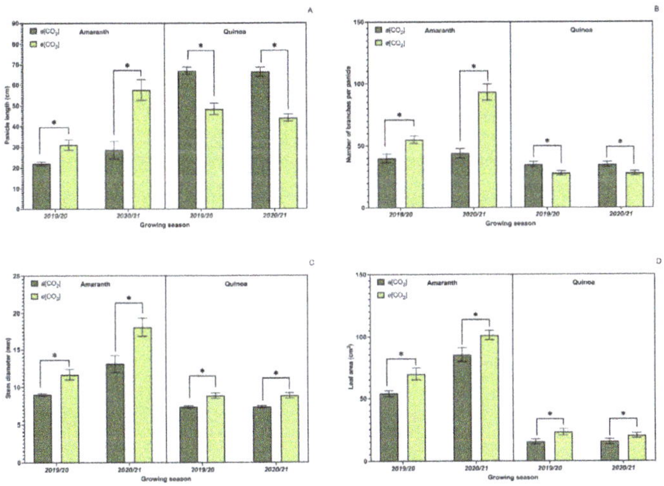

Figure 2. Effect of CO_2 on growth parameters. Panicle length (**A**); number of branches per panicle (**B**); stem diameter (**C**); leaf area (**D**) of amaranth and quinoa plants. $a[CO_2]$ = plants grown in OTC with 400 ± 50 µmol mol^{-1} CO_2; $e[CO_2]$ = plants grown in OTC with 700 ± 50 µmol mol^{-1} CO_2. The experiment was conducted during the 2019/2020 and 2020/2021 growing seasons. Error bars correspond to the 95% confidence interval. * Indicates significant difference between CO_2 concentration (t test $p \leq 0.05$, n = 10).

3.2. Photosynthetic Pigments

In general, photosynthetic pigments (Figure 3) were not affected by CO_2 treatments. Significant changes were observed only for chlorophyll-*a* (Figure 3A), chlorophyll-*b* (Figure 3B), and carotenoids (Figure 3C) in amaranth plants in the agricultural year 2020/2021.

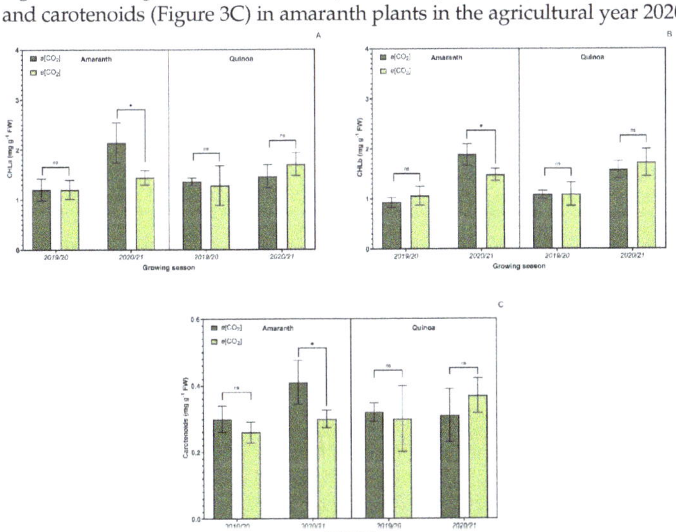

Figure 3. Effect of CO_2 concentration on photosynthetic pigments. Chlorophyll-*a* (CHLa) (**A**); chlorophyll-*b* (CHLb) (**B**); carotenoids (**C**) of amaranth and quinoa plants in transition stadium between vegetative and flowering. $a[CO_2]$ = plants grown in OTC with 400 ± 50 µmol mol^{-1} CO_2; $e[CO_2]$ = plants grown in OTC with 700 ± 50 µmol mol^{-1} CO_2. The experiment was conducted during the 2019/2020 and 2020/2021 growing seasons. Error bars correspond to the 95% confidence interval. * Indicates significant difference between CO_2 concentration (t test $p \leq 0.05$, n = 10). ns indicates non-significant differences (t test $p \leq 0.05$, n = 10).

3.3. Leaf Gas Exchange

Overall, $e[CO_2]$ produced changes in the leaf gas exchange (Figure 4), demonstrating an expressive difference between the treatments ($p \leq 0.05$) to all measured parameters. CO_2 assimilation (Figure 4A) was higher for both amaranth and quinoa at $e[CO_2]$. High CO_2 increases the net carbon assimilation rate, which is also evidenced by Ci (Figure 4C) for both crops. Therefore, a decrease in stomatal conductance and transpiration rate was observed under $e[CO_2]$. Thus, the A/E ratio was higher at $e[CO_2]$. Water use efficiency (Figure 4E) also showed a significant difference between the two treatments for both species, higher at $e[CO_2]$. However, the results for quinoa and amaranth point to the same trend regarding water use efficiency and its influence on the photosynthesis rate.

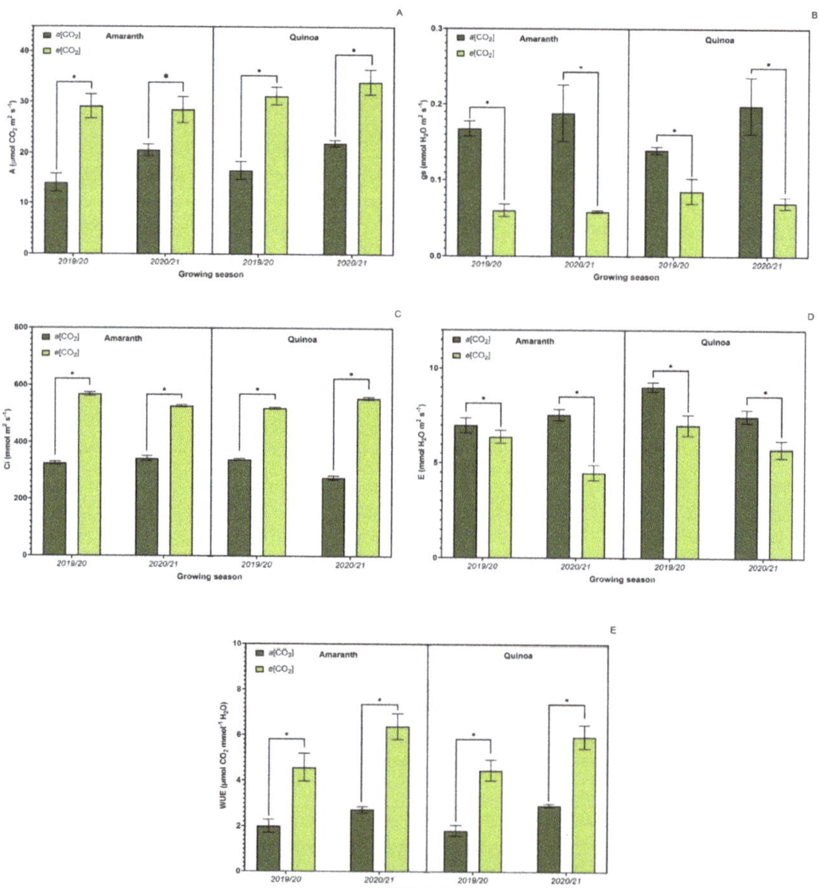

Figure 4. Effect of CO_2 on leaf gas exchange. Net CO_2 assimilation (**A**); stomatal conductance (**B**); internal concentration of CO_2 (**C**); transpiration rate (**D**); water use efficiency (WUE) (**E**) of amaranth and quinoa plants in transition stadium between vegetative and flowering. $a[CO_2]$ = plants grown in OTC with 400 ± 50 µmol mol^{-1} CO_2; $e[CO_2]$ = plants grown in OTC with 700 ± 50 µmol mol^{-1} CO_2. The experiment was conducted during the 2019/2020 and 2020/2021 growing seasons. Error bars correspond to the 95% confidence interval. * Indicates significant difference between CO_2 concentration (t test $p \leq 0.05$, n = 10).

3.4. *Sucrose Metabolism-Related Enzyme Activity*

Analysis of enzymatic activity related to sucrose metabolism (Figure 5) showed a significant increase in neutral invertase activity (Figure 5A), except for amaranth in the first agricultural year, in quinoa and amaranth plants grown under *e*[CO_2]. Acidic invertases such as soluble acid invertases of the vacuole (Figure 5B) and cell wall acid invertase (Figure 5C) showed similar trends with higher values under *e*[CO_2], except for amaranth plants in 2020/2021, when no difference was observed. Sucrose synthase activity (Figure 5D) was higher in *e*[CO_2] for both species in agricultural years.

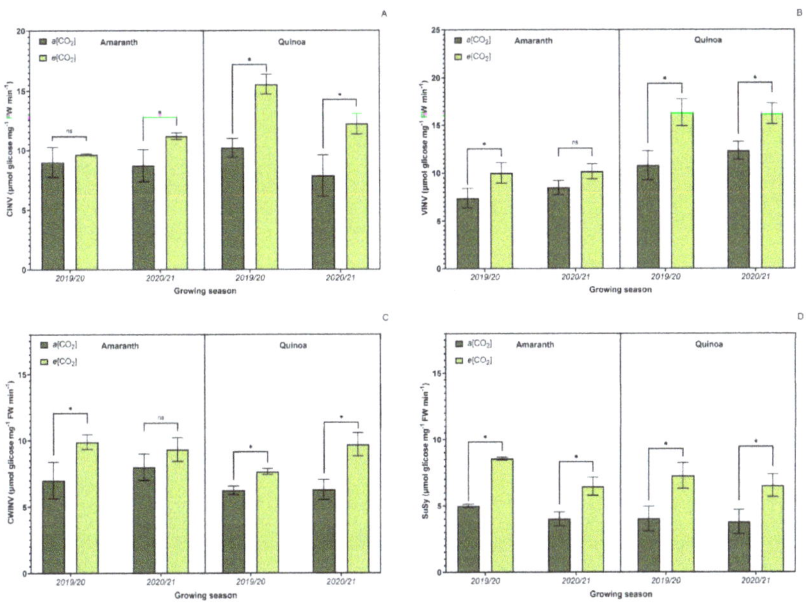

Figure 5. Effect of CO_2 on sucrose metabolism-related enzyme activity in leaves: soluble neutral invertases of cytosol (CINV) (**A**); soluble acid invertases of the vacuole (VINV) (**B**); cell wall acid invertase (CWINV) (**C**); and sucrose synthase (SuSy) (**D**) activity. *a*[CO_2] = plants grown in OTC with 400 $\pm$ 50 µmol mol^{-1} CO_2; *e*[CO_2] = plants grown in OTC with 700 $\pm$ 50 µmol mol^{-1} CO_2. The experiment was conducted during the 2019/2020 and 2020/2021 growing seasons. Error bars correspond to the 95% confidence interval. *Indicates significant difference between CO_2 concentration (*t* test $p \leq 0.05$, n = 4). ns indicates non-significant.

3.5. *Carbohydrate Metabolism*

Significant changes ($p \leq 0.05$) were observed in carbohydrate metabolism (Figure 6). The total content of soluble sugars (Figure 6A) and sucrose (Figure 6B) was higher for both cultures under *e*[CO_2], while the entire starch content (Figure 6C) showed a unique trend for each culture. Amaranth plants had a decrease in these parameters under *e*[CO_2], while quinoa plants increased, although the total soluble amino acid content (Figure 6D) was higher in both cultures under *a*[CO_2].

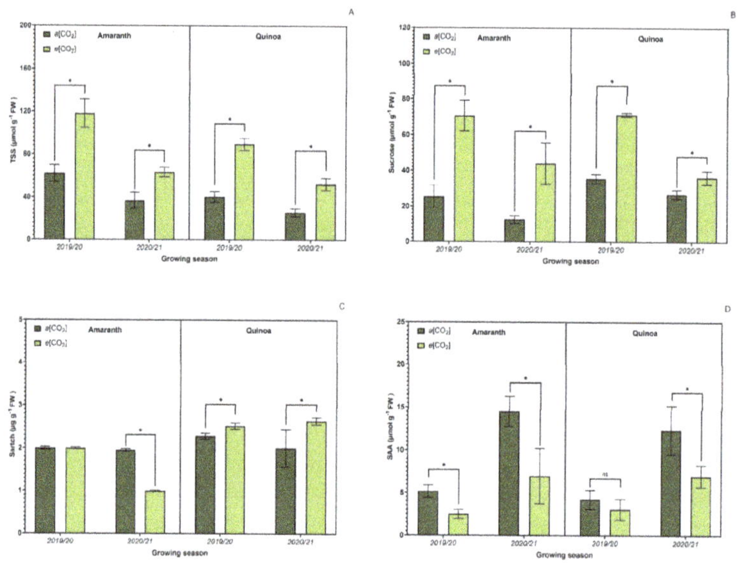

Figure 6. Effect of CO_2 on carbohydrate metabolism: total content of soluble sugars (**A**), sucrose-SUC (**B**), starch (**C**), and total soluble amino acids (**D**) on amaranth and quinoa leaves in transition stadium between vegetative and flowering. $a[CO_2]$ = plants grown in OTC with 400 ± 50 µmol mol^{-1} CO_2; $e[CO_2]$ = plants grown in OTC with 700 ± 50 µmol mol^{-1} CO_2. The experiment was conducted during the 2019/2020 and 2020/2021 growing seasons. Error bars correspond to the 95% confidence interval. * Indicates significant difference between CO_2 concentration (t test $p \leq 0.05$, n = 4). ns indicates non-significant.

3.6. Antioxidant Activity

The antioxidant capacity evaluated by the DPPH scavenging assay (Figure 7) showed that amaranth and quinoa under $e[CO_2]$ significantly increased antioxidant capacity for both crops. Data obtained for amaranth and quinoa under $e[CO_2]$ corresponded to 8.12 (±0.67) and 9.10 (±1.04) µg caffeic acid eq g^{-1} DW, respectively, indicating the deactivation of free radicals.

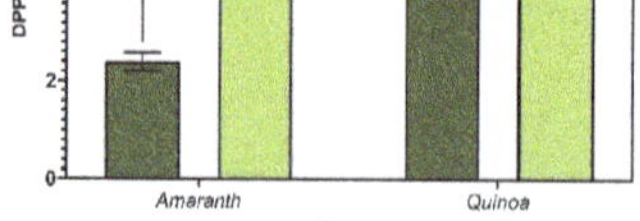

Figure 7. Antioxidant capacity evaluated by DPPH from amaranth and quinoa in transition stadium between vegetative and flowering, grown in the presence of different CO_2 concentrations (400 and 700 µmol mol^{-1} CO_2). The extracts evaluated were obtained using the growing season 19/20 and 2020/2021. $a[CO_2]$ = plants grown in OTC with 400 ± 50 µmol mol^{-1} CO_2; $e[CO_2]$ = plants grown in OTC with 700 ± 50 µmol mol^{-1} CO_2. The experiment was conducted during the 2019/2020 and 2020/2021 growing seasons. Error bars correspond to the 95% confidence interval. * Indicates significant difference between CO_2 concentration (t test $p \leq 0.05$, n = 10).

3.7. Nutrient Contents in Leaves

Overall, *e*[CO_2] decreased some macronutrients (Table 1) and micronutrients (Table 2) in amaranth and quinoa leaves. Here, we had different results for crops and agricultural years. Amaranth plants in *e*[CO_2] revealed a decline in leaf contents of N; K; Ca; Zn; Mn and Cu, whereas quinoa plants presented a decrease in N; P (quinoa first year); K; Ca (quinoa second year); Zn; and Mn in *e*[CO_2] (amaranth and quinoa two agricultural years). In general, magnesium leaf contents were non-significant. Fe and P levels were higher in *e*[CO_2] for both crops and agricultural years.

Table 1. Macronutrient contents in leaves of amaranth and quinoa plants in transition stadium between vegetative and flowering as affected by CO_2 concentration in two growing seasons.

Treatment [CO_2]	Amaranth (First Year)	Amaranth (Second Year)	Quinoa (First Year)	Quinoa (Second Year)
		N (g kg^{-1})		
a[CO_2]	52.0 ± (1.79) *	53.0 ± (1.66) *	54.9 ± (4.57) *	53.2 ± (0.54) *
e[CO_2]	49.2 ± (0.98)	36.0 ± (1.79)	48.4 ± (0.27)	38.5 ± (1.71)
CV%	1.15	1.57	2.52	1.12
		P (g kg^{-1})		
a[CO_2]	3.4 ± (0.13) *	3.1 ± (0.08) *	5.1 ± (0.42) *	2.7 ± (0.08) *
e[CO_2]	4.3 ± (0.17)	3.7 ± (0.17)	4.0 ± (0.37)	3.0 ± (0.18)
CV%	1.65	1.56	3.54	2.03
		K (g kg^{-1})		
a[CO_2]	46.5 ± (2.33) *	46.5 ± (3.87) *	50.8 ± (2.07) *	50.9 ± (1.97) *
e[CO_2]	30.7 ± (4.20)	33.9 ± (1.62)	46.3 ± (1.86)	45.7 ± (0.64)
CV%	3.54	2.97	1.63	1.22
		Ca (g kg^{-1})		
a[CO_2]	16.9 ± (1.88) *	27.3 ± (2.75) *	11.3 ± (0.46) ns	21.7 ± (1.63) *
e[CO_2]	12.9 ± (1.63)	16.9 ± (1.88)	11.2 ± (2.58)	15.3 ± (2.97)
CV%	4.74	4.29	6.66	5.23
		Mg (g kg^{-1})		
a[CO_2]	8.1 ± (0.88) *	8.8 ± (1.07) ns	6.3 ± (1.13) ns	8.9 ± (0.78) ns
e[CO_2]	9.9 ± (0.21)	7.8 ± (1.34)	6.0 ± (1.03)	8.8 ± (0.71)
CV%	2.99	5.87	7.06	3.38

* Indicates significant difference by *t*-test ($p \leq 0.05$, n = 4). ns indicates non-significant. CV: coefficient of variation. Values in parentheses correspond to the 95% confidence interval. *a*[CO_2] = plants grown in OTC with 400 ± 50 µmol mol^{-1} CO_2; *e*[CO_2] = plants grown in OTC with 700 ± 50 µmol mol^{-1} CO_2.

Table 2. Micronutrient contents in leaves of amaranth and quinoa plants in transition stadium between vegetative and flowering in response to CO_2 concentration in two growing seasons.

Treatment [CO_2]	Amaranth (First Year)	Amaranth (Second Year)	Quinoa (First Year)	Quinoa (Second Year)
		Fe (mg kg^{-1})		
a[CO_2]	164.4 ± (4.50) *	174.5 ± (1.33) *	80.1 ± (7.83) *	180.0 ± (3.63) *
e[CO_2]	185.4 ± (2.37)	207.8 ± (3.60)	135.1 ± (8.64)	235.5 ± (2.66)
CV%	0.89	2.04	3.09	
		Zn (mg kg^{-1})		
a[CO_2]	22.2 ± (2.10) *	36.1 ± (2.99) *	11.3 ± (2.31) *	49.0 ± (1.03) *
e[CO_2]	15.5 ± (4.45)	26.0 ± (2.92)	6.4 ± (4.55)	20.8 ± (0.88)
CV%	7.41	3.82	16.37	1.11

Table 2. *Cont.*

Treatment $[CO_2]$	Amaranth (First Year)	Amaranth (Second Year)	Quinoa (First Year)	Quinoa (Second Year)
		Mn (mg kg^{-1})		
$a[CO_2]$ $e[CO_2]$	554.7 ± (9.19) * 424.8 ± (2.60)	254.2 ± (4.18) * 193.7 ± (8.29)	348.4(5.42) * 284.0 ± (3.86)	755.0 ± (8.93) * 585.1 ± (9.43)
CV%	0.56	1.18	0.6	
		Cu (mg kg^{-1})		
$a[CO_2]$ $e[CO_2]$	17.6 ± (1.72) * 15.2 ± (1.72)	19.6 ± (1.72) * 17.6 ± (1.72)	12.4 ± (1.72) * 31.4 ± (1.72)	14.0 ± (1.72) * 29.2 ± (1.72)
CV%	4.21	3.73	3.15	3.2

* Indicates significant difference by *t*-test ($p \leq 0.05$, n = 4). CV: coefficient of variation. Values in parentheses correspond to the 95% confidence interval. n = 4. $a[CO_2]$ = plants grown in OTC with 400 ± 50 μmol mol^{-1} CO_2; $e[CO_2]$ = plants grown in OTC with 700 ± 50 μmol mol^{-1} CO_2.

3.8. Yield Components

Regarding the yield components, under $e[CO_2]$, there was a reduction in the 1000 grain weight for both crops (Table 3). However, there was a different performance regarding the number of grains per panicle. While the values for amaranth were non-significative, $e[CO_2]$ decreased the number of grains per panicle in quinoa plants. The grain yield per pot was reduced for quinoa under $e[CO_2]$, with 50.34% and 61.72% reduction in the first and second growing seasons, respectively.

Table 3. Yield components in grains of amaranth and quinoa plants in transition stadium between vegetative and flowering in response to CO_2 concentration.

Treatment $[CO_2]$	Amaranth (First Year)	Amaranth (Second Year)	Quinoa (First Year)	Quinoa (Second Year)
		Weight of 1000 grains (g)		
$a[CO_2]$ $e[CO_2]$	1.11 ± (0.01) * 0.86 ± (0.03)	1.10 ± (0.00) * 0.81 ± (0.01)	2.41 ± (0.07) * 1.71 ± (0.04)	2.48 ± (0.01) * 1.70 ± (0.00)
CV%	1.87	0.74	1.89	0.28
		Number of grains per panicle		
$a[CO_2]$ $e[CO_2]$	10.921 ± (11.28) ^{ns} 9.876 ± (10.52)	9.879 ± (7.5) ^{ns} 10.922 ± (9.14) ^{ns}	2.598 ± (2.66) * 2.174 ± (3.48)	2.772 ± (18.45) * 1.830 ± (14.0)
CV%	13.12	11.44	18.2	9.99
		Grain yield per pot (g)		
$a[CO_2]$ $e[CO_2]$	9.37 ± (0.96) ^{ns} 9.30 ± (0.73)	8.40 ± (0.66) ^{ns} 8.41 ± (0.70)	5.72 ± (0.69) * 2.84 ± (0.62)	5.67 ± (0.78) * 2.17 ± (0.47)
CV%	12.83	11.38	21.58	23.08

* Indicates significant difference by *t* test ($p \leq 0.05$, n = 4); ns indicates non-significant. CV: coefficient of variation. Values in parentheses correspond to the 95% confidence interval (n = 10). $a[CO_2]$ = plants grown in OTC with 400 ± 50 μmol mol^{-1} CO_2; $e[CO_2]$ = plants grown in OTC with 700 ± 50 μmol mol^{-1} CO_2.

3.9. Crude Protein

For the grain protein content (SPC), it was possible to observe that $e[CO_2]$ reduced this component (Table 4). The protein levels in grains were reduced in both species under $e[CO_2]$. On average, during two growing seasons, protein content was reduced by 32% in amaranth and 50.5% in quinoa seeds.

Table 4. Effect of CO_2 concentration during the plant growing season on seed/grain crude protein content of amaranth and quinoa plants.

Treatment [CO_2]	Crude Protein (%)			
	Amaranth (First Year)	Amaranth (Second Year)	Quinoa (First Year)	Quinoa (Second Year)
a[CO_2]	18.28 ± (1.35) *	17.99 ± (0.37) *	15.97 ± (0.97) *	15.94 ± (0.92) *
e[CO_2]	12.74 ± (0.62)	11.87 ± (0.30)	7.97 ± (0.89)	7.79 ± (0.41)
CV%	4.29	1.45	4.92	3.81

* Indicates significant difference by *t* test ($p \leq 0.05$, n = 4); CV: coefficient of variation. Values in parentheses correspond to standard deviation. *a*[CO_2] = plants grown in OTC with 400 ± 50 µmol mol^{-1} CO_2; *e*[CO_2] = plants grown in OTC with 700 ± 50 µmol mol^{-1} CO_2.

4. Discussion

The hypothesis that C4 plants can respond to atmospheric CO_2 concentrations as much as C3 plants has proven useful, though not always entirely predictive. This variability is influenced by genetic improvements [38], which result in differing responses depending on genotype, variety, and cultivar [39]. Consequently, the long-held assumption, deeply embedded in models of past vegetation–climate interactions, may require a fresh perspective [40].

Plant growth is a complex process influenced by numerous factors, including increased CO_2 levels. This is what makes growth parameters important since they are related to greater light absorption [41,42]. Here, all of those results were increased by *e*[CO_2]. Both an increase in shoot and root length (consequently, shoot and root dry matter) are major traits that can be attributed to the initial effects of *e*[CO_2] in plants. Similar results were presented by Song et al. [42] for amaranth cultivated in 500 ppm and Bunce [43] for quinoa in 600 ppm of CO_2, respectively. These experiments also were conducted in OTC.

The root system not only takes up soil nutrients and water for sustainable plant production but also pumps photosynthetically fixed C to soil organic matter (SOM) pools. It plays a crucial role in terrestrial C cycling. The *e*[CO_2] exerts a substantial impact on these systems by influencing the morphology of root length and distribution. In addition, secondary roots control ecosystem C and N cycling as plants obtain water and nutrients and release exudates. In this study, it was possible to observe that the root system of amaranth and quinoa (Figure 8) plants from both species collected after grain maturation presented more secondary roots when in *e*[CO_2], possibly as a strategy to maintain the nutritional and water status of the plants.

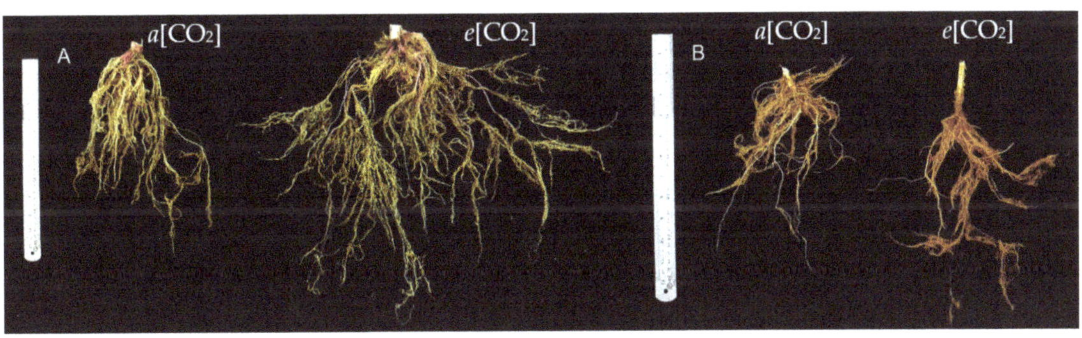

Figure 8. Effects of CO_2 concentration on roots of amaranth (**A**) and quinoa (**B**) collected after grain maturation. *a*[CO_2] = plants grown in OTC with 400 ± 50 µmol mol^{-1} CO_2; *e*[CO_2] = plants grown in OTC with 700 ± 50 µmol mol^{-1} CO_2.

An increase in root elongation and branching was evidenced in *Sedum alfredii*, $e[CO_2]$ [44]. However, other studies found that a variety of plant species show just an increased fine root production [45,46]. Corroborating the results presented here, a meta-analysis carried out by Nie et al. [47] showed that the fine root biomass of plants had a significantly stronger response to $e[CO_2]$ in OTC experiments (+35.8%).

According to Ferreira et al. [48], the stem diameter is important because it is related to the falling of the plants. To be maintained upright and support it, the $e[CO_2]$ significantly increased the stem diameter of amaranth and quinoa. Similar changes in plant morphology have been reported elsewhere, for example, $e[CO_2]$ significantly increased the shoot of C4 plants like maize [49], sugarcane [50], foxtail millet [51] and C3 plants like quinoa [52] coffee tree [53], *Stylosanthes capitata* Vogel. [54]. While to the panicle length, quinoa plants had a controversial result. The control treatment presented a higher PL than $e[CO_2]$. This phenomenon has been reported by some authors [55,56] as a downregulation or acclimatization of species to the increase in CO_2 during the stadiums. Physiological processes often develop mechanisms of compensation that reduce or minimize the long-term effects of CO_2; sometimes, a limitation of its regeneration capacity can be observed [57].

Chlorophylls and carotenoids are pigments capable of absorbing visible radiation and triggering photochemical reactions of photosynthesis [58]. In general, plants grown in $e[CO_2]$ show alteration in photosynthetic pigments. Some authors have mentioned the reduction of chlorophyll and carotenoids in leaves due to the increase in biomass under $e[CO_2]$, called the N dilution effect [59,60]. In some cases, changes in the C:N ratio cause an effect of nitrate (NO_3) assimilation inhibition by the roots [61–63]. However, it was possible to observe two important factors: (1) change in pigments for amaranth plants (2020/2021), and (2) non-significant change but mostly higher levels in the control treatment, evidencing the importance of chlorophyll breakdown. This a highly coordinated and integral process of the plant development stages that are programmed to facilitate the dynamic remobilization of nutrients from organs/tissues to parts of the plant that are still growing, in particular to reproductive/storage organs during the transition of plants to reproductive growth [64,65].

Growth, photosynthetic pigments, and leaf gas exchange are closely linked. C4 plant species possess a "distinct pathway" of photosynthesis from C3 species, which in low atmospheric CO_2 concentration could be concentrated to enable a more efficient carboxylation reaction due to changes in the ratio of CO_2:O_2 [66]. This does not mean that C4 species do not change the Calvin–Benson cycle's fundamental machinery; rather, they have functionalized structural and biochemical additions around C3 photosynthesis to improve its efficiency. It is necessary to consider that most C4 plants fix CO_2 in mesophyll cells with phosphoenolpyruvate carboxylase (PEPC), an enzyme that, unlike Rubisco, is insensitive to O_2. Subsequently, CO_2 is released in the bundle sheath cells where Rubisco is localized, and the Calvin–Benson cycle occurs. This additional step increases the availability of CO_2 around Rubisco and minimizes its chance of catalyzing the oxygenation reaction [67].

Concentration of carbon dioxide is an important regulator in the dynamics of opening and closing of stomata. Through these mechanisms, plants exchange gas with the external environment [68]. Opening the stomata allows the diffusion of CO_2 for photosynthesis, in addition to providing a path for water to diffuse from the leaves to the atmosphere [69]. Therefore, plants regulate the level of stomatal opening (stomatal conductance), seeking to maintain high rates of photosynthesis and reduce water loss. In the current study, $e[CO_2]$ increased CO_2 assimilation (Figure 6A) and Ci (Figure 6C) in quinoa plants. Partial closure of the stomata could have been responsible for reducing stomatal conductance and transpiration rate. Thus, it was possible to maintain high rates of photosynthesis without compromising the internal concentration of CO_2 since the greater difference in CO_2 concentration between the atmosphere and the interior of the leaf compensates for the increase in stomatal resistance [70].

This compensation is relatively less understood for C4 species [71]. Theoretically, when it comes to C4 species in general, researchers consider C4 plants saturated at $a[CO_2]$. Also, they might not be stimulated by $e[CO_2]$ [55]. We found significant stimulation rates, like Zhang et al. [71] for Broomcorn millet; Li et al. [51] for Foxtail millet; Davis and Ainsworth [72] for *Amaranthus rudis*, even though Santos [73] and Leakey et al. [74] had opposite results for *Amaranthus viridis* and maize, respectively.

C4 plants such as amaranth have increased vein densities during the evolution process, causing a reduction in intercellular air spaces and enhancement of bundle sheath organelles [75,76]. In addition, the increasing vein densities may not only increase structural integrity or enhance leaf water status [76] but may also allow xylem-transported CO_2 to be utilized for photosynthesis by reducing the distance between vascular bundles transporting xylem-transported CO_2 and photosynthetic cells [77].

The results of changes in C4 photosynthesis have not yet reached a consensus among researchers. Some authors mention the fact that increases observed in A at $e[CO_2]$, such as our results, are noticeable during the transition stadium between vegetative and flowering or apparent only during the early stage in some crops because of the carbohydrate metabolism. At this stage, plants are preparing to export macro and micronutrients to produce grains/seeds [78,79]. In addition, studies related to gene regulation have increasingly reported the particularity of genotypes, showing different responses to $e[CO_2]$ for the same species, mainly for C4 plants [80].

Carbon assimilated in photosynthesis is stored through carbohydrates, which are compounds generated in high quantities by plants and have high proportions of carbon [81]. In general, increases in carbohydrate production resulting from the increment in photosynthesis by $e[CO_2]$ can result in alterations in the production and partition of carbohydrates, as it raises the activity of enzymes that hydrolyze sucrose into sink organs [82,83].

Among the sugars synthesized in a plant, only a few of them are transported in the phloem over a long distance [84]. Upon arriving at sink tissues, sucrose can follow different pathways that can modulate sink strength and carbon flux [85]. Sucrose might be unloaded from the phloem to the apoplast by transporters or be hydrolyzed by invertases (CINV, CWINV, and VINV) to yield glucose and fructose, which can enter the sink cells via hexose transporters [86]. Besides that, as a reversible cleavage, SuSy might catalyze sucrose using UDP to yield fructose and UDP-G [87] and utilize other nucleotide phosphates for the cleavage, especially ADP, but usually with a lower affinity [88].

In the present study, the activity of sucrolytic enzymes was altered in leaves of amaranth and quinoa in the transition stage between vegetative and flowering by $e[CO_2]$. At this stage, considering that grain yield is dependent on the plant source/sink relationship, the top two leaves are the primary source, and the florets are the primary sink for photosynthesis. This might also change plant carbon and nitrogen metabolism [64,89]. In addition to this direct effect on photosynthesis, many physiological processes are indirectly regulated, mainly through sugar detection and signaling pathways. Sugar signaling plays an important role in the plant's response to $e[CO_2]$; however, this is not well understood concerning the plant's nutritional quality [90].

Accumulation of soluble sugars is a direct effect of $e[CO_2]$ due to the growth of triose phosphate synthesis in leaves, which can be further transformed into other carbohydrates, e.g., glucose, fructose, and sucrose. A meta-analysis made with publications between 1990 and 2018 showed that $e[CO_2]$ increased the concentrations of sucrose by 3.7% and total soluble sugar by 17.5% in leaves. This data shows a non-integration of newly fixed carbohydrates to growth, accumulating them in the leaves [91] and confirms the results presented here. Dong et al. [92] still proposed that the synthesized carbohydrates in leaves cannot be fully translocated to fruits as well as to roots, although one needs to be cautious regarding species variation.

Regarding sucrose, it is important to remember its relationship to plant growth and development. If photosynthetic capacity exceeds demand, excess photoassimilates remain in the chloroplast and can be stored in the form of starch. Thus, it is believed that the sink can control the activity of the source. Furthermore, there is an intricate relationship between source and sink, as both activities are controlled by environmental factors such as CO_2 [85,93].

As we have seen so far, $e[CO_2]$ plays a crucial role in the physiology of plants. The plant photosynthesis, stomatal aperture, biomass production, yield, and water use efficiency could be modulated by the CO_2 environment [94]. As a result, more carbohydrates could be transferred into the grains due to the increased photosynthesis in plants grown under a CO_2-enriched environment. Nevertheless, there is generally a reduction in mineral contents, particularly N concentrations (Table 3), in grown plants at $e[CO_2]$, probably due to restricted root nutrient uptake caused by reduced mass flow and dilution effect, which is reflected directly on the total soluble amino acids content in leaves [51].

Re-translocation of nutrients from leaves could be a strategy to retain P efficiently. However, this is more observed in senescence [95–97]. Re-translocation by resorption serves to withdraw nutrients from leaves before abscission for later redeployment in developing tissues. The extent to which P is re-translocated and re-used depends on the plant's nutrient status. Generally, with decreasing nutrient availability in an ecosystem, the amount of resorption of both N and P tends to increase [98,99]. This ratio could also be an indicator of which nutrient is most limiting in the ecosystem [100]. Considering that both nitrogen and phosphorus are involved in the photosynthetic machinery, most studies have reported that these nutrients also decrease due to increased carbon assimilation in several crops [101]. However, in this experiment, amaranth and quinoa (2020/2021) plants demonstrated an increase in P (Table 3).

Given that N and P availability control the global carbon cycle response to environmental changes, aspects like stoichiometric balance are important to consider. Nevertheless, the importance of P dynamics under $e[CO_2]$ is less clear. The assumption of a homeostatic N:P ratio in elevated CO_2 due to similar proportional N and P responses need to be clarified more [102,103]. Xu et al. [104] also reinforce the idea that C3 and C4 plants also have different nutritional changes under $e[CO_2]$. In a study where nitrogen availability was analyzed, Sudderth et al. [105] observed that N availability in the presence of $e[CO_2]$ increased foliar N content in *Amaranthus viridis* (C4 plant).

Several studies reported that $e[CO_2]$ decreased the mineral concentration by a dilution effect, suggesting that the decrease in mineral concentration is not specifically regulated by certain metabolic processes but by a dilution effect due to the increased biomass, as it was demonstrated earlier. It is well elucidated that the higher growth rate under high CO_2 also increases the activity of anabolic processes that require nutrients, including osmoregulation (K), cell elongation and nucleic acid metabolism (B), metabolic pathways that require nutrients as cofactors (Ca, Mg, and Mn) and redox reactions (Fe, Zn, and Cu). Furthermore, to maintain homeostasis, plants can alter the absorption of these nutrients [104]. However, Table 4 shows that iron content increased significantly in $e[CO_2]$. So far, we do not have a specific explanation for that. Perhaps more studies about the nutritional theme are necessary, especially when we talk about food security and food supply.

During the grain formation, the panicles act as nitrogen sinks in the plant. The process of accumulating storage during grain development in the middle and late maturation stages is crucial to determine crude protein [106]. These reserves are essential for human nutrition, with a better balance of essential amino acids than most cereals and legumes seed germination and seedling establishment [107–109]. Despite that, when plants grow under $e[CO_2]$, photosynthetic machinery is faster, causing a reduction in the nitrogen concentration and protein content of the grains [110]. Our result shows us a decrease in CSP; however, the mechanism of how every plant responds to the crude protein content is not totally clarified.

In theory, increased photosynthesis increases the availability of carbohydrates, which results in a gain in biomass and consequently in grain yield [5,55,111]. However, for the yield components presented, we must also consider the increase in average temperature (Table 4) in both agricultural years. Numerous studies have reported that the weight of 1000 grains and crude protein is reduced in $e[CO_2]$ when compared to $a[CO_2]$ conditions. These are fundamental characteristics when it comes to selecting species for a future climate change scenario. In this study, it was possible to observe that amaranth plants showed a smaller decrease in productivity when compared to quinoa plants [71,72].

5. Conclusions

Despite their different biochemical pathways (C3 in quinoa and C4 in amaranth), projected rise in atmospheric [CO_2] significantly increases photosynthesis and growth in both crops. Such changes are associated with various metabolic processes, ultimately affecting plant nutritional quality, carbohydrate production, antioxidant capacity, and grain yield.

Elevated [CO_2] alters the metabolism of amaranth and quinoa plants, as evidenced by the changes in growth patterns and photosynthetic pigments that reflect changes in gas exchange, carbohydrate metabolism, and nutritional status. As a result, plants may experience a reduction on essential nutrients like nitrogen. However, a notable increase in secondary metabolites was observed in both crops, and together with the enhanced enzymatic antioxidant system, these compounds may help protect the plants against stress.

For quinoa plants, the *e*[CO_2] led to reduction in grain yield and protein content. In contrast, there was no reduction in grain yield for amaranth, but there was a considerable reduction in grain protein content. This reduction is particularly significant in these crops renowned for its high protein content and quality.

Future research should focus on evaluating the protein quality and other nutritional aspects in these pseudocereals under *e*[CO_2]. Our findings ultimately underscore the critical need to breed plants that can adapt to *e*[CO_2] as means to mitigate its negative effects and to ensure sustainable and nutritious crop production in future environmental conditions.

Supplementary Materials: The following supporting information can be downloaded at: https://www.mdpi.com/article/10.3390/plants13233453/s1.

Author Contributions: Conceptualization, B.E.P.S., L.A.d.A. and S.D.; methodology, B.E.P.S., S.N.P., N.d.S.F., L.H.C., S.R.L., F.S.C. and S.D.; software, B.E.P.S., S.N.P. and S.D.; validation, B.E.P.S., L.A.d.A. and S.D.; formal analysis, B.E.P.S., S.N.P., S.B.T., S.R.L. and S.D.; investigation, B.E.P.S., S.N.P., S.B.T., N.d.S.F., L.H.C., F.S.C. and S.D.; resources, B.E.P.S., S.N.P., S.B.T. and S.D.; data curation, B.E.P.S., S.N.P., and S.D.; writing—original draft preparation, B.E.P.S., F.R.d.S., L.A.d.A. and S.D.; writing—review and editing, B.E.P.S., F.R.d.S., L.A.d.A. and S.D.; visualization, B.E.P.S., F.R.d.S. and S.D.; supervision, B.E.P.S., L.A.d.A. and S.D.; project administration, L.A.d.A. and S.D.; funding acquisition, L.A.d.A. and S.D. All authors have read and agreed to the published version of the manuscript.

Funding: This research was funded by Coordination for the Improvement of Higher Education Personnel (CAPES)—financing code 001 for S.N. Pires. The Brazilian Council for Scientific and Technological Development (CNPq) provided research fellowship to S. Deuner (grant number 429531/2016-8) and to Luis Avila (N.Proc. 310830/2019-2).

Data Availability Statement: The original contributions presented in this study are included in the article/supplementary material. Further inquiries can be directed to the corresponding author.

Acknowledgments: The present work was carried out with the support of the Coordination for the Improvement of Higher Education Personnel (CAPES), supported by resources from the Brazilian Council for Scientific and Technological Development (CNPq), Luis Avila thanks CNPq for the his Research Fellowship received during the period (N.Proc. 310830/2019-2). The authors acknowledge the assistance of artificial intelligence tools in proofreading and grammar-checking this manuscript.

Conflicts of Interest: The authors declare no conflicts of interest.

References

1. Hannah, L. *Climate Change Biology*, 3rd ed.; Academic Press: London, UK, 2021.
2. IPCC. *Climate Change: Impacts, Adaptation, and Vulnerability*; IPCC: Geneva, Switzerland, 2022.
3. Letcher, T.M. Global warming-a complex situation. *Clim. Change Obs. Impacts Planet Earth Third Ed.* **2021**, 3–17. [CrossRef]
4. Leakey, A.D.B.; Ainsworth, E.A.; Bernacchi, C.J.; Rogers, A.; Long, S.P.; Ort, D.R. Elevated CO_2 effects on plant carbon, nitrogen, and water relations: Six important lessons from FACE. *J. Exp. Bot.* **2009**, *60*, 2859–2876. [CrossRef] [PubMed]

5. Kimball, B.A. Crop responses to elevated CO_2 and interactions with H_2O, N, and temperature. *Curr. Opin. Plant Biol.* **2016**, *3[illegible]*, 36–43. [CrossRef] [PubMed]
6. Escribano, J.; Cabanes, J.; Jiménez-Atiénzar, M.; Ibañez-Tremolada, M.; Gómez-Pando, L.R.; García-Carmona, F.; Gandía-Herrero, F. Characterization of betalains, saponins and antioxidant power in differently colored quinoa (*Chenopodium quinoa*) varieties. *Food Chem.* **2017**, *234*, 285–294. [CrossRef]
7. Antolin, L.A.; Heinemann, A.B.; Marin, F.R. Impact assessment of common bean availability in Brazil under climate change scenarios. *Agric. Syst.* **2021**, *191*, 103174. [CrossRef]
8. Omar, M.E.D.M.; Moussa, A.M.A.; Hinkelmann, R. Impacts of climate change on water quantity, water salinity, food security, and socioeconomy in Egypt. *Water Sci. Eng.* **2021**, *14*, 17–27. [CrossRef]
9. Manners, R.; Varela-Ortega, C.; van Etten, J. Protein-rich legume and pseudo-cereal crop suitability under present and future European climates. *Eur. J. Agron.* **2019**, *113*, 125974. [CrossRef]
10. Mir, N.A.; Riar, C.S.; Singh, S. Nutritional constituents of pseudo cereals and their potential use in food systems: A review. *Trends Food Sci. Technol.* **2018**, *75*, 170–180. [CrossRef]
11. Schoenlechner, R. Properties of pseudocereals, selected specialty cereals and legumes for food processing with special attention to gluten-free products/Verarbeitungseigenschaften von Pseudogetreide, ausgewählten Spezialitätengetreide und Leguminosen mit speziellem Fokuss auf glutenfreie Produkte. *Die Bodenkultur J. Land Manag. Food Environ.* **2016**, *67*, 239–248. [CrossRef]
12. Chandra, A.K.; Chandora, R.; Sood, S.; Malhotra, N. Global production, demand, and supply. In *Millets and Pseudo Cereals: Genetic Resources and Breeding Advancements*; Woodhead Publishing: Cambridge, UK, 2020. [CrossRef]
13. Hatfield, J.L.; Walthall, C.L. Climate Change: Cropping System Changes and Adaptations. *Encycl. Agric. Food Syst.* **2014**, 256–265. [CrossRef]
14. Hofstra, G.; Hesketh, J.D. The Effects of Temperature and CO_2 Enrichment on Photosynthesis in Soybean. In *Environmental and Biological Control of Photosynthesis*; Springer: Dordrecht, The Netherlands, 1975; pp. 71–80. [CrossRef]
15. Sionit, N.; Hellmers, H.; Strain, B.R. Interaction of Atmospheric CO_2 Enrichment and Irradiance on Plant Growth. *Agron. J.* **1982**, *74*, 721–725. [CrossRef]
16. Long, S.P.; Zhu, X.; Naidu, S.L.; Ort, D.R. Can improvement in photosynthesis increase crop yields? *Plant Cell Environ.* **2006**, *29*, 315–330. [CrossRef]
17. Zong, Y.-Z.; Shangguan, Z.-P. Increased sink capacity enhances C and N assimilation under drought and elevated CO_2 conditions in maize. *J. Integr. Agric.* **2016**, *15*, 2775–2785. [CrossRef]
18. Münch, E. Movement of Solutes within the Plant. *Nature* **1931**, *127*, 550–551. [CrossRef]
19. Herbers, K.; Sonnewald, U. Molecular determinants of sink strength. *Curr. Opin. Plant Biol.* **1998**, *1*, 207–216. [CrossRef] [PubMed]
20. Ho, L.C. Metabolism and Compartmentation of Imported Sugars in Sink Organs in Relation to Sink Strength. *Annu. Rev. Plant Physiol. Plant Mol. Biol.* **1988**, *39*, 355–378. [CrossRef]
21. Liu, J.; Hu, T.; Fang, L.; Peng, X.; Liu, F. CO2 elevation modulates the response of leaf gas exchange to progressive soil drying in tomato plants. *Agric. For. Meteorol.* **2019**, *268*, 181–188. [CrossRef]
22. Khanam, U.K.S.; Oba, S. Bioactive substances in leaves of two amaranth species, *Amaranthus tricolor* and *A. hypochondriacus*. *Can. J. Plant Sci.* **2013**, *93*, 47–58. [CrossRef]
23. Hariadi, Y.; Marandon, K.; Tian, Y.; Jacobsen, S.-E.; Shabala, S. Ionic and osmotic relations in quinoa (*Chenopodium quinoa* Willd.) plants grown at various salinity levels. *J. Exp. Bot.* **2010**, *62*, 185–193. [CrossRef]
24. James, L.E.A. Chapter 1 Quinoa (*Chenopodium quinoa* Willd.): Composition, chemistry, nutritional, and functional properties. In *Advances in Food and Nutrition Research*; Academic Press: Cambridge, MA, USA, 2009; Volume 58, pp. 1–31. [CrossRef]
25. Ruiz, K.B.; Biondi, S.; Oses, R.; Acuña-Rodríguez, I.S.; Antognoni, F.; Martinez-Mosqueira, E.A.; Coulibaly, A.; Canahua-Murillo, A.; Pinto, M.; Zurita-Silva, A.; et al. Quinoa biodiversity and sustainability for food security under climate change. A review. *Agron. Sustain. Dev.* **2013**, *34*, 349–359. [CrossRef]
26. Monteiro, J.E.B.A.; Sentelhas, P.C.; Chiavegato, E.J.; Guiselini, C.; Santiago, A.V.; Prela, A. Estimação da área foliar do algodoeiro por meio de dimensões e massa das folhas. *Bragantia* **2005**, *64*, 15–24. [CrossRef]
27. Benincasa, M.M. *Análise de Crescimento de Plantas: Noções Básicas*, 2nd ed.; Jaboticabal Funep: São Paulo, Brazil, 2003.
28. Wellburn, A.R. The Spectral Determination of Chlorophylls a and b, as well as Total Carotenoids, Using Various Solvents with Spectrophotometers of Different Resolution. *J. Plant Physiol.* **1994**, *144*, 307–313. [CrossRef]
29. Graham, D.; Smydzuk, J. Use of anthrone in the quantitative determination of hexose phosphates. *Anal. Biochem.* **1965**, *11*, 246–255. [CrossRef] [PubMed]
30. van Handel, E. Direct microdetermination of sucrose. *Anal. Biochem.* **1968**, *22*, 280–283. [CrossRef]
31. Cocking, E.C.; Yemm, E.W. Estimation of amino acids by ninhydrin. *Process Biochem.* **1954**, *58*.
32. Zeng, Y.; Wu, Y.; Avigne, W.T.; Koch, K.E. Rapid Repression of Maize Invertases by Low Oxygen. Invertase/Sucrose Synthase Balance, Sugar Signaling Potential, and Seedling Survival. *Plant Physiol.* **1999**, *121*, 599–608. [CrossRef] [PubMed]
33. Miller, G.L. Use of Dinitrosalicylic Acid Reagent for Determination of Reducing Sugar. *Anal. Chem.* **1959**, *31*, 426–428. [CrossRef]
34. Lowell, C.A.; Tomlinson, P.T.; Koch, K.E. Sucrose-Metabolizing Enzymes in Transport Tissues and Adjacent Sink Structures in Developing Citrus Fruit. *Plant Physiol.* **1989**, *90*, 1394–1402. [CrossRef]
35. Pérez-Tortosa, V.; López-Orenes, A.; Martínez-Pérez, A.; Ferrer, M.A.; Calderón, A.A. Antioxidant activity and rosmarinic acid changes in salicylic acid-treated *Thymus membranaceus* shoots. *Food Chem.* **2012**, *130*, 362–369. [CrossRef]

36. Tedesco, M.; Gianello, C.; Bissiani, C.A.; Bohnem, H.; Volkweiss, S.J. *Análise de Solo, Plantas e Outros Materiais*; Universidade Federal do Rio Grande do Sul: Porto Alegre, Brazil, 1995.
37. Kirk, P.L. Kjeldahl Method for Total Nitrogen. *Anal. Chem.* **1950**, *22*, 354–358. [CrossRef]
38. Reich, P.B.; Hobbie, S.E.; Lee, T.D.; Pastore, M.A. Unexpected reversal of C_3 versus C_4 grass response to elevated CO_2 during a 20-year field experiment. *Science* **2018**, *360*, 317–320. [CrossRef] [PubMed]
39. Rocha, M.D.M. *Características da Planta (Árvore do Conhecimento-Feijão Cauipi)*; Agência Embrapa de Informação Tecnologica (Ageitec): Agência, Brazil, 2019.
40. Xu, C.; McDowell, N.G.; Sevanto, S.; Fisher, R.A. Our limited ability to predict vegetation dynamics under water stress. *New Phytol.* **2013**, *200*, 298–300. [CrossRef] [PubMed]
41. Westoby, M.; Falster, D.S.; Moles, A.T.; Vesk, P.A.; Wright, I.J. Plant ecological strategies: Some leading dimensions of variation between species. *Annu. Rev. Ecol. Syst.* **2002**, *33*, 125–159. [CrossRef]
42. Song, N.; Zhang, X.; Wang, F.; Zhang, C.; Tang, S. Elevated CO_2 increases Cs uptake and alters microbial communities and biomass in the rhizosphere of *Phytolacca americana* Linn (pokeweed) and *Amaranthus cruentus* L. (purple amaranth) grown on soils spiked with various levels of Cs. *J. Environ. Radioact.* **2012**, *112*, 29–37. [CrossRef]
43. Bunce, J.A. Variation in yield responses to elevated CO_2 and a brief high temperature treatment in quinoa. *Plants* **2017**, *6*, 26. [CrossRef]
44. Li, T.; Di, Z.; Han, X.; Yang, X. Elevated CO_2 improves root growth and cadmium accumulation in the hyperaccumulator *Sedum alfredii*. *Plant Soil* **2011**, *354*, 325–334. [CrossRef]
45. Pritchard, S.G.; Rogers, H.H. Spatial and temporal deployment of crop roots in CO_2-enriched environments. *New Phytol.* **2000**, *147*, 55–71. [CrossRef]
46. Tingey, D.T.; Phillips, D.L.; Johnson, M.G. Elevated CO_2 and conifer roots: Effects on growth, life span and turnover. *New Phytol.* **2000**, *147*, 87–103. [CrossRef]
47. Nie, M.; Lu, M.; Bell, J.; Raut, S.; Pendall, E. Altered root traits due to elevated CO_2: A meta-analysis. *Glob. Ecol. Biogeogr.* **2013**, *22*, 1095–1105. [CrossRef]
48. Jayme-Oliveira, A.; Júnior, W.Q.R.; Ramos, M.L.G.; Ziviani, A.C.; Jakelaitis, A. Amaranth, quinoa, and millet growth and development under different water regimes in the Brazilian Cerrado. *Pesqui. Agropecu. Bras.* **2017**, *52*, 561–571. [CrossRef]
49. Xie, H.; Liu, K.; Sun, D.; Wang, Z.; Lu, X.; He, K. A field experiment with elevated atmospheric CO_2-mediated changes to C_4 crop-herbivore interactions. *Sci. Rep.* **2015**, *5*, 13923. [CrossRef] [PubMed]
50. DE Souza, A.P.; Gaspar, M.; DA Silva, E.A.; Ulian, E.C.; Waclawovsky, A.J.; Nishiyama, M.Y., Jr.; DOS Santos, R.V.; Teixeira, M.M.; Souza, G.M.; Buckeridge, M.S. Elevated CO_2 increases photosynthesis, biomass and productivity, and modifies gene expression in sugarcane. *Plant Cell Environ.* **2008**, *31*, 1116–1127. [CrossRef] [PubMed]
51. Li, P.; Li, B.; Seneweera, S.; Zong, Y.; Li, F.Y.; Han, Y.; Hao, X. Photosynthesis and yield response to elevated CO_2, C_4 plant foxtail millet behaves similarly to C_3 species. *Plant Sci.* **2019**, *285*, 239–247. [CrossRef] [PubMed]
52. Oliveira Filho, A.F.D. Produção e Qualidade de Sementes de Quinoa em Função do Arranjo Espacial. Ph.D. Thesis, Universidade Federal do Ceará, Fortaleza, Brazil, 2017.
53. Tozzi, F.R.O.; Ghini, R. Impacto do aumento da concentração atmosférica de dióxido de carbono sobre a ferrugem e o crescimento do cafeeiro. *Pesqui. Agropecu. Bras.* **2016**, *51*, 933–941. [CrossRef]
54. Gonzalez-Meler, M.A.; Silva, L.B.C.; Dias-De-Oliveira, E.; Flower, C.E.; Martinez, C.A. Experimental air warming of a *Stylosanthes capitata*, vogel dominated tropical pasture affects soil respiration and nitrogen dynamics. *Front. Plant Sci.* **2017**, *8*, 46. [CrossRef] [PubMed]
55. Ainsworth, E.A.; Long, S.P. What have we learned from 15 years of free-air CO_2 enrichment (FACE)? A meta-analytic review of the responses of photosynthesis, canopy properties and plant production to rising CO_2. *New Phytol.* **2005**, *165*, 351–372. [CrossRef]
56. Dorneles, K.; Posso, D.A.; Rebhahn, I.; Deuner, S.; Pazdiora, P.; Avila, L.; Dallagnol, L.J. Respostas morfofisiológicas e rendimento de grãos do trigo mediados pelo aumento da concentração de CO_2 atmosférico. *Rev. Bras. De Cienc. Agrar.* **2019**, *14*, 1–7. [CrossRef]
57. Busch, F.A.; Sage, R.F. The sensitivity of photosynthesis to O_2 and CO_2 concentration identifies strong Rubisco control above the thermal optimum. *New Phytol.* **2016**, *213*, 1036–1051. [CrossRef]
58. Siefermann-Harms, D. The light-harvesting and protective functions of carotenoids in photosynthetic membranes. *Physiol. Plant.* **1987**, *69*, 561–568. [CrossRef]
59. Conroy, J.; Hocking, P. Nitrogen nutrition of C_3 plants at elevated atmospheric CO_2 concentrations. *Physiol. Plant.* **1993**, *89*, 570–576. [CrossRef]
60. Wang, L., Deng, F.; Ren, W.-J. Shading tolerance in rice is related to better light harvesting and use efficiency and grain filling rate during grain filling period. *Field Crop. Res.* **2015**, *180*, 54–62. [CrossRef]
61. Bloom, A.J.; Asensio, J.S.R.; Randall, L.; Rachmilevitch, S.; Cousins, A.B.; Carlisle, E.A. CO_2 enrichment inhibits shoot nitrate assimilation in C_3 but not C_4 plants and slows growth under nitrate in C_3 plants. *Ecology* **2012**, *93*, 355–367. [CrossRef] [PubMed]
62. Bhargava, S.; Mitra, S. Elevated atmospheric CO_2 and the future of crop plants. *Plant Breed.* **2020**, *140*, 1–11. [CrossRef]
63. Krämer, K.; Kepp, G.; Brock, J.; Stutz, S.; Heyer, A.G. Acclimation to elevated CO_2 affects the C/N balance by reducing de novo N-assimilation. *Physiol. Plant.* **2022**, *174*, e13615. [CrossRef] [PubMed]

64. White, A.C.; Rogers, A.; Rees, M.; Osborne, C.P. How can we make plants grow faster? A source–sink perspective on growth rat *J. Exp. Bot.* **2015**, *67*, 31–45. [CrossRef]
65. Kuai, B.; Chen, J.; Hörtensteiner, S. The biochemistry and molecular biology of chlorophyll breakdown. *J. Exp. Bot.* **2017**, *6* 751–767. [CrossRef]
66. Makino, A.; Mae, T. Photosynthesis and plant growth at elevated levels of CO_2. *Plant Cell Physiol.* **1999**, *40*, 999–1006. [CrossRe
67. Turkan, I.; Uzilday, B.; Dietz, K.-J.; Bräutigam, A.; Ozgur, R. Reactive oxygen species and redox regulation in mesophyll an bundle sheath cells of C_4 plants. *J. Exp. Bot.* **2018**, *69*, 3321–3331. [CrossRef]
68. Taub, D. Effects of rising atmospheric concentrations of carbon dioxide on plants. *Nat. Educ. Knowl.* **2010**, *3*, 21.
69. Xu, Y.; Ibrahim, I.M.; Harvey, P.J. The influence of photoperiod and light intensity on the growth and photosynthesis of *Dunaliell salina* (chlorophyta) CCAP 19/30. *Plant Physiol. Biochem.* **2016**, *106*, 305–315. [CrossRef]
70. Wang, T.; Wei, Q.; Wang, Z.; Liu, W.; Zhao, X.; Ma, C.; Gao, J.; Xu, Y.; Hong, B. CmNF-YB8 affects drought resistance i chrysanthemum by altering stomatal status and leaf cuticle thickness. *J. Integr. Plant Biol.* **2021**, *64*, 741–755. [CrossRef] [PubMed
71. Zhang, D.; Li, A.; Lam, S.K.; Li, P.; Zong, Y.; Gao, Z.; Hao, X. Increased carbon uptake under elevated CO_2 concentration enhance water-use efficiency of C_4 broomcorn millet under drought. *Agric. Water Manag.* **2020**, *245*, 106631. [CrossRef]
72. Davis, A.S.; Ainsworth, E.A. Weed interference with field-grown soyabean decreases under elevated [CO_2] in a FACE experimen *Weed Res.* **2012**, *52*, 277–285. [CrossRef]
73. Dos Santos, L.A.; Pinto, C.C.; Szareski, V.J.; Carvalho, I.R.; Koch, F.; Pimentel, J.R.; Troyjack, C.; Dubal, T.P.; Reolon, F.; Da Rosa T.C.; et al. Respiratory activity and physiological performance of maize seeds classified according to their shapes and sizes. *Aus J. Crop Sci.* **2018**, *12*, 1882–1889. [CrossRef]
74. Leakey, A.D.; Uribelarrea, M.; Ainsworth, E.A.; Naidu, S.L.; Rogers, A.; Ort, D.R.; Long, S.P. Photosynthesis, productivity, anc yield of maize are not affected by open-air elevation of CO_2 concentration in the absence of drought. *Plant Physiol.* **2006**, *14C* 779–790. [CrossRef]
75. Monson, R.K. The Origins of C_4 Genes and Evolutionary Pattern in the C_4 Metabolic Phenotype. *C4 Plant Biol.* **1999**, 377–41C [CrossRef]
76. Sage, R.F. The evolution of C_4 photosynthesis. *New Phytol.* **2003**, *161*, 341–370. [CrossRef]
77. Stutz, S.S.; Hanson, D.T. Contribution and consequences of xylem-transported CO_2 assimilation for C_3 plants. *New Phytol.* **2019** *223*, 1230–1240. [CrossRef]
78. de Souza Osório, C.R.W.; Teixeira, G.C.M.; Barreto, R.F.; Campos, C.N.S.; Leal, A.J.F.; Teodoro, P.E.; Prado, R.D.M. Macronutrien deficiency in snap bean considering physiological, nutritional, and growth aspects. *PLoS ONE* **2020**, *15*, e0234512. [CrossRef]
79. Wand, S.J.E.; Midgley, G.F.; Jones, M.H.; Curtis, P.S. Responses of wild C_4 and C_3 grass (Poaceae) species to elevated atmospheric CO_2 concentration: A meta-analytic test of current theories and perceptions. *Glob. Chang. Biol.* **1999**, *5*, 723–741. [CrossRef]
80. da Silva, R.G.; Alves, R.d.C.; Zingaretti, S.M. Increased [CO_2] causes changes in physiological and genetic responses in C_4 crops A brief review. *Plants* **2020**, *9*, 1567. [CrossRef] [PubMed]
81. Leegood, R.C. C_4 photosynthesis: Principles of CO_2 concentration and prospects for its introduction into C_3 plants. *J. Exp. Bot* **2002**, *53*, 581–590. [CrossRef] [PubMed]
82. Roitsch, T.; González, M.-C. Function and regulation of plant invertases: Sweet sensations. *Trends Plant Sci.* **2004**, *9*, 606–613 [CrossRef] [PubMed]
83. Li, L.; Wang, X.; Manning, W.J. Effects of Elevated CO_2 on Leaf Senescence, Leaf Nitrogen Resorption, and Late-Season Photosynthesis in *Tilia americana* L. *Front. Plant Sci.* **2019**, *10*, 1217. [CrossRef]
84. Lemoine, R.; La Camera, S.; Atanassova, R.; Dédaldéchamp, F.; Allario, T.; Pourtau, N.; Bonnemain, J.L.; Laloi, M.; Coutos-Thévenot, P.; Maurousset, L.; et al. Source-to-sink transport of sugar and regulation by environmental factors. *Front. Plant Sci* **2013**, *4*, 272. [CrossRef]
85. Ma, N.; Ma, C.; Liu, Y.; Shahid, M.O.; Wang, C.; Gao, J. Petal senescence: A hormone view. *J. Exp. Bot.* **2018**, *69*, 719–732 [CrossRef]
86. Ruan, Y.-L. Sucrose metabolism: Gateway to diverse carbon use and sugar signaling. *Annu. Rev. Plant Biol.* **2014**, *65*, 33–67 [CrossRef]
87. Lal, M.K.; Sharma, N.; Adavi, S.B.; Sharma, E.; Altaf, M.A.; Tiwari, R.K.; Kumar, R.; Kumar, A.; Dey, A.; Paul, V.; et al. From source to sink: Mechanistic insight of photoassimilates synthesis and partitioning under high temperature and elevated [CO_2] *Plant Mol. Biol.* **2022**, *110*, 305–324. [CrossRef]
88. Stein, O.; Granot, D. An overview of sucrose synthases in plants. *Front. Plant Sci.* **2019**, *10*, 95. [CrossRef]
89. Peleg, Z.; Blumwald, E. Hormone balance and abiotic stress tolerance in crop plants. *Curr. Opin. Plant Biol.* **2011**, *14*, 290–295 [CrossRef]
90. Thompson, M.; Gamage, D.; Hirotsu, N.; Martin, A.; Seneweera, S. Effects of elevated carbon dioxide on photosynthesis and carbon partitioning: A perspective on root sugar sensing and hormonal crosstalk. *Front. Physiol.* **2017**, *8*, 578. [CrossRef] [PubMed]
91. Poorter, H.; Bühler, J.; van Dusschoten, D.; Climent, J.; Postma, J.A. Pot size matters: A meta-analysis of the effects of rooting volume on plant growth. *Funct. Plant Biol.* **2012**, *39*, 839–850. [CrossRef] [PubMed]
92. Dong, J.; Gruda, N.; Lam, S.K.; Li, X.; Duan, Z. Effects of elevated CO_2 on nutritional quality of vegetables: A review. *Front. Plant Sci.* **2018**, *9*, 924. [CrossRef] [PubMed]
93. Taiz, L.; Zeiger, E. *Fisiologia e Desenvolvimento Vegetal*, 6th ed.; Artmed: Porto Alegre, Brazil, 2017.

94. Pazzagli, P.T.; Weiner, J.; Liu, F. Effects of CO_2 elevation and irrigation regimes on leaf gas exchange, plant water relations, and water use efficiency of two tomato cultivars. *Agric. Water Manag.* **2016**, *169*, 26–33. [CrossRef]
95. Escudero, A.; del Arco, J.M.; Sanz, I.C.; Ayala, J. Effects of leaf longevity and retranslocation efficiency on the retention time of nutrients in the leaf biomass of different woody species. *Oecologia* **1992**, *90*, 80–87. [CrossRef] [PubMed]
96. Aerts, R. Nutrient Resorption from Senescing Leaves of Perennials: Are there General Patterns? *J. Ecol.* **1996**, *84*, 597. [CrossRef]
97. Killingbeck, K.T. Nutrients in senesced leaves: Keys to the search for potential resorption and resorption proficiency. *Ecology* **1996**, *77*, 1716–1727. [CrossRef]
98. Reed, S.C.; Townsend, A.R.; Davidson, E.A.; Cleveland, C.C. Stoichiometric patterns in foliar nutrient resorption across multiple scales. *New Phytol.* **2012**, *196*, 173–180. [CrossRef]
99. Vergutz, L.; Manzoni, S.; Porporato, A.; Novais, R.F.; Jackson, R.B. Global resorption efficiencies and concentrations of carbon and nutrients in leaves of terrestrial plants. *Ecol. Monogr.* **2012**, *82*, 205–220. [CrossRef]
100. de Campos, M.C.R.; Pearse, S.J.; Oliveira, R.S.; Lambers, H. Downregulation of net phosphorus-uptake capacity is inversely related to leaf phosphorus-resorption proficiency in four species from a phosphorus-impoverished environment. *Ann. Bot.* **2013**, *111*, 445–454. [CrossRef]
101. Pang, J.; Zhu, J.-G.; Xie, Z.-B.; Liu, G.; Zhang, Y.-L.; Chen, G.-P.; Zeng, Q.; Cheng, L. A new explanation of the N concentration decrease in tissues of rice (*Oryza sativa* L.) exposed to elevated atmospheric pCO2. *Environ. Exp. Bot.* **2005**, *57*, 98–105. [CrossRef]
102. Crous, K.Y.; Wujeska-Klause, A.; Jiang, M.; Medlyn, B.E.; Ellsworth, D.S. Nitrogen and phosphorus retranslocation of leaves and stemwood in a mature eucalyptus forest exposed to 5 years of elevated CO_2. *Front. Plant Sci.* **2019**, *10*, 664. [CrossRef] [PubMed]
103. Zhang, Y.; Chen, X.M.; Zhang, C.C.; Pan, G.X.; Zhang, X.H. Availability of soil nitrogen and phosphorus under elevated [CO_2] and temperature in the Taihu Lake region, China. *J. Plant Nutr. Soil Sci.* **2014**, *177*, 343–348. [CrossRef]
104. Xu, H.; Xie, H.; Wu, S.; Wang, Z.; He, K. Effects of elevated CO_2 and increased n fertilization on plant secondary metabolites and chewing insect fitness. *Front. Plant Sci.* **2019**, *10*, 739. [CrossRef]
105. Sudderth, E.A.; Stinson, K.A.; Bazzaz, F.A. Host-specific aphid population responses to elevated CO_2 and increased N availability. *Glob. Chang. Biol.* **2005**, *11*, 1997–2008. [CrossRef]
106. Müntz, K.; Belozersky, M.; Dunaevsky, Y.; Schlereth, A.; Tiedemann, J. Stored proteinases and the initiation of storage protein mobilization in seeds during germination and seedling growth. *J. Exp. Bot.* **2001**, *52*, 1741–1752. [CrossRef]
107. Goyoaga, C.; Burbano, C.; Cuadrado, C.; Romero, C.; Guillamón, E.; Varela, A.; Pedrosa, M.M.; Muzquiz, M. Content and distribution of protein, sugars and inositol phosphates during the germination and seedling growth of two cultivars of Vicia faba. *J. Food Compos. Anal.* **2011**, *24*, 391–397. [CrossRef]
108. Erbaş, S.; Tonguç, M.; Şanli, A. Mobilization of seed reserves during germination and early seedling growth of two sunflower cultivars. *J. Appl. Bot. Food Qual.* **2016**, *89*. [CrossRef]
109. Gu, J.; Chao, H.; Gan, L.; Guo, L.; Zhang, K.; Li, Y.; Wang, H.; Raboanatahiry, N.; Li, M. Proteomic dissection of seed germination and seedling establishment in *Brassica napus*. *Front. Plant Sci.* **2016**, *7*, 1482. [CrossRef]
110. Chen, C.; Jiang, Q.; Ziska, L.H.; Zhu, J.; Liu, G.; Zhang, J.; Ni, K.; Seneweera, S.; Zhu, C. Seed vigor of contrasting rice cultivars in response to elevated carbon dioxide. *Field Crop. Res.* **2015**, *178*, 63–68. [CrossRef]
111. Högy, P.; Wieser, H.; Köhler, P.; Schwadorf, K.; Breuer, J.; Franzaring, J.; Muntifering, R.; Fangmeier, A. Effects of elevated CO_2 on grain yield and quality of wheat: Results from a 3-year free-air CO_2 enrichment experiment. *Plant Biol.* **2009**, *11*, 60–69. [CrossRef] [PubMed]

plants

MDPI

Article

Geographical Distribution Dynamics of *Acorus calamus* in China Under Climate Change

Chunlei Yue, Hepeng Li and Xiaodeng Shi *

Zhejiang Academy of Forestry, Hangzhou 310023, China
* Correspondence: sxdpaper@163.com

Abstract: *Acorus calamus*, a perennial emergent herb, is highly valued for its ornamental appeal, water purification ability, and medicinal properties. However, there is a significant contradiction between the rapidly increasing demand for *A. calamus* and the diminishing wild resources. Understanding its geographical distribution and the influence of global climate change on its geographical distribution is imperative for establishing a theoretical framework for the conservation of natural resources and the expansion of its cultivation. In this study, 266 distribution records of *A. calamus* and 18 selected key environmental factors were utilized to construct an optimal MaxEnt model via the ENMeval package. We simulated the potential geographical distributions under current conditions and under three different climate scenarios (SSP126, SSP370, and SSP585) in the 2050s, 2070s, and 2090s. Additionally, we employed the jackknife method and response curves to identify the environmental factors with the greatest influence on the distribution of *A. calamus*, and their response intervals. The results indicate that the regularization multiplier (RM) of 3.5 and the feature combinations (FC) of linear (L), quadratic (Q), hinge (H), and product (P) are the optimal model parameter combinations. With these parameters, the model predictions are highly accurate, and the consistency of the results is significant. The dominant environmental factors and their thresholds affecting the distribution of *A. calamus* are the precipitation of the wettest month (≥109.87 mm), human footprint (≥5.39), annual precipitation (≥388.56 mm), and mean diurnal range (≤12.83 °C). The primary land use types include rivers and channels, reservoirs and ponds, lakes, urban areas, marshes, other constructed lands, rice fields, forested areas, and shrublands. Under current climate conditions, the suitable geographical distribution of *A. calamus* in China is clearly located east of the 400 mm precipitation line, with high- and low-suitability areas covering 121.12×10^4 km^2, and 164.20×10^4 km^2, respectively. Under future climate conditions, both high- and low- suitability areas are projected to increase significantly, whereas unsuitable areas are expected to decrease, with the centroid of each suitability zone shifting northward. This study provides a theoretical foundation for sustainable utilization, future production planning, and the development of conservation strategies for wild germplasm resources of *A. calamus*.

Keywords: *Acorus calamus*; MaxEnt; ENMeval package; land use; human

1. Introduction

The geographical distribution patterns of plant populations are predominantly influenced by climate, which plays a critical role in shaping plant growth and survival [1]. Global warming has now become a widely accepted international consensus [2]. According to recent studies, the average global surface temperature increase may have already exceeded 1.5 °C and could surpass 2 °C by the end of this decade [3]. The impact of global climate change is becoming increasingly pronounced. The increase in climate anomalies as a result of global warming has resulted in the increased occurrence and severity of extreme weather events, notably high temperatures and droughts, which are becoming more prominent each year [4]. This situation has severely impacted the global ecological environment and significantly disrupted the structure and function of terrestrial ecosystems, community composition, and species distribution patterns [5]. Global climate change may lead to shifts and alterations in the distribution patterns of most species, habitat loss, and fragmentation, and the extinction of endangered species with narrow natural distribution ranges [6]. Therefore, understanding the impacts of global climate change on the potential geographical distribution of plants is vital for assessing their responses to climate variations [7]. It is vital to evaluate the vulnerability of species and develop appropriate adaptive management strategies under changing climate conditions [8]. This knowledge is highly important for the conservation, utilization, and sustainable development of plant species.

Species distribution models (SDMs) have demonstrated an excellent ability to predict changes in the potential distribution areas of various species under global climate change [9]. These methods are widely applied in climate change research, ecosystem functional community analysis, plant diversity conservation, and biological invasion assessment, among other fields [10]. SDMs can be used to systematically evaluate the probability of species suitability in different regions by correlating species distribution information with corresponding environmental variables [11]. In theory, species distribution models can predict the distribution of species on a global scale and are expected to be used to guide species introduction efforts to suitable areas. Commonly used SDMs include the Maximum Entropy Model (MaxEnt), Genetic Algorithm for Rule-Set Production (GARP), Bioclim Domain, Climex, and Ecological Niche Factor Analysis (ENFA), among others [12]. Different models require varying data and algorithms for species distribution simulations, leading to potential discrepancies in the calculated results [6]. Among these, the MaxEnt model has become one of the most widely used SDMs due to its stable simulation results and high accuracy [13]. This model can avoid overfitting and maintain the stability and reliability of prediction accuracy even under conditions of incomplete data or small sample sizes [14]. Additionally, this model can directly generate spatial maps of potentially suitable habitats and assess the contributions of environmental variables through jackknife testing [15]. It has been widely used in various research fields to simulate the potential distribution areas of species and evaluate their responses to climate change, biodiversity conservation, and biological evolution [16].

Acorus is a unique genus within monocotyledons and belongs to the family Acoraceae [17]. It was previously classified under the Araceae family, but modern taxonomic studies indicate that it should form its own family and be sister to all other monocots [18]. It has garnered significant attention due to its traditional medicinal value. According to the widely accepted classifications by the Flora of China (FOC), the Plant List, and the International Plant Names Index (IPNI), only two species within the *Acorus* genus are recognized globally: *Acorus calamus* and *Acorus gramineus*. However, several studies have shown that this genus contains many more species that have been presumed before, especially in Eurasia [17,18]. *Acorus calamus* contains one species (*A. calamus* L. var. *calamus*) and one variety (*A. calamus* L. var. *verus* L.), both of which belong to the genus *Acorus* of Acoraceae and are important perennial emergent herbs; this study does not distinguish between them [19]. It is widely distributed from the northern temperate zone to subtropical regions, and China serves as the primary distribution center for the *Acorus* genus [20].

A. calamus has an attractive appearance, with most of its stems and leaves standing upright above the water surface, while the lower stems and roots grow underwater. It has high ornamental value and is often used as a feature in water gardens or as a plant in constructed wetlands [21]. Recent studies have shown that, in addition to its great ornamental value, *A. calamus* has significant applications in water remediation due to its excellent water purification capabilities; thus, it is commonly used to remediate polluted water bodies, such as domestic sewage [22], and heavy-metal-contaminated water sources, including wastewater containing cadmium [23] and zinc [24].

Moreover, *A. calamus* is known for its aromatic properties, with its rhizomes being used in traditional medicine [25]. The plant is characterized by its pungent, bitter, and warm nature, offering various pharmacological benefits, such as insect repellent, analgesic, antibacterial, antifungal, antianxiety, antioxidant, and renoprotective effects [26]. It holds significant potential for development and application in the healthcare sector [27]. The essential oil extracted from *A. calamus* contains monoterpenes, sesquiterpenes, alkaloids, anthraquinones, aromatic essential oils, and flavonoids [28], making it an excellent ingredient for food flavoring and cosmetic fragrances. In many Asian countries, *A. calamus* is used as a substitute for other spices, such as ginger and cinnamon [29]. In recent years, the demand for products derived from *A. calamus* has increased due to the continuous development of related medicinal formulations [30].

Although *A. calamus* is widely distributed, natural populations are often confined to small areas along riverbanks, streams, or marshy wetlands [31]. The natural seedling rate is extremely low, and rhizome branches grow slowly, making them vulnerable to extensive root harvesting [20]. As a result, the natural resources of *A. calamus* are highly susceptible to damage. More critically, the current exploitation of *A. calamus* heavily depends on wild germplasm resources. With the increasing demand for *A. calamus*, wild populations are being subjected to unrestrained harvesting, leading to a sharp decline in resource reserves [32]. Currently, research on *A. calamus* has mainly focused on areas such as water pollution control, endophytic fungal inhibition, heavy metal accumulation, and water stress [33]. However, there are few reports on the suitable geographical distribution areas for *A. calamus*, despite its importance as a key ornamental and medicinal plant.

Therefore, there is an urgent need to study the suitable geographical distribution of *A. calamus* to understand its precise growth range and expand artificial cultivation areas to meet market demand. Additionally, considering the substantial pace of global climate change, it is crucial to examine how these changes affect its geographical distribution. This undertaking will facilitate the development of effective strategies for *A. calamus* introduction, cultivation, resource conservation, and adaptation to future global climate change. Moreover, this study offers a theoretical foundation for the sustainable use of *A. calamus*, aiding in future production planning and the development of strategies for protecting wild germplasm resources.

Based on the characteristics of China's climate under global climate change, temperature and precipitation are expected to increase to varying degrees in the northeastern and southwestern regions of China [34]. We hypothesize that the future suitable geographical distribution of *A. calamus* will expand but become more fragmented. To verify our hypothesis, we utilized an optimized MaxEnt model based on the ENMeval 0.3.0 package to focus on the following objectives: (1) to investigate the potential geographic distribution of *A. calamus* in China under current climate conditions; (2) to identify the dominant environmental factors influencing the geographic distribution and determine the optimal conditions for *A. calamus*; and (3) to predict and compare the potential geographic distribution and migration trends under different climate scenarios in the 2050s (2041–2060), 2070s (2061–2080), and 2090s (2081–2100) against the backdrop of global warming.

2. Materials and Methods

2.1. Occurrence Data

The study area was located in China (3.85–53.55° N, 73.55–135.08° E), where *A. calamus* commonly grows in wetland environments such as riverbanks, marshes, or lakeshores, typically at altitudes below 2600 m [31]. It thrives in warm, humid climates with abundant sunlight, with an optimal growth temperature range of 20–25 °C. Growth halts when temperatures fall below 10 °C [31]. During winter, its rhizomes retreat into the soil to overwinter. It prefers loose, well-aerated soils with good water retention and nutrient-holding capacity and can tolerate periods of waterlogging [31]. In shallow water environments, *A. calamus* grows well, with lush foliage and thick rhizomes; however, under flooded conditions, its growth slows or even halts [35]. In this study, the distribution data of *A. calamus* were sourced mainly from the Global Biodiversity Information Facility (GBIF, http://www.gbif.org/), the Chinese Virtual Herbarium (CVH, http://www.cvh.ac.cn/), relevant literature, and field surveys conducted from May 2022 to May 2023. A total of 339 distribution points of *A. calamus* were collected. Duplicate coordinates and points with unclear geographic coordinates were removed. Additionally, for points with distances less than 5 km between them, only one point was retained to reduce the potential clustering effects that could lead to model overfitting [36]. This process resulted in a final dataset of 266 distribution points of *A. calamus* for modeling (Figure 1).

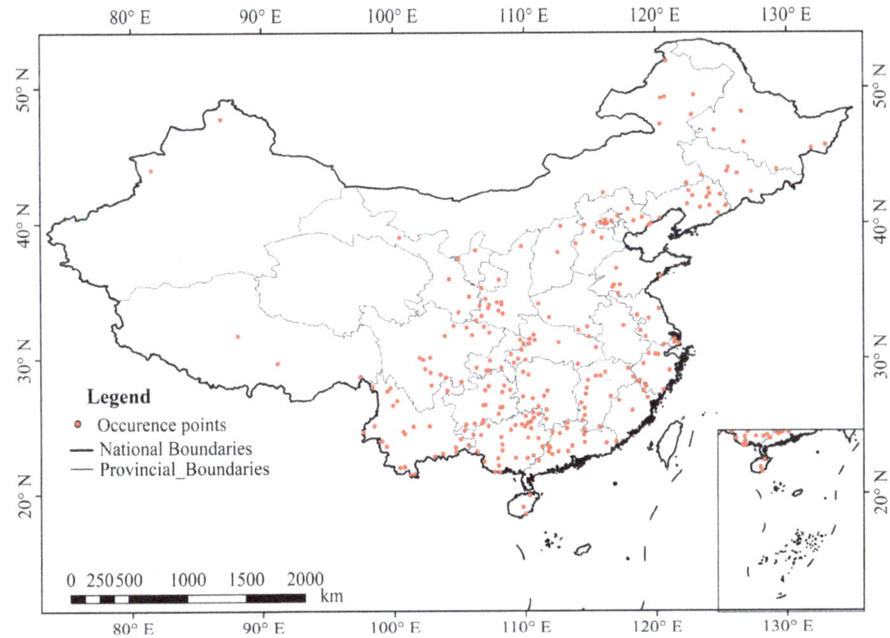

Figure 1. Distributions of occurrence points of *A. calamus*.

2.2. Filtering and Data Processing of Environmental Variables

In this study, the selected environmental variables related to the geographical distribution of *A. calamus* included climatic, topographic, soil, human, and land variables (as listed in Table S1). The 19 climatic variables and elevation data were obtained from the WorldClim database (www.worldclim.org). The slope and aspect factors were derived from the elevation data. The periods included the current (1970–2000) and three future scenarios (2050s (2041–2060), 2070s (2061–2080), and 2090s (2081–2100)). In this study, three different SSP scenarios were used: SSP1-2.6 (SSP126), SSP3-7.0 (SSP370), and SSP5–8.5 (SSP585). SSPs are reference pathways that describe plausible scenarios in the evolution of society and ecosystems [37]. For future climate scenarios, we utilized the BCC-CSM2-MR climate change model from the Coupled Model Intercomparison Project Phase 6 (CMIP6) dataset [38].

Soil factor data were obtained from the Harmonized World Soil Database (HWSD) (http://webarchive.iiasa.ac.at/Research/LUC/External-World-soil-database/HTML/index.html?sb=1, accessed on 10 December 2023).

Human footprint data were collected from the International Geoscience Information-Network Center (https://sedac.ciesin.columbia.edu, accessed on 5 August 2024).

The land use dataset used was provided by the National Tibetan Plateau/Third Pole Environment Data Center (http://data.tpdc.ac.cn, accessed on 21 August 2024).

To avoid multicollinearity among the environmental variables leading to model overfitting, pre-modeling experiments were initially conducted using MaxEnt 3.4.1 with existing distribution data for *A. calamus* and 35 environmental variables. The objective of the experiment was to assess the predictive contributions of individual environmental factors to the geographical distribution of the species. SPSS 25.0 software (IBM Corp. Released 2017. IBM SPSS Statistics for Windows, Version 25.0. Armonk, NY, USA: IBM Corp.) was used to analyze the correlations among the environmental factors. Finally, considering both the contribution rates and correlation coefficient matrices, environmental factors with low contribution rates and correlation coefficients $|r| > 0.8$ were removed. Ultimately, 18 environmental factors with significant impacts on the potential geographical distribution of *A. calamus* were identified, as shown in Table S2.

The spatial resolution of all the ecological factor data is 2.5′. The administrative zoning map and map of China are from the National Geomatics Center of China (http://www.ngcc.cn/, accessed on 1 December 2023).

2.3. Model Optimization

The MaxEnt model, which is a sophisticated machine learning algorithm, is sensitive to sampling bias and susceptible to overfitting [39]. A review and assessment of recent MaxEnt-related literature revealed that using the default parameters of the model for data simulation may lead to inaccurate data analysis results and difficulties in evaluating the distributions of corresponding species [40]. The regularization multiplier (RM) and feature combination (FC) are important factors affecting the accuracy of MaxEnt model construction. In this study, the ENMeval 0.3.0 package [41] in R 3.6.1 software was utilized to adjust the RM and FC parameters. The RM was set from 0.5 to 4.0 with 8 levels, with adjacent levels spaced by 0.5. Six types of FCs were tested, namely, L, LQ, H, LQH, LQHP, and LQHPT, where L is linear, Q is quadratic, H is hinge, P is product, and T is threshold. The model performance was evaluated based on various criteria, including delta.AICc, avg.test.AUC (average test Area Under the Curve), the difference between the training and testing AUC values (AUC.DIFF), and a 10% omission rate (OR10) [42].

2.4. Model Construction

Upon importing the collected data points of the *A. calamus* distribution and the chosen environmental factors into the MaxEnt 3.4.1 modeling software, 25% of the data points were reserved as a test dataset to validate the model. The system convergence threshold was established at 10^{-5}, and the maximum number of iterations was capped at 1000. The number of bootstrap replicates was set to 10. Random seeds were selected. Response curves and jackknife functions were chosen to analyze the impact of environmental factors on the *A. calamus* distribution. The RM and FC were set to the optimized parameters, while the rest were kept at the default settings.

2.5. Model Accuracy Evaluation

The model's accuracy was evaluated using the area under the receiver operating characteristic (ROC) curve (AUC) and Cohen's kappa values [39]. The AUC ranges from 0 to 1, with a higher AUC indicating greater model accuracy [43]. The kappa coefficient ranges from −1 to 1, where −1 signifies complete disagreement between the model predictions and sample data, and 1 signifies perfect agreement. Higher kappa values indicate better consistency and more reliable model predictions [44].

2.6. Geographical Distribution Classification

The suitability of a species' geographical distribution is commonly expressed on a scale from 0 to 1, where higher values denote greater suitability for the species' survival in the specified area [45]. In our study, the ASCII files produced by MaxEnt were imported into ArcGIS and transformed into raster files (TIFF format). With the "Reclassify" function, the potentially suitable geographical distribution areas for the species were classified into three levels: unsuitable ($p < 0.2$), low suitability ($0.2 \leq p < 0.4$), and high suitability ($p \geq 0.4$)

3. Results

3.1. Model Optimization and Accuracy Evaluation

In this study, we utilized the mature ENMeval package to optimize the parameters of the MaxEnt model to simulate the geographical distribution of *A. calamus*. The optimization results are shown in Table 1. The default settings included: feature combinations of linear (L), quadratic (Q), hinge (H), product (P), and threshold (T), with a regularization multiplier of 1. Under these settings, the average diff.AUC was 0.0422, the delta AICc was 130.7619, the avg.OR10 was 0.1884, and the avg.test.AUC was 0.8444. After optimization using the ENMeval package, the MaxEnt model settings were adjusted to feature combinations of linear (L), quadratic (Q), hinge (H), and product (P), with a regularization multiplier of 3.5. This optimization resulted in decreases in the avg.diff.AUC and avg.OR10 of 45.26% and 23.25%, respectively, and a delta AICc of 0, whereas the avg.test.AUC significantly increased compared with that before optimization. This finding indicates that the degree of overfitting in the model decreased by 45.26%, the complexity improved markedly, and the model's transferability was enhanced. With the optimal parameters, the MaxEnt model was run 10 times, achieving an average training AUC of 0.8958 and an average testing AUC of 0.8849, both above 0.8 and higher than the random test AUC value (0.5). The kappa coefficient was 0.7733, demonstrating that the model's predictive results were accurate and consistent. These results validate the high performance of the optimized MaxEnt model in accurately predicting the geographical distribution of *A. calamus*.

Table 1. Predictive performance of the MaxEnt model with default and optimized parameter settings.

Settings	Regularization Multiplier	Feature Combination	Avg.Test.AUC	Avg.Diff.AUC	Avg.OR10	Delta.AICc
Default	1	LQHPT	0.8444	0.0422	0.1884	130.7619
Optimized	3.5	LQPH	0.8546	0.0231	0.1466	0

3.2. Environmental Factor Analysis

Through the optimized MaxEnt model simulation, the percent contribution and permutation importance values reveal that the six most significant environmental factors affecting the distribution of *A. calamus* are the precipitation of the wettest month (bio13), human footprint (humanfoot), land use (lucc), annual precipitation (bio12), mean diurnal range (bio2), and elevation. The cumulative values of the percent contribution and permutation importance for these factors are 80.47% and 84.87%, respectively (Table S3). According to the jackknife test results (Figure S1), the factors that significantly increase the regularized training gain of the model are humanfoot, bio13, and bio12. In summary, the distribution of *A. calamus* is influenced primarily by temperature and precipitation factors, including bio13, humanfoot, lucc, bio12, bio2, and elevation. Soil factors have minimal impacts on the current distribution of *A. calamus* (Table S3 and Figure S1).

An examination of the response curves of *A. calamus* to the six dominant environmental factors revealed a trend where the probability of distribution initially increased and eventually stabilized as the concentrations of bio13, humanfoot, and bio12 increased (Figure 2). Conversely, an increase in bio2 and elevation led to a decrease in the distribution probability (Figure 2). The primary land use types for the distribution of *A. calamus* are rivers and channels, reservoirs and ponds, and lakes. When the distribution probability of *A. calamus* in a suitable geographical distribution exceeds the threshold (0.2), the dominant environmental factors that restrict its distribution fall within the following ranges: bio13 $\geq$ 109.87 mm, humanfoot $\geq$ 5.39, bio12 $\geq$ 391.25 mm, bio2 $\leq$ 12.83 °C, elevation $\leq$ 2469.24 m and land use types include rivers and channels, reservoirs and ponds, lakes, urban areas, marshes, other constructed lands, rice fields, forested areas, and shrublands, as shown in Figure 2.

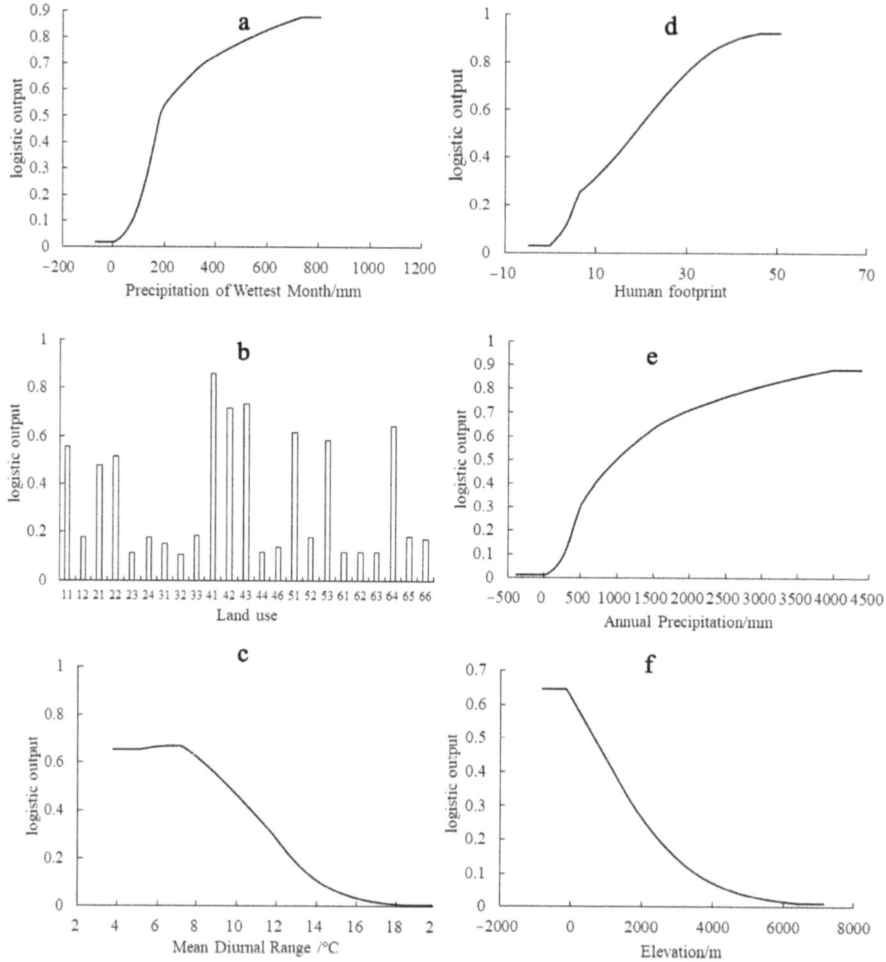

Figure 2. Response curves of the six major environmental factors. (**a**) Precipitation of the wettest month; (**b**) Land use; (**c**) Mean diurnal range; (**d**) Human footprint; (**e**) Annual precipitation; (**f**) Elevation.

3.3. Current Potential Geographical Distribution

The results of the simulated current distribution of *A. calamus* via the optimized MaxEnt model are shown in Figure 3. In this figure, the green areas represent high-suitability distribution zones, the red areas represent low-suitability distribution zones, and the yellow areas represent unsuitable distribution zones (similarly indicated in subsequent figures). The figure shows that the suitable geographical distribution areas for *A. calamus* in China are distinctly located east of the 400 mm isohyet line. However, its distribution is highly fragmented. High-suitability distribution areas are found primarily in southeastern Tibet, eastern and southern Yunnan, Guizhou, Guangxi, Guangdong, Hunan, Jiangxi, Fujian, Zhejiang, and most of Taiwan. There are also scattered distributions in central Hainan, central Sichuan, southern Jiangsu, Henan, Shandong, Shanxi, Shaanxi, Beijing, Liaoning, Jilin, and Heilongjiang, covering 121.12×10^4 km^2, which accounts for 12.61% of the study area. The low-suitability distribution areas include most of Hainan, northern Yunnan, eastern Sichuan, northern Chongqing, southern Gansu, southern Shaanxi, southern Ningxia,

Shanxi, Hebei, Hubei, Henan, Anhui, and Shandong. Eastern Inner Mongolia, northern Liaoning, Jilin, Heilongjiang, and southeastern Tibet (above the medium-suitability areas) also have notable distributions. These regions cover 164.20×10^4 km^2 accounting for 17.09% of the study area. Additionally, regions where $p < 0.2$ are considered unsuitable for the growth of *A. calamus*; these regions cover 675.27×10^4 km^2, which accounts for 70.30% of the area in China.

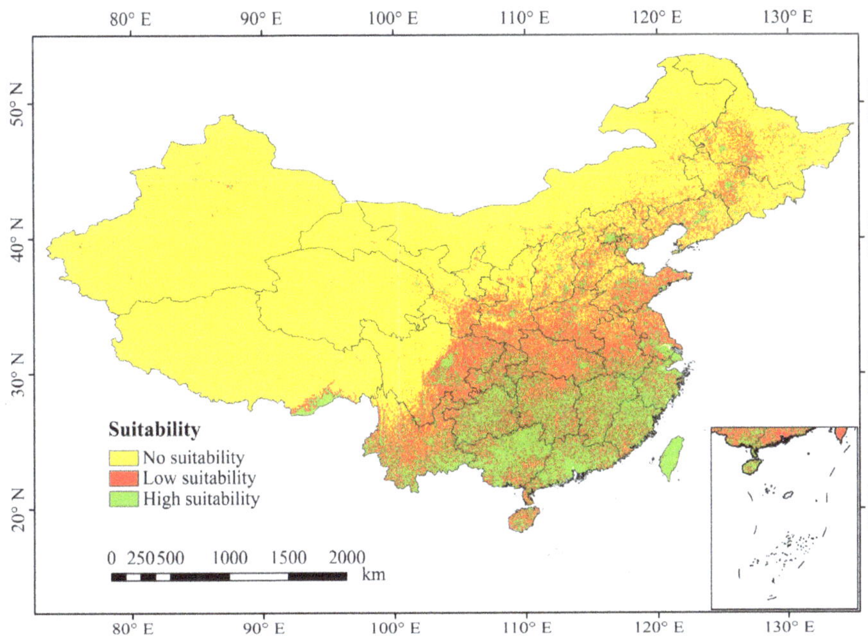

Figure 3. Potential geographical distribution pattern of *A. calamus* in China under current climatic conditions.

3.4. Future Geographical Distribution

Following the same standards as in the previous sections, predictions were made for the distribution areas of *A. calamus* for three future periods under three target scenario pathways. The resulting future geographical distribution maps are shown in Figure 4. Dynamic change maps for each distribution grade are shown in Figure 5, and the areas and variation ranges of each suitability zone are presented in Table 2. The comprehensive results indicate that under the three shared socioeconomic pathways and representative concentration pathway climate scenarios for the 2050s, 2070s, and 2090s, the locations and areas of the *A. calamus* distribution at the different suitability levels will undergo varying degrees of change. Both the high-suitability and low-suitability areas are projected to significantly expand, whereas unsuitable areas are expected to decrease.

In the future, the area of high-suitability distribution zones for *A. calamus* will first increase and then decrease over time under the SSP126 and SSP585 scenarios (Table 2). However, under the SSP370 scenario, the area will expand continuously, showing an overall increasing trend with decreasing fragmentation. The growth rate will range from 12.07% to 40.46% (Table 2). Notably, under the SSP585 scenario in 2090, the high-suitability distribution area will reach its maximum of 170.13×10^4 km^2, representing an increase of 40.46% compared with that under the current climatic conditions (Table 2, Figures 4i and 5i). This expansion will mainly result from the transformation of low-suitability zones in most of Yunnan, eastern Sichuan, northern Guizhou, southern Hubei and Anhui, and the eastern Shandong provinces (Figure 5i).

In the future, under the SSP126 scenario, the low-suitability distribution zones for *A. calamus* will initially increase and then decrease over time. Under the SSP370 and SSP585 scenarios, the increase will initially decrease and then increase, with an overall increase ranging from 0.90% to 14.13% (Table 2). Notably, under the SSP585 scenario in 2050, the low-suitability distribution area will reach its maximum at 187.39×10^4 km^2 (Table 2). Compared with the current climatic conditions, the increase in area will come from the conversion of primarily the unsuitable zones in Liaoning, central Jilin, northern Heilongjiang, eastern Inner Mongolia, northern Hebei, northern Shanxi, central Ningxia, and southern Gansu, as well as sporadic unsuitable regions in eastern Heilongjiang, eastern Qinghai, western Sichuan, and eastern Tibet (Figures 4c and 5c). In the future, the unsuitable zones for *A. calamus* will show a continuous decreasing trend with increasing radiative forcing over time, with an overall reduction ranging from 3.89% to 10.15% (Table 2).

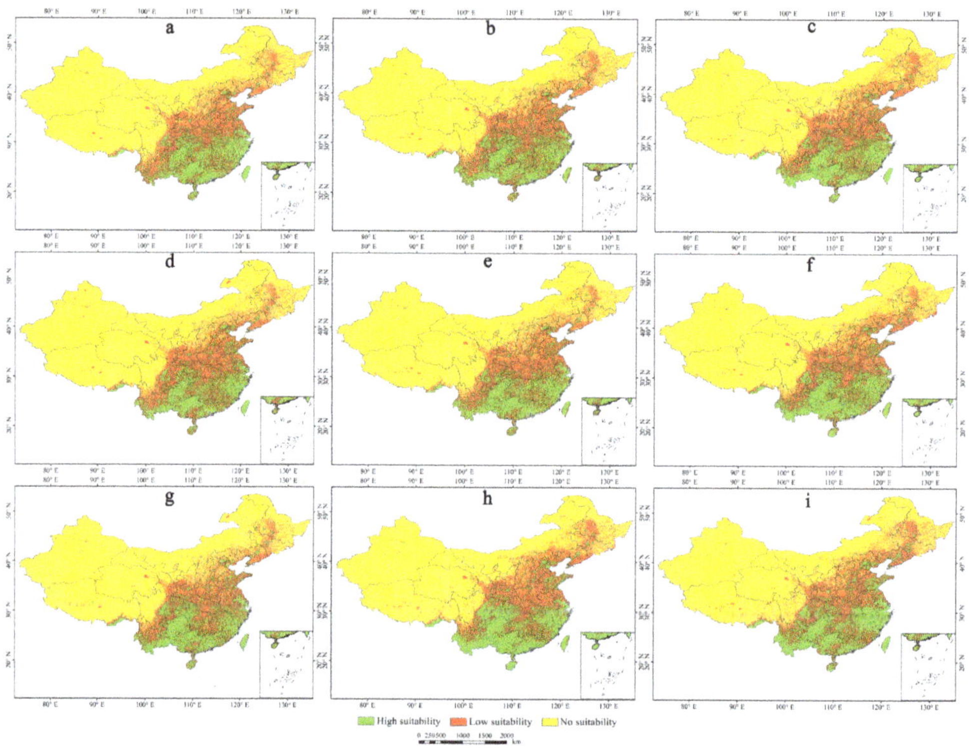

Figure 4. Future geographical distribution maps of *A. calamus.* (**a**) 2050s-SSP126; (**b**) 2050s-SSP370; (**c**) 2050s-SSP585; (**d**) 2070s-SSP126; (**e**) 2070s-SSP370; (**f**) 2070s-SSP585; (**g**) 2090s-SSP126; (**h**) 2090s-SSP370; (**i**) 2090s-SSP585.

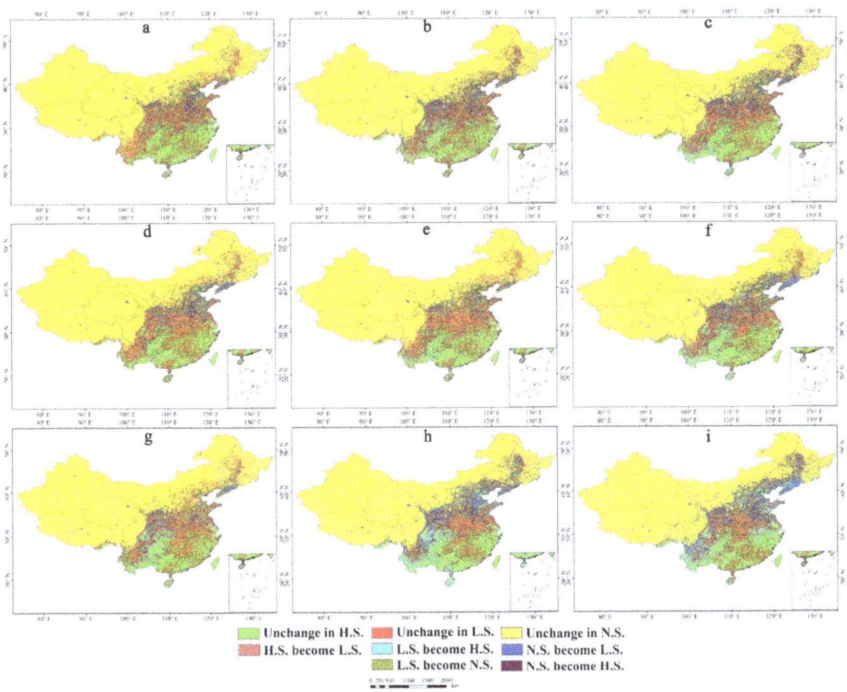

Figure 5. Future dynamic changes of different grade distribution areas. (**a**) 2050s-SSP126; (**b**) 2050s-SSP370; (**c**) 2050s-SSP585; (**d**) 2070s-SSP126; (**e**) 2070s-SSP370; (**f**) 2070s-SSP585; (**g**) 2090s-SSP126; (**h**) 2090s-SSP370; (**i**) 2090s-SSP585.

Table 2. Areas and variation ranges of the distribution area.

Scenario	Period	High Suitability Area		Low Suitability Area		Total Area($\times 10^4$/km^2)	Total Change
		Area($\times 10^4$/km^2)	Change	Area($\times 10^4$/km^2)	Change		
current		121.12		164.20		675.25	
SSP126	2050s	137.93	13.87%	178.41	8.65%	644.23	−4.59%
	2070s	135.74	12.07%	181.63	10.62%	643.20	−4.75%
	2090s	144.21	19.06%	165.68	0.90%	650.68	−3.64%
SSP370	2050s	142.92	17.99%	184.65	12.46%	633.00	−6.26%
	2070s	139.46	15.14%	172.14	4.84%	648.97	−3.89%
	2090s	167.55	38.33%	180.90	10.17%	612.12	−9.35%
SSP585	2050s	147.86	22.07%	187.39	14.13%	625.32	−7.39%
	2070s	154.90	27.89%	177.58	8.15%	628.08	−6.98%
	2090s	170.13	40.46%	183.72	11.89%	606.72	−10.15%

3.5. Centroid Changes in the A. calamus Distribution Pattern

Figure 6 depicts the projected shifts in centroid positions within the potential geographical distribution areas of *A. calamus* across various climate change scenarios. Specifically, Figure 6(a1,b1) depict the movement distances of the distribution area centroids in the east-west, and north-south directions relative to the current centroid across the three periods and three climate change scenarios. As depicted in the figures, the centroid of the potential geographical distribution area for *A. calamus* is predominantly situated within Hubei Province, with a noticeable trend of northward migration across all future climate scenarios. In contrast, the centroids of unsuitable areas tend to move northwest.

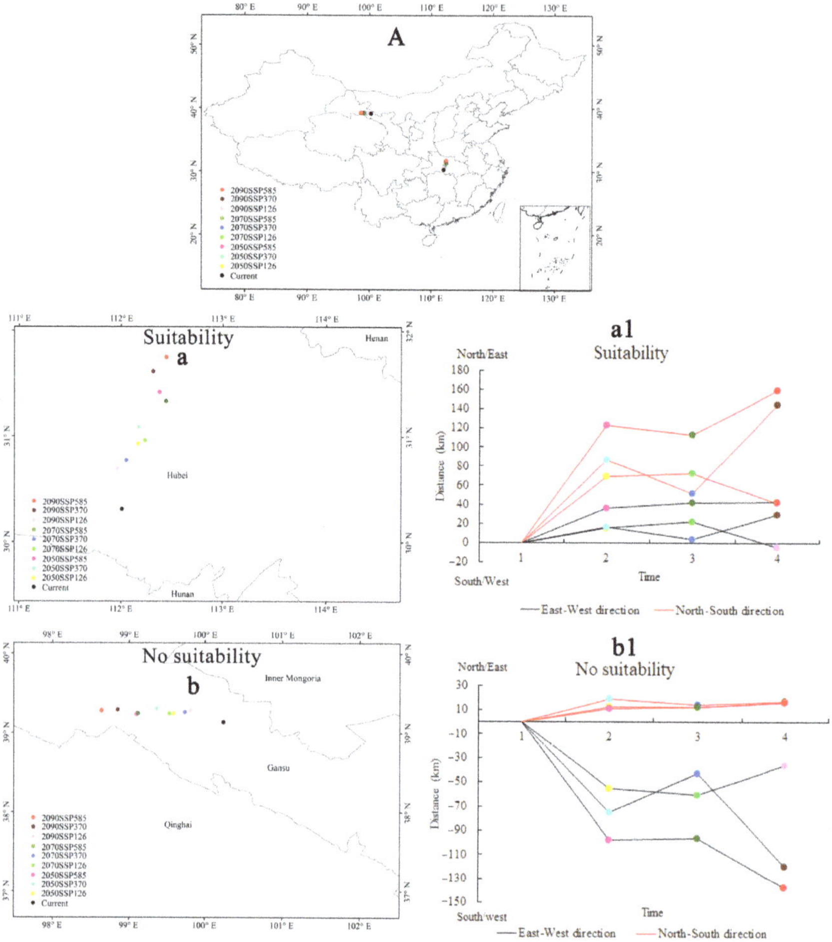

Figure 6. Future centroid movement of the potential geographical distribution area of *A. calamus* under different climate change scenarios. (**A**) Centroids in China; (**a**) Centroids in suitable area. (**b**) Centroids in unsuitable area; (**a1**,**b1**) indicates the distances that centroids of different grades of suitable distribution area migrate in two directions (north–south and east–west) under future climate change.

4. Discussion

4.1. Model Accuracy

This study highlights the inaugural utilization of an optimized MaxEnt model to project the potential geographical distribution of the herbaceous plant *A. calamus* across China under present and future climate change scenarios. Among the various species distribution models, the MaxEnt model stands out for its robust predictive accuracy, even with limited sample sizes, narrow geographic ranges, and constrained environmental tolerances [46]. Consequently, it ranks among the most widely employed species distribution models [47]. Nevertheless, in practical applications of the MaxEnt model, optimizing model parameters is imperative to increase the precision and dependability of species distribution forecasts [48]. Compared with alternative data packages employed for optimizing the MaxEnt model, leveraging the ENMeval package to fine-tune the regularization multiplier and feature combination parameters yielded a model that demonstrated superior predictive performance [40]. The response curves were smoother, and the transferability of the model was greater, allowing for a more reasonable reflection of the response of a species to environmental factors and accurate simulations of potential species distribution. In this study, after the model parameters were optimized via the ENMeval package, the FC was set to LQPH, and the RM was set to 3.5. The optimized model achieved an AUC above 0.8 and a kappa coefficient above 0.7, with smooth and extended response curves. These findings indicate that the model constructed in this study has high predictive sensitivity and meets accuracy standards, resulting in highly accurate prediction outcomes. Relevant studies indicate that the optimal regularization multiplier is generally greater than the default value in MaxEnt software, with values typically between 2 and 4 yielding the best overall performance in predicting species distributions [49,50]. According to the results obtained from the MaxEnt model simulation in this study, the current distribution of *A. calamus* is distinctly located east of the 400 mm isohyet in China. This region almost entirely covers all the concentrated sample points and closely aligns with the dominant environmental factor of annual precipitation of ≥391.25 mm. This alignment further demonstrated that the model constructed in this study is highly precise and reliable. This finding also indicates that under current climatic conditions, the cultivation of *A. calamus* can be simply guided by selecting cultivation areas along the 400 mm isohyet.

4.2. Response of *A. calamus* to Dominant Environmental Factors

Climate and human factors are considered the most important driving factors controlling plant geographical distribution [51], and their influence on the geographical distribution of *A. calamus* is no exception. This study revealed that the dominant environmental factors influencing the distribution of *A. calamus* are the precipitation of the wettest month, the human footprint, land use, annual precipitation, the mean diurnal range, and elevation. Thus, the primary climatic factors affecting the geographical distribution of *A. calamus* are precipitation and temperature, which is consistent with the dominant environmental factors of the co-distributed wetland plant *Carex doniana* [14]. *A. calamus* is a typical wetland plant that has high water requirements. In this study, both of the top two dominant environmental factors influencing its distribution were related to precipitation. The rainfall during the wettest month reflects extreme moisture conditions, which typically occur in summer in China. Summer is the peak growing season for herbaceous plants, including the perennial *A. calamus* [45]. Additionally, annual precipitation indicates that *A. calamus* requires a substantial water supply for optimal growth. Summer is also the critical period for *A. calamus* to flower and fruit [51]. Adequate water during this time can increase atmospheric humidity and soil moisture, thereby promoting the growth of *A. calamus*.

Temperature plays an important role in regulating plant life activities and physiological processes, thereby shaping plant distribution [46]. Within certain thresholds, temperature can increase a plant's adaptability to geographical distribution conditions and environmental fluctuations. However, exceeding a specific temperature threshold can cause the geographical distribution range of a plant to shrink or even disappear [52]. In this study, bio2 was ≤12.83 °C indicating that *A. calamus* thrives in regions with relatively small daily temperature fluctuations. Research has shown that *A. calamus* requires a certain amount of heat for growth, ceases to grow below 10 °C, and survives winter by submerging its rhizomes into the mud [27]. Elevation also constrains the potential geographic distribution of *A. calamus*. In the model, the elevation threshold for its potential distribution is 2469.24 m, which aligns with its natural distribution typically found below 2600 m [31], indicating *A. calamus*' ability to adapt to a specific range of altitude variations.

Based on the aforementioned climate factors and thresholds, *A. calamus* thrives in warm and humid climates with abundant rainfall and minimal diurnal temperature variation. This finding aligns with those of previous studies [19]. *A. calamus* is predominantly found in the Yangtze River Basin and regions to the south, which exhibit significant seasonal changes but relatively mild temperature variations [27]. Consequently, the fundamental growth characteristics of *A. calamus* can be preliminarily analyzed based on the variations in these 6 key environmental factors across different regions.

The human footprint and land use have become the second and third most significant factors influencing *A. calamus* distribution, highlighting the substantial impact of human activities on its geographic range. Field surveys revealed that infrastructure projects such as waterway construction, road construction, farmland expansion, and urban development all disrupt *A. calamus* habitats, with some populations facing severe challenges such as habitat shrinkage and fragmentation (Figure 3). Among the primary land use types within the *A. calamus* distribution areas, urban and other constructed lands ranked the highest, further confirming that human activities are critical factors affecting and even constraining suitable *A. calamus* habitats. In this study, the model identified key land types in potential *A. calamus* distribution areas, including rivers, reservoirs, ponds, lakes, urban areas, marshes, other constructed lands, paddy fields, forested areas, and shrublands—many of which contain water sources. Naturally, Acorus species grow around wetlands such as mountain streams, ponds, and shallow lake areas [31], indicating distinctive habitat associations.

Therefore, some basic growth characteristics of *A. calamus* can be preliminarily analyzed based on the variations in six primary environmental factors across different regions in this study.

4.3. Changes in the Geographical Distribution Pattern of A. calamus

The results of this study suggest that the total geographical distribution area of *A. calamus* will markedly expand under various future climate scenarios, likely due to the sensitivity of this species to moisture. In the context of future climate warming and increased atmospheric moisture content, predictions indicate that precipitation in southern China will likely increase [53]. Additionally, the northward shift of the rain belt may lead to increased rainfall in previously drier northern regions [54], thereby expanding the geographical distribution of *A. calamus*. The largest increase in area is observed under the SSP58.5 climate scenario, suggesting that high-concentration greenhouse gas scenarios are favorable for the geographic expansion of *A. calamus*. Similar results have been reported in studies on other plants, such as *Prunus mume Siebold & Zucc.* [44], *Ageratina adenophora* R.M. King & H. Rob. [55], *Fritillaria cirrhosa* D. Don [56], and *Actinidia chinensis* Planch. [57], where the suitable geographical distribution areas were found to increase under scenarios with increased greenhouse gas emissions.

However, the changes in the suitable geographical distributions of different suitability levels for *A. calamus* under future climate scenarios vary. Both high-suitability and low-suitability distribution areas exhibit a northward expansion trend under future climate change. The most significant changes occur mainly in North China and Northeast China, where unsuitable areas transform into low-suitability areas, and low-suitability areas gradually convert into high-suitability areas. Notably, the high-suitability distribution area shows not only an increasing trend but also decreasing fragmentation. The fragmentation of plant habitats can directly impact plant reproduction, gene flow, and population connectivity [58]. Research suggests that future precipitation in northeastern China will show a significant increasing trend, with greater precipitation increases in high-emission scenarios than in low-emission scenarios and in drier areas than in wetter areas. These regions will meet the growth requirements of *A. calamus* under these scenarios, facilitating its growth and reproduction. This also accounts for the significant northward movement of the centroid of the suitable geographical distribution for *A. calamus* across all three projected climate scenarios. Recent studies have shown that most species in China are gradually migrating northward in response to global climate change [59].

In conclusion, the geographical distribution of *A. calamus*, a valuable plant resource, is expected to expand in China under the influence of global climate change. Policies and regulations should be established to protect and utilize this resource effectively, ensuring the continuity of its habitats. These results provide a foundation for formulating conservation strategies for *A. calamus* in response to global climate change.

5. Conclusions

This study is the first to utilize an optimized MaxEnt model to simulate the potential geographical distribution pattern of *A. calamus* in China under both current and future climate change scenarios. The dominant environmental factors affecting the geographical distribution of *A. calamus* in China are bio13, humanfoot, lucc, bio12, bio2, and elevation. Under current climate conditions, the suitable geographical distribution areas for *A. calamus* in China are located primarily east of the 400 mm isohyet. In future climate scenarios, both the high- and low-suitability distribution areas of *A. calamus* will significantly increase over time, whereas the unsuitable areas will decrease. The centroid of the potential geographical distribution area of *A. calamus* is projected to shift overall toward higher latitudes in the north. Our findings provide a theoretical basis for future efforts focused on the introduction, cultivation, and conservation of *A. calamus*, as well as for addressing the impacts of global climate change. We recommend large-scale artificial cultivation in regions such as Guangdong, Guangxi, and Fujian to meet market demand while mitigating the effects of human activities and climate change on its wild populations.

Supplementary Materials: The following supporting information can be downloaded at: https://www.mdpi.com/article/10.3390/plants13233352/s1, Table S1: Environmental variables used in this study; Table S2: The environmental variables used in this study; Table S3: Percent contribution and permutation importance of each environmental variable of *A. calamus*; Figure S1: Jackknife test for the environmental variables.

Author Contributions: C.Y.: Investigation, Methodology, Supervision, Project administration, Funding Acquisition. H.L.: Conceptualization, Investigation, Data Curation, Methodology, Formal Analysis; X.S.: Investigation, Methodology, Formal Analysis, Writing—Original Draft Preparation, Writing—Review & Editing, Funding Acquisition. All authors have read and agreed to the published version of the manuscript.

Funding: This study was funded by Zhejiang Forestry Science and Technology Project (2022SY06), "Pioneer" and "Leading Goose" R&D Program of Zhejiang (2022C02038), and Zhejiang Provincial Scientific Research Institute Special Project (2023F1068-4).

Data Availability Statement: The data presented in this study are available as Supplementary Materials.

Conflicts of Interest: The authors declare no conflict of interest.

References

1. Skendžić, S.; Zovko, M.; Živković, I.P.; Lešić, V.; Lemić, D. The impact of climate change on agricultural insect pests. *Insects* **2021**, *12*, 440. [CrossRef] [PubMed]
2. IPCC. *Climate Change 2023: Synthesis Report*; Core Writing Team, Lee, H., Romero, J., Eds.; IPCC: Geneva, Switzerland, 2023.
3. McCulloch, M.T.; Winter, A.; Sherman, C.E.; Trotter, J.A. 300 years of sclerosponge thermometry shows global warming has exceeded 1.5 °C. *Nat. Clim. Chang.* **2024**, *14*, 171–177. [CrossRef]
4. Li, E.; Zhao, J.; Pullens, J.W.; Yang, X. The compound effects of drought and high temperature stresses will be the main constraints on maize yield in Northeast China. *Sci. Total Environ.* **2022**, *812*, 152461. [CrossRef]
5. Ostad-Ali-Askari, K.; Shayannejad, M.; Ghorbanizadeh Kharazi, H.; Zareian, M.J. Effect of Climate Change on Precipitation Patterns in an Arid Region Using GCM Models: Case Study of Isfahan-Borkhar Plain. *Nat. Hazards Rev.* **2020**, *21*, 04020006. [CrossRef]
6. Shi, X.D.; Wang, J.W.; Zhang, L.; Chen, S.; Zhao, A.L.; Ning, X.; Fan, G.; Wu, N.; Zhang, L.; Wang, Z. Prediction of the potentially suitable areas of Litsea cubeba in China based on future climate change using the optimized MaxEnt model. *Ecol. Indic.* **2023**, *148*, 110093. [CrossRef]
7. Yan, X.; Wang, S.; Duan, Y.; Han, J.; Huang, D.; Zhou, J. Current and future distribution of the deciduous shrub Hydrangea macrophylla in China estimated by MaxEnt. *Ecol. Evol.* **2021**, *11*, 16099–16112. [CrossRef]
8. Cao, B.; Bai, C.; Xue, Y.; Yang, J.; Gao, P.; Liang, H.; Li, G. Wetlands rise and fall: Six endangered wetland species showed different patterns of habitat shift under future climate change. *Sci. Total Environ.* **2020**, *731*, 138518. [CrossRef]
9. Zhan, P.; Wang, F.; Xia, P.; Zhao, G.; Wei, M.; Wei, F.; Han, R. Assessment of suitable cultivation region for Panax notoginseng under different climatic conditions using MaxEnt model and high-performance liquid chromatography in China. *Ind. Crops Prod.* **2022**, *176*, 114416. [CrossRef]

10. Darrah, S.E.; Bland, L.M.; Bachman, S.P.; Clubbe, C.P.; Trias-Blasi, A. Using coarse-scale species distribution data to predic extinction risk in plants. *Divers. Distrib.* **2017**, *23*, 435–447. [CrossRef]
11. Liu, D.; Lei, X.; Gao, W.; Guo, H.; Xie, Y.; Fu, L.; Lei, Y.; Li, Y.; Zhang, Z.; Tang, S. Mapping the potential distribution suitability c 16 tree species under climate change in northeastern China using MaxEnt modelling. *J. For. Res.* **2022**, *33*, 1739–1750. [CrossRef]
12. Phillips, S.J.; Anderson, R.P.; Schapire, R.E. Maximum entropy modeling of species geographic distributions. *Ecol. Model.* **200** *190*, 231–259. [CrossRef]
13. Grimmett, L.; Whitsed, R.; Horta, A. Presence-only species distribution models are sensitive to sample prevalence: Evaluatin models using spatial prediction stability and accuracy metrics. *Ecol. Model.* **2020**, *431*, 109194. [CrossRef]
14. Rathore, M.K.; Sharma, L.K. Efficacy of species distribution models (SDMs) for ecological realms to ascertain biological conserva tion and practices. *Biodivers. Conserv.* **2023**, *32*, 3053–3087. [CrossRef]
15. Merow, C.; Smith, M.J.; Silander, J.A., Jr. A practical guide to MaxEnt for modeling species' distributions: What it does, and wh inputs and settings matter. *Ecography* **2013**, *36*, 1058–1069. [CrossRef]
16. Wang, F.; Yuan, X.; Sun, Y.; Liu, Y. Species distribution modeling based on MaxEnt to inform biodiversity conservation in th Central Urban Area of Chongqing Municipality. *Ecol. Indic.* **2024**, *158*, 111491. [CrossRef]
17. Sokoloff, D.D.; Degtjareva, G.V.; Skaptsov, M.V.; Vislobokov, N.A.; Kirejtshuk, A.G.; Sennikov, A.N.; Severova, E.E.; Chepinoga V.V.; Samigullin, T.H.; Valiejo-Roman, C.M.; et al. Diploids and tetraploids of *Acorus* (Acoraceae) in temperate Asia are pseu docryptic species with clear differences in micromorphology, DNA sequences and distribution patterns, but shared pollinatior biology. *Taxon* **2024**, *73*, 718–761. [CrossRef]
18. Sokoloff, D.D.; Remizowa, M.V.; Nuraliev, M.S.; Averyanov, L.V.; Sennikov, A.N. The first genome from the basal monoco family has been misnamed: Taxonomic identity of *Acorus tatarinowii* (Acoraceae), a source of numerous chemical compounds o pharmaceutical importance. *Diversity* **2023**, *15*, 176. [CrossRef]
19. Shou, J.X.; Wu, X.; Guo, C.Y.; Ying, J.; Chen, M.S. Anatomical features of vegetative organs of *Acorus calamus* and their ecologica adaptability. *Subtrop. Plant Sci.* **2018**, *47*, 23–27.
20. Wu, J.L. Study on resources of Acorus calamus medicinal materials. *J. West China Phar.* **1992**, *7*, 123–126.
21. Lai, W.L.; Zhang, X.Y.; Fang, L.; Lai, S.; Zhang, M. Response of Protecting Enzymes of Antioxidant System in *Acorus calamus* L. tc Naphthalene Pollution. *North. Hortic.* **2016**, *40*, 79–82.
22. Zhou, S.L.; Fang, Y.; Sun, G.; Du, X.L.; Hu, Z.Y. Effects of *Acorus calamus* on Nitrogen and Phosphorus Purification from Water *North. Hortic.* **2013**, *26*, 51–53.
23. Lin, H.; Liu, J.F.; Liu, L.L.; Dong, Y.B. Physiological responses of *Acorus calamus* and reed under composite heavy metal stress and their enrichment ability. *Chin. J. Eng.* **2017**, *39*, 1123–1128.
24. Ren, J.; Fu, Z.W.; Tao, L.; Yang, Q. Accumulation Effect of Phragmites australis, *Acorus calamus* and *Scirpus tabernaemontani* on Zn^{2+} in Water Body. *Wetl. Sci.* **2011**, *9*, 322–327.
25. Joshi, R.K. *Acorus calamus* Linn. phytoconstituents and bactericidal property. *World J. Microbiol. Biotechnol.* **2016**, *32*, 164 [CrossRef]
26. Xiong, Q.; Luo, S.J.C.; Yang, C.Y.; Wang, X.C.; Song, W.X.; Yan, Z.H.; Zhu, G.H. Study on GC/Q-TOF MS analysis of essential oil components of *Acorus calamus*. *Tibet Sci. Technol.* **2023**, *359*, 10–13+35.
27. Zhang, Y.L.; Sun, B.; Deng, Z.C.; Zhang, Y.S.; Liu, C.M.; Yang, X.J. High yield and high quality selenium rich cultivation technology of *Acorus calamus*. *Crop Res.* **2015**, *29*, 787–788.
28. Qiao, D.; Gan, L.S.; Mo, J.X.; Zhou, C. Chemical constituents of *Acorus calamus*. *China J. Chin. Mater. Med.* **2012**, *37*, 3430–3433.
29. Wei, G.M.; Li, D.W.; Wang, S.M.; Luo, L.S.; Leng, J.; Hou, Y.; Li, J. Genetic Diversity of Morphological Characters in *Acorus calamus* Germplasm Resources in Guizhou. *Guizhou Agric. Sci.* **2020**, *48*, 111–113.
30. Gu, S.Y.; Yang, F.; Zhang, Y.M.; Zhang, Z.; Xie, K.; Guan, X. Effects of Flooding Stress on Physiological Characteristics and Rhizosphere Bacterial Community of *Acorus calamus*. *J. Ecol. Rural Environ.* **2020**, *36*, 488–498.
31. Mu, M.J.; Zhao, H.; Wu, Q.L. Research Progress on Chemical Constituents and Physiological Activities of *Acorus calamus* L. *J. Anhui Agric. Sci.* **2020**, *48*, 9–12.
32. Liu, L.; Nie, P.; Xiao, B.Y.; Luo, H.M.; Li, L.; Li, W.L. Identification of 5 species of medicinal plants for *Acorus* L. based on DNA barcoding sequences. *Cent. South Pharm.* **2015**, *13*, 388–392.
33. Xia, M.H.; Liu, M.Y.; Ye, L.; Li, X.; Li, Y.Y.; Wu, J.S. Rhizosphere effect of three types of macrophytes during nitrogen removal process of high loaded swine wastewater. *J. Agro-Environ. Sci.* **2024**, *43*, 896–905.
34. Yin, D.Q.; Cheng, D.X.; Li, X.; Li, L.; Zheng, C.; Huang, Y.; Wei, J. Future Projection of Spatial-temporal Changes of Hydro-Solar-Wind Power Generation Factors in Nine Major Basins of China. *J. Basic Sci. Eng.* **2024**, *3*, 754–772.
35. Yuan, G.X.; Wu, A.P.; Ge, D.B.; Chu, Z. Effects of water depth on the growth of four emergent macrophytes. *Acta Sci. Circumstantiae* **2011**, *31*, 2690–2697.
36. Song, H.J.; Zhang, X.Z.; Wang, X.Y.; Wang, Y.; Li, S.; Xu, Y.D. Not the expected poleward migration: Impact of climate change scenarios on the distribution of two endemic evergreen broad-leaved Quercus species in China. *Sci. Total Environ.* **2023**, *889*, 164273. [CrossRef] [PubMed]
37. Guisan, A.; Zimmermann, N.E.; Elith, J.; Graham, C.H.; Phillips, S.; Peterson, A.T. What matters for predicting the occurrences of trees: Techniques, data, or species' characteristics? *Ecol. Monogr.* **2007**, *77*, 615–630. [CrossRef]

3. Zhang, L.; Liu, Q.X.; Wu, T.W.; Shi, X.L.; Zhang, F.; Li, J.L.; Chu, M.; Yan, J.H.; Ma, Q.; Wei, M. BCC BCC-CSM2MR model output prepared for CMIP6 CFMIP. *Earth Syst. Grid Fed.* **2019**, 1725. [CrossRef]
9. Peterson, A.T.; Papeş, M.; Soberón, J. Rethinking receiver operating characteristic analysis applications in ecological niche modeling. *Ecol. Model.* **2008**, *213*, 63–72. [CrossRef]
0. Zhu, G.P.; Qiao, H.J. Effect of the Maxent model's complexity on the prediction of species potential distributions. *Biodivers. Sci.* **2016**, *24*, 1189–1196. [CrossRef]
1. Kass, J.M.; Muscarella, R.; Galante, P.J.; Bohl, C.L.; Pinilla-Buitrago, G.E.; Boria, R.A.; Soley-Guardia, M.; Anderson, R.P. ENMeval 2.0: Redesigned for customizable and reproducible modeling of species' niches and distributions. *Methods Ecol. Evol.* **2021**, *12*, 1602–1608. [CrossRef]
2. Muscarella, R.; Galante, P.J.; Soley-Guardia, M.; Boria, R.A.; Kass, J.M.; Uriarte, M.; Anderson, R.P. ENMeval: An R package for conducting spatially independent evaluations and estimating optimal model complexity for Maxent ecological niche models. *Methods Ecol. Evol.* **2014**, *5*, 1198–1205. [CrossRef]
3. Assefa, A.; Tibebu, A.; Bihon, A.; Yimana, M. Global ecological niche modelling of current and future distribution of peste des petits ruminants virus (PPRv) with an ensemble modelling algorithm. *Transbound. Emerg. Dis.* **2021**, *68*, 3601–3610. [CrossRef] [PubMed]
4. Chen, Y.N.; Liu, Z.G.; Yu, T.; Liao, H.; Zhou, J.Y. Prediction of Potential Distribution of Prunus mume Based on MaxEnt Model. *Chin. Wild Plant Resour.* **2024**, *43*, 107–113+126.
5. Li, Z.S.; Ma, Y.S.; Li, Y.X.; Liu, Y.; Wang, Y.L.; Wang, X.Y. Spatial and temporal dynamics of the potential habitat of Asterothamnus centraliasiaticus on the Qinghai-Tibet Plateau under climate change. *Chin. J. Ecol.* **2024**, *43*, 1566.
6. Wang, W.; Li, Z.J.; Zhang, Y.L.; Xu, X.Q. Current Situation, Global Potential Distribution and Evolution of Six Almond Species in China. *Front. Plant Sci.* **2021**, *12*, 619883. [CrossRef]
7. Hao, Z.; Fang, G.; Huang, W.; Ye, H.; Zhang, B.; Li, X. Risk prediction and variable analysis of pine wilt disease by a maximum entropy model. *Forests* **2022**, *13*, 342. [CrossRef]
8. Morales, N.S.; Fernández, I.C.; Baca-González, V. MaxEnt's parameter configuration and small samples: Are we paying attention to recommendations? A systematic review. *Peer J.* **2017**, *5*, e3093. [CrossRef]
9. Zou, X.; Peng, Y.; Wang, L.; Li, Y.; Zhang, W.X.; Liu, X. Impact of climate change on the distribution pattern of *Malus baccata* (L.) Borkh. in China since the Last Glacial Maximum. *Plant Sci. J.* **2018**, *36*, 676–686.
50. Shi, X.D.; Yin, Q.; Sang, Z.Y.; Zhu, Z.L.; Jia, Z.K.; Ma, L.Y. Prediction of potentially suitable areas for the introduction of *Magnolia wufengensis* under climate change. *Ecol. Indic.* **2021**, *127*, 107762. [CrossRef]
51. Zhang, K.; Zhang, Y.; Zhou, C.; Meng, J.; Sun, J.; Zhou, T.; Tao, J. Impact of climate factors on future distributions of *Paeonia ostii* across China estimated by MaxEnt. *Ecol. Inform.* **2019**, *50*, 62–67. [CrossRef]
52. Corlett, R.T.; Westcott, D.A. Will plant movements keep up with climate change? *Trends Ecol. Evol.* **2013**, *28*, 482–488. [CrossRef] [PubMed]
53. Huang, X.Y.; Li, X.H. Future Projection of Rainstorm and Flood Disaster Risk in Southwest China Based on CMIP6 Models. *J. Appl. Meteorol. Sci.* **2022**, *33*, 231–243.
54. Yang, S.L.; Ding, Z.L.; Li, Y.Y.; Huang, X.F. Warming-induced northwestward migration of the East Asian monsoon rain belt from the Last Glacial Maximum to the mid-Holocene. *Proc. Natl. Acad. Sci. USA* **2015**, *112*, 13178–13183. [CrossRef] [PubMed]
55. Wei, B.; Liu, L.S.; Gu, C.J.; Yu, H.B.; Zhang, Y.L.; Zhang, B.; Cui, B.; Gong, D.; Tu, Y. The climate niche is stable and the distribution area of *Ageratina adenophora* is predicted to expand in China. *Biodivers. Sci.* **2022**, *30*, 21443. [CrossRef]
56. Jiang, R.; Zou, M.; Qin, Y.; Tan, G.; Huang, S.; Quan, H.; Zhou, J.; Liao, H. Modeling of the potential geographical distribution of three fritillaria species under climate change. *Front. Plant Sci.* **2022**, *12*, 749838. [CrossRef]
57. Wang, R.L.; Wen, G.; Li, Q.; Wang, M.T.; Guo, X.; Lin, S.; Jiang, G.; Shen, Z.H. Geographical Distribution Simulation of *Actinidia deliciosa* in China and Influence of Climate Change. *J. Trop. Subtrop. Bot.* **2018**, *26*, 335–345.
58. Nong, X.X.; Yu, H.Y.; Xiang, Y.Y.; Yang, P.; Zhang, Q. Analysis of potential suitable habitat change of *Pseudotsuga sinensis* based on Maxent model. *J. Guangxi Norm. Univ. Nat. Sci. Ed.* **2024**, 42. [CrossRef]
59. You, J.; Qin, X.; Ranjitkar, S.; Lougheed, S.C.; Wang, M.; Zhou, W.; Ouyang, D.; Zhou, Y.; Xu, J.; Zhang, W.; et al. Response to climate change of montane herbaceous plants in the genus Rhodiola predicted by ecological niche modelling. *Sci. Rep.* **2018**, *8*, 5879. [CrossRef]

plants

MDPI

Article

Impact of Climate Change on Distribution of Endemic Plant Section *Tuberculata* (*Camellia* L.) in China: MaxEnt Model-Based Projection

Xu Xiao [1], Zhi Li [1,*], Zhaohui Ran [1], Chao Yan [1] and Juyan Chen [2]

1 College of Forestry, Guizhou University, Guiyang 550025, China; xiaoxu199801@163.com (X.X.); ranzhaohui1998@outlook.com (Z.R.); godblessuuus@126.com (C.Y.)
2 Key Laboratory of National Forestry and Grassland Administration on Biodiversity Conservation in Karst Mountainous Areas of Southwestern China, Guizhou Academy of Forestry, Guiyang 550005, China; jychen2008@163.com
* Correspondence: zli7@gzu.edu.cn

Abstract: Sect. *Tuberculata*, as one of the endemic plant groups in China, belongs to the genus *Camellia* of the Theaceae family and possesses significant economic and ecological value. Nevertheless, the characteristics of habitat distribution and the major eco-environmental variables affecting its suitability are poorly understood. In this study, using 65 occurrence records, along with 60 environmental factors, historical, present and future suitable habitats were estimated using MaxEnt modeling, and the important environmental variables affecting the geographical distribution of sect. *Tuberculata* were analyzed. The results indicate that the size of the its potential habitat area in the current climate was 1.05×10^5 km^2, and the highly suitable habitats were located in Guizhou, central-southern Sichuan, the Wuling Mountains in Chongqing, the Panjiang Basin, and southwestern Hunan. The highest probability of presence for it occurs at mean diurnal range (bio2) $\leq$ 7.83 °C, basic saturation (s_bs) $\leq$ 53.36%, temperature annual range (bio7) $\leq$ 27.49 °C, −7.75 °C < mean temperature of driest quarter (bio9) < 7.75 °C, annual UV-B seasonality (uvb2) $\leq 1.31 \times 10^5$ W/m^2, and mean UV-B of highest month (uvb3) $\leq$ 5089.61 W/m^2. In particular, bio2 is its most important environmental factor. During the historical period, the potential habitat area for sect. *Tuberculata* was severely fragmented; in contrast, the current period has a more concentrated habitat area. In the three future periods, the potential habitat area will change by varying degrees, depending on the aggressiveness of emissions reductions, and the increase in the potential habitat area was the largest in the SSP2.6 (Low-concentration greenhouse gas emissions) scenario. Although the SSP8.5 (High-concentration greenhouse gas emissions) scenario indicated an expansion in its habitat in the short term, its growth and development would be adversely affected in the long term. In the centroid analysis, the centroid of its potential habitat will shift from lower to higher latitudes in the northwest direction. The findings of our study will aid efforts to uncover its originsand geographic differentiation, conservation of unique germplasms, and forestry development and utilization.

Keywords: MaxEnt model; sect. *Tuberculata*; potential habitat area; climate change

1. Introduction

Global climate change is the greatest challenge facing mankind and the earth's ecosystems in the 21st century [1]. As climate strongly influences plant growth and proliferation, it is the determining factor for the geographic range of plant species [2]. Some studies have found that the ranges of species have shifted toward higher altitudes or latitudes in response to climate change [3,4]. However, changes in the climate niches of species are usually outpaced by climate change. Consequently, climate change has caused habitat loss and created geographic barriers to the dispersal of species. This is especially pertinent to species with narrow habitat ranges, which have been forced to change their geographic range to adapt to new climatic regimes [5].

Species distribution models (SDMs) are important tools for ecological and biogeographic research and are typically used to match a species to their ecological niche factors [6]. SDMs also play an important role in the prediction of potential ranges of species based on the changes in environmental factors. The maximum entropy (MaxEnt) model is intuitive, accurate, and relatively easy to use. Over the past few decades, numerous studies have been conducted on the prediction of species distributions, detection of habitat fragmentation, novel ecological modeling methodologies, biodiversity prediction, and the effects of climate change on terrestrial biodiversity [7,8].

Sect. *Tuberculata* H.T. Chang belongs to the genus *Camellia* L. and is named after its "ovary and pericarp, both of which are markedly tuberculate". *Callima tuberculata* Chien was discovered by the famous botanist Prof. Qian Chongping in Sichuan, China and first reported in 1939 [9]. Subsequently, in 1981, Chang Hongda first proposed the group of *Callima tuberculata* Chien based on its original traits, and categorized it into the six species discovered. To date, 18 species have been reported from this group [10,11]. Sect. *Tuberculata* is endemic to the southwest and parts of southern China, mainly in the karst landscape of southwest China, with Guizhou as the center of distribution and spreading to neighboring provinces. It grows mostly in limestone zones, showing strong regionalization, and is one of the representative taxa of typical karst regions [12–14]. Therefore, the evolution, distribution, and dispersal of sect. *Tuberculata* have been of interest to researchers, including phylogeny, community characteristics, and other aspects [15–19]. However, little has been reported on the potential distribution areas of sect. *Tuberculata* under climate change, with some reports focusing more on resource surveys and zonation studies of some of the important taxa of this group of plants at provincial scales, within nature reserves, or at smaller regional scales such as mountain systems [20,21]. As an endemic taxon of the genus *Camellia* that maintains its original traits, sect. *Tuberculata* has gone through a long evolutionary history. Predicting the spatial change pattern of its potential release area based on the species distribution model and analyzing the important environmental factors affecting its geographic distribution are of great scientific significance for the protection and conservation of reconstructing the endemic vegetation in this region.

The geographical distribution ranges of the species and their trends in different periods under different climatic conditions show similarities and differences due to environmental changes. The prediction model of a suitable area can reveal the future trend of sect. *Tuberculata* and its main environmental influence factors. Therefore, in this study, 18 sect. *Tuberculata* plants that have been reported were selected, and, based on the optimized MaxEnt model and ArcGIS v10.8 software, sect. *Tuberculata* plants were simulated from the Last interglacial (LIG), Last glacial maximum (LGM), Mid-Holocene (MH), Contemporary (Current), 2020–2040 (2030), 2040–2060 (2050), 2060–2080 (2070), and 2080–2100 (2090) geographic distributions and spatial change patterns. This will reveal the environmental impact factors affecting the distribution of sect. *Tuberculata* plants and their possible habitat areas, with a view to providing a theoretical basis for the in-depth field investigation, conservation, development, and utilization of sect. *Tuberculata*, and, at the same time, providing an auxiliary basis of geographical distribution for the study of this group of plants in the phylogenetic classification.

2. Results

2.1. Model Optimization and Accuracy Validation

The default settings of the MaxEnt model were generated by conducting trials to fit the 65 occurrence points of sect. *Tuberculata* plants [22]. We found that RM = 2 and FC = QP minimized the complexity and overfitting and maximized the accuracy. The accuracy of the model was then evaluated through its ROC curve. The average AUC was 0.951 ± 0.15 for all historical, current, and future climate scenarios (Figure 1), indicating that the predictions of the optimized MaxEnt model produce a smooth ROC curve with an AUC of ≥ 0.9. Hence, the fine-tuned model can produce reasonable predictions of the response of sect. *Tuberculata* to climate change (in terms of the geographic range), and accurately predict its potential distribution. This result also shows that the MaxEnt model can produce highly credible predictions of the potential range of sect. *Tuberculata* plants.

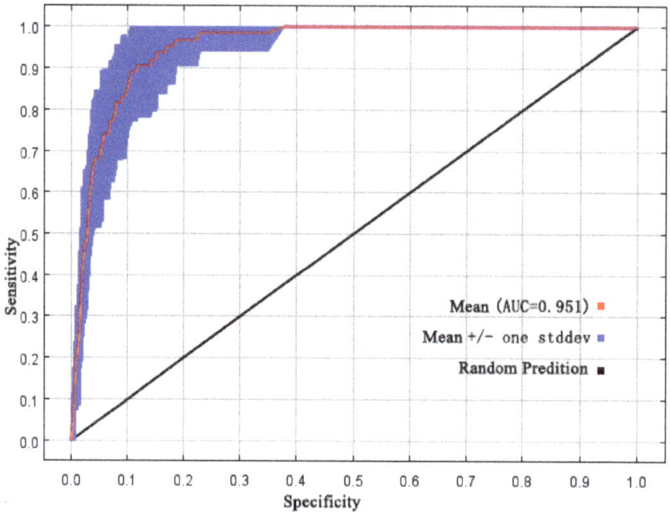

Figure 1. The verification of the ROC curve of prediction for sect. *Tuberculata* by the MaxEnt model.

2.2. The Most Influential Climatic Factors for the Geographic Range of Sect. Tuberculata Plants

The most important climatic factors for the geographic range of sect. *Tuberculata* plants were determined from the output of the MaxEnt model, jackknife test of regularized training gain contribution, permutation importance, and univariate response curves (Table 1, Figures 2 and 3). As presented in Table 1 and shown in Figure 2, the gain contributions of the mean diurnal temperature range (bio2), subsoil base saturation (S_bs), annual temperature range (bio7), mean temperature of the driest quarter (bio9), seasonality of UV-B (uvb2), and mean UV-B of the highest month (uvb3) were 46.9%, 10.4%, 10.1%, 8.9%, 7.1%, and 3.4%, respectively (86.8% cumulatively). Therefore, these six factors are the most important environmental factors for the range of sect. *Tuberculata*.

Table 1. Importance of each dominant environment variable in the MaxEnt model.

Environmental Variables	Description	Contribution (%)	Suitable Range
Bio2	Mean diurnal range	46.9	<7.83 °C
S_bs	Subsoil base saturation	10.4	<53.36%
Bio7	Annual temperature range	10.1	<27.49 °C
Bio9	Mean temperature of the driest quarter	8.9	−7.75–7.75 °C
Uvb2	Seasonality of UV-B	7.1	$<1.31 \times 10^5$ W/m^2
Uvb3	Mean UV-B of the highest month	3.4	<5089.61 W/m^2

Based on the above six dominant environmental factors, single-factor simulations were carried out, and the single-factor response curves characterized the relationship between the predicted species' pure presence probability and each variable, which can clearly reveal the correlation between the presence probability of the suitable area of the plants in sect. *Tuberculata* and the dominant factors. It is generally considered that a presence probability greater than 5% can be regarded as the most favorable range for species growth [23]. From the single-factor response curves (Figure 3), it can be seen that the probability of the existence of sect. *Tuberculata* is 0 when the average daily difference in temperature (bio2) is greater than 13.2 °C, and its existence probability increases with the decrease in bio2, and the probability of the sect. *Tuberculata* distribution reaches its peak when bio2 equals to

5.5 °C, which is the optimal fitness for submission. Taking the existence probability greater than 0.5 as the optimal fitness range, the optimal fitness range of sect. *Tuberculata* in bio2 is less than 7.83 °C. Similarly to bio2, the sect. *Tuberculata* existence probability increased with the decrease in bio7, S_bs, uvb2, and uvb3, with the optimal peaks of 13.1 °C, 0, 0.89×10^5 W/m^2, and 3350 W/m^2, respectively, and the optimal ranges of bio7 less than 27.49 °C, S_bs less than 53.36%, uvb2 less than 1.31×10^5 W/m^2, S_bs less than 53.36%, uvb3 less than 1.31×10^5 W/m^2, and uvb3 less than 5089.61 W/m^2. When bio9 is equal to −33 °C, the distribution probability of sect. *Tuberculata* is infinitely close to 0, and its existence probability is improved with the increase in bio9, etc. The distribution probability of sect. *Tuberculata* reaches its peak when bio9 is equal to 0 °C, which is the optimal condition for survival, and then, with the continued increase in bio9, the existence probability is reduced gradually, and it is extremely low when it reaches 22.5 °C. The existence probability is extremely low, so the optimal survival range of sect. *Tuberculata* in bio9 is −7.75 to 7.75 °C.

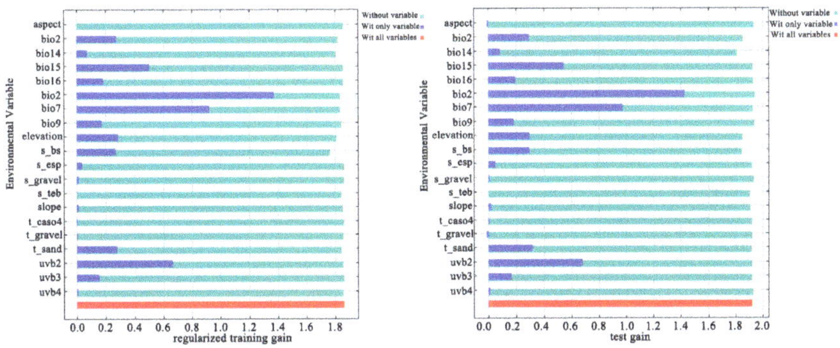

Figure 2. Evaluation of major environmental factors by the jackknife method.

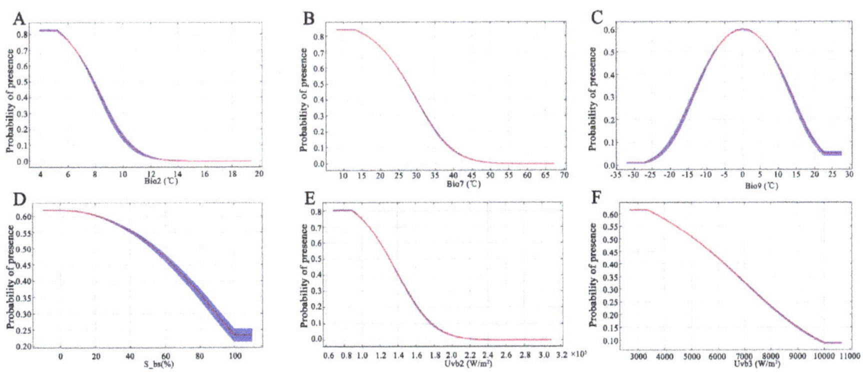

Figure 3. Relationship between potential suitable areas and single factor response variables. (**A**), Mean diurnal range; (**B**), Subsoil base saturation; (**C**), Annual temperature range; (**D**), Mean temperature of the driest quarter; (**E**), Seasonality of UV-B; (**F**), Mean UV-B of the highest month.

Table 2. Potential habitat areas for sect. *Tuberculata* in different periods.

Climate Scenarios		High ($\times 10^5$ km^2)	Medium ($\times 10^5$ km^2)	Low ($\times 10^5$ km^2)
Last Interglacial		0.01	0.12	0.5
Last Glacial Maximum		0.01	0.05	0.1
Mid Holocene		0.01	0.06	0.09
Current		0.21	0.35	0.49
2030s	SSP2.6	0.21	0.35	0.53
	SSP4.5	0.12	0.26	0.49
	SSP8.5	0.16	0.3	0.51
2050s	SSP2.6	0.18	0.38	0.58
	SSP4.5	0.19	0.43	0.58
	SSP8.5	0.22	0.41	0.6
2070s	SSP2.6	0.17	0.31	0.54
	SSP4.5	0.16	0.32	0.57
	SSP8.5	0.07	0.2	0.48
2090s	SSP2.6	0.21	0.41	0.54
	SSP4.5	0.12	0.33	0.66
	SSP8.5	0.1	0.3	0.62

2.3. Potential Habitat of Sect. Tuberculata Species in the Current Climate

The results of our study (Figure 4 and Table 2) showed that sect. *Tuberculata* plants are largely distributed between 22.54°–30.96° N and 103.83°–111.76° E. In the current climate (Figure 4), their total habitat area is 1.05 $\times$ 10^5 km^2, and their main area of distribution is southwestern China and some parts of China's coastline.

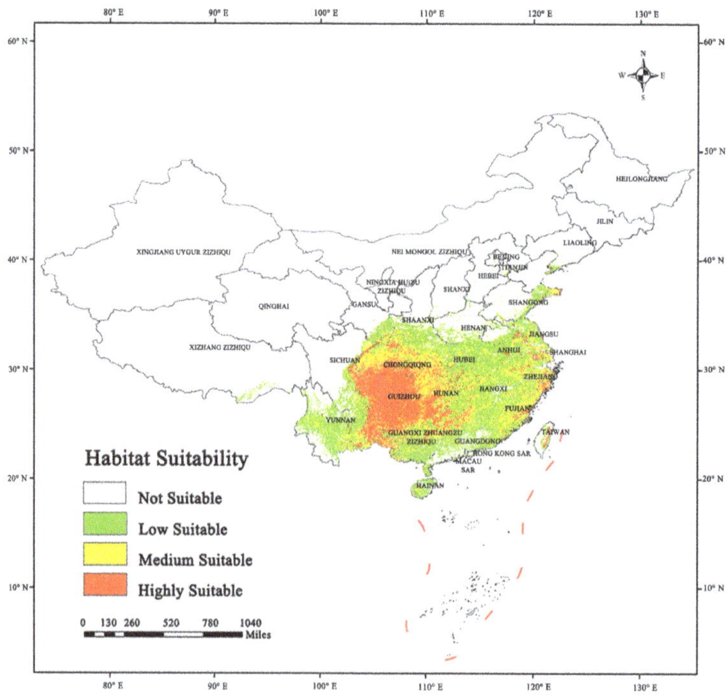

Figure 4. The potential habitat areas of sect. *Tuberculata* plants in China under the current climate.

High-suitability areas: these areas are mainly located in Guizhou, central and southern Sichuan, central and southern Chongqing, located at the junction of Guangxi, Yunnan and Guizhou provinces and a small part of Hunan China. The total area of high-suitability areas is 0.21×10^5 km^2, and they account for 20% of the total habitat area. Medium-suitability areas: these areas are generally located around high-suitability areas, and include eastern Yunnan, northeastern Sichuan, northern Chongqing, the border between Hunan and Hubei, central and northern Guangxi, central Fujian, and eastern-central Zhejiang. The total area of medium-suitability areas is 0.35×10^5 km^2, and they account for 33.33% of the total habitat area. Low-suitability areas: these areas are located in central Yunnan, Hainan, southern Guangxi, spots in northern Guangdong, the Qin Mountains in southern Shaanxi, the Dabie Mountains in eastern-central Hubei, eastern Hunan, the entirety of Jiangxi, central Anhui, southern Henan, and southern Xizang. The total area of low-suitability areas is 0.49×10^5 km^2, and they account for 46.67% of the total habitat area.

2.4. Simulation of Potential Habitat Areas for Sect. Tuberculata Plants in the Past and Future

The potential distribution of sect. *Tuberculata* plants was predicted for eight different periods. Based on the results (Table 2 and Figure 5), we found that the total habitat area of sect. *Tuberculata* decreased from 0.63×10^5 km^2 in the LIG to 0.16×105^5 km^2 in the LGM, remained at 0.16×10^5 km^2 in the MH and pre-modern period, and subsequently increased to 1.05×10^5 km^2 in the current period. Therefore, the potential habitat area decreased after the LIG, stabilized during the LGM and MH, and subsequently rapidly increased in the current period. The area of highly suitable habitats did not change throughout these periods, but the areas of the medium- and low-suitability habitats gradually decreased from the LIG to the MH.

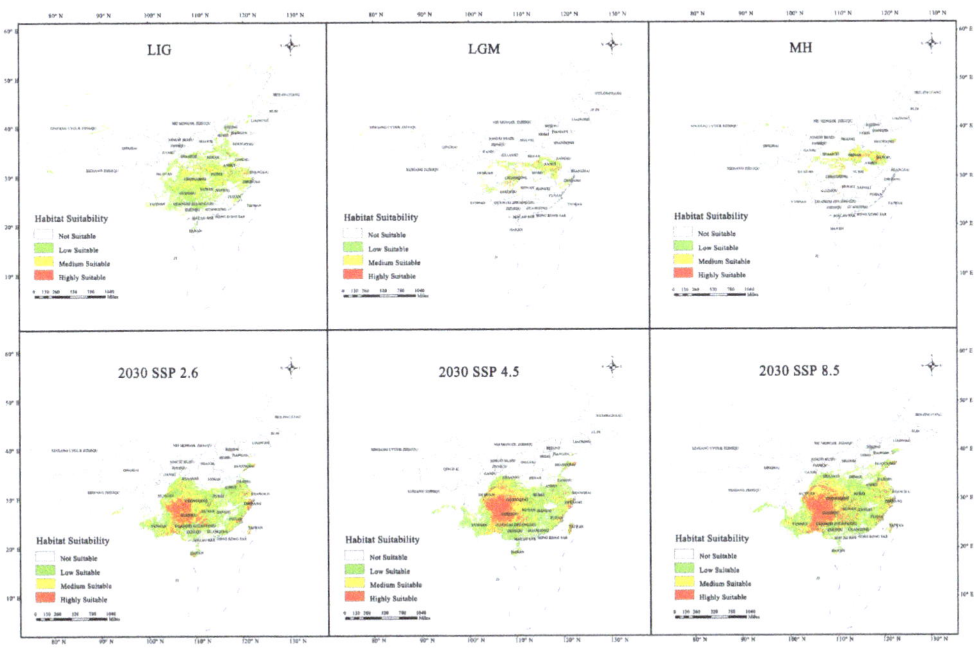

Figure 5. Distribution changes of the suitability of sect. *Tuberculata* in historical period and 2030. (LIG, Last Interglacial; LGM, Last Glacial Maximum; MH, Mid Holocene; SSP2.6, Low-concentration greenhouse gas emissions; SSP4.5, Medium-concentration greenhouse gas emissions; SSP8.5, High-concentration greenhouse gas emissions).

As presented in Table 3 and shown in Figure 6, the total habitat area was predicted to decrease in the future. In all three scenarios for 2030, the total habitat area will stay greater than 0.87×10^5 km^2. In the SSP2.6 scenario, the total habitat and high-suitability areas were predicted to be 1.09×10^5 and 0.21×10^5 km^2, respectively, which are similar to those in the current period. In the SSP4.5 scenario, the total habitat and high-suitability areas will be 0.87 $\times$ 10^5 and 0.12 $\times$ 10^5 km^2, respectively, which are 0.18 $\times$ 10^5 and 0.09 $\times$ 10^5 km^2 less than those in the current period. In the SSP8.5 scenario, the total habitat and high-suitability areas will decrease by 0.05 $\times$ 10^5 and 0.08 $\times$ 10^5 km^2 compared with those in the current period to 0.97 $\times$ 10^5 and 0.16 $\times$ 10^5 km^2, respectively.

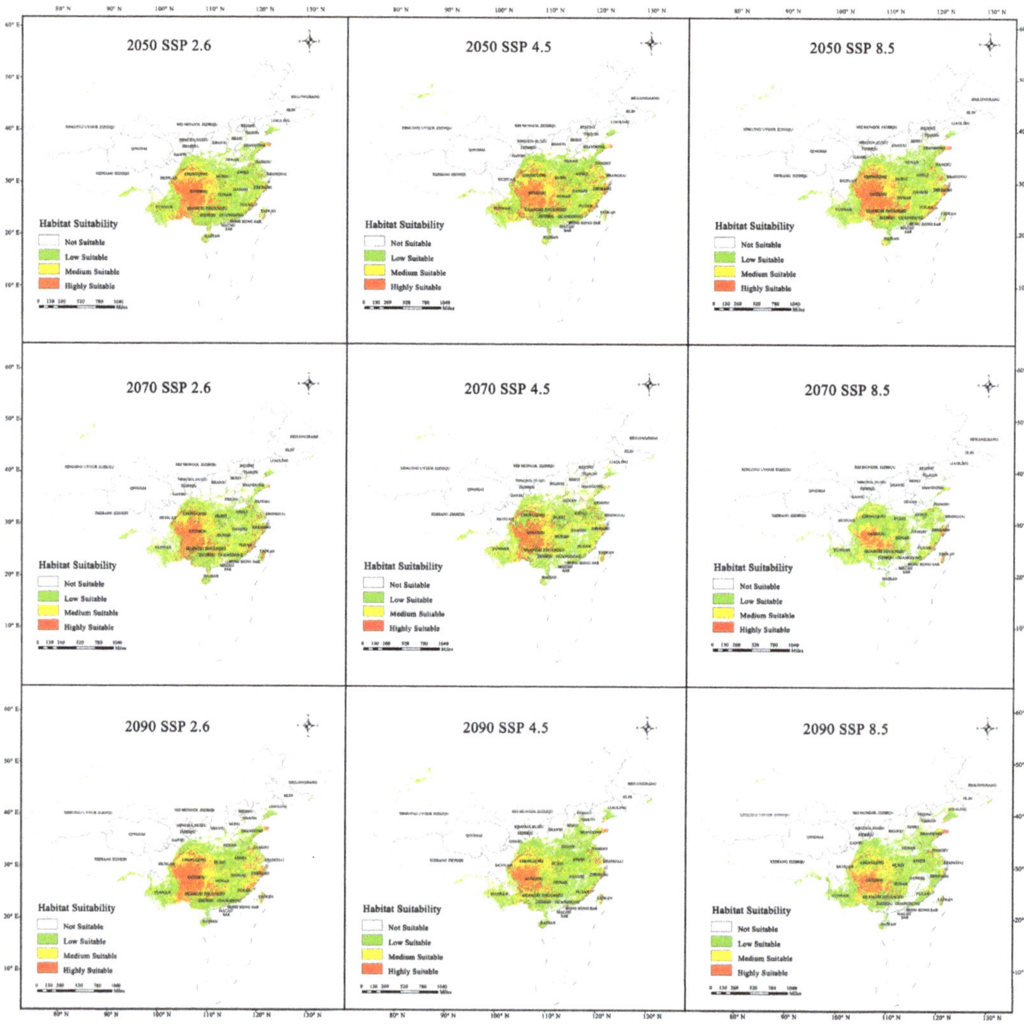

Figure 6. Distribution changes of the suitability of sect. *Tuberculata* in 2050, 2070, and 2090. (SSP2.6, Low-concentration greenhouse gas emissions scenario; SSP4.5, Medium-concentration greenhouse gas emissions scenario; SSP8.5, High-concentration greenhouse gas emissions).

In all three scenarios for 2050 (Figure 6), the total habitat area was predicted to remain above 1.14 $\times$ 10^5 km^2. In the SSP2.6 scenario, the total habitat area will be 1.14 $\times$ 10^5 km^2, which is 0.09 $\times$ 10^5 km^2 larger than that of the current period. The high-suitability area will be 0.18 $\times$ 10^5 km^2, which is 0.03 $\times$ 10^5 km^2 less than that of the current period. In the SSP4.5 scenario, the total habitat area will be 1.2 $\times$ 10^5 km^2 (0.15 $\times$ 10^5 km^2 higher than that of the current period), whereas the high-suitability area will be 0.19 $\times$ 10^5 km^2 (0.02 $\times$ 10^5 km^2 less than that of the current period). In the SSP8.5 scenario, the total habitat and high-suitability areas will be 1.23×10^5 and 0.22×10^5 km^2, namely, 0.18×10^5 and 0.01×10^5 km^2 higher than those of the current period, respectively.

In all three scenarios for 2070 (Figure 6), the total habitat area will remain above 0.75×10^5 km^2. In the SSP2.6 scenario, the total habitat and high-suitability areas will be 1.02×10^5 and 0.17×10^5 km^2, which are 0.03×10^5 and 0.04×10^5 km^2 less than those of the current period. In the SSP4.5 scenario, the total habitat area will be 1.05×10^5 km^2 (unchanged from that of the current period), whereas the high-suitability area will be 0.16×10^5 km^2 (0.05×10^5 km^2 less than that of the current period). In the SSP8.5 scenario, the total habitat and high-suitability areas will be 0.75×10^5 and 0.07×10^5 km^2, namely, 0.3×10^5 and 0.14×10^5 km^2 less than those of the current period, respectively.

In all three scenarios for 2090 (Figure 6), the total habitat area will remain above 1.02×10^5 km^2. In the SSP2.6 scenario, the total habitat area will be 1.16×10^5 km^2 (0.11×10^5 km^2 higher than that of the current period), while the high-suitability area will be 0.21×10^5 km^2 (unchanged from that of the current period). In the SSP4.5 scenario, the total habitat area will be 1.11×10^5 km^2 (0.06×10^5 km^2 higher than that of the current period), whereas the high-suitability area will be 0.12×10^5 km^2 (0.09×10^5 km^2 less than that of the current period). In the SSP8.5 scenario, the total habitat and high-suitability areas will be 1.02×10^5 and 0.1×10^5 km^2, namely, 0.3×10^5 and 0.11×10^5 km^2 less than those of the current period, respectively.

In summary, the potential high-suitability area for sect. *Tuberculata* will be less than those of the current period in all 12 future climate scenarios. However, the habitat areas predicted for the SSP8.5 scenarios were generally lower than those predicted for the SSP2.6 scenarios. As shown in Figure 6, the most suitable habitats for it varied from one period to another by differing degrees. Nonetheless, the geographical range was always located in the karst region in southwestern China and centered on Guizhou, from which it dispersed to other adjacent provinces. Additionally, although the potential habitat areas were relatively contiguous in the current period, they could become fragmented owing to human activity or climate change. Nonetheless, in the four future periods, habitat fragmentation appeared to be relatively mild.

2.5. Changes in the Spatial Pattern of Sect. Tuberculata Habitats

From the LIG to the MH, the size of the habitat of sect. *Tuberculata* initially decreased and subsequently increased (Table 3 and Figure 7). During the LIG, a habitat area of 12.0×10^5 km^2 (22.93%) was added, whereas that of 1.86×10^5 km^2 (12.45%) was lost. The newly added habitats were in central and eastern Yunnan, eastern Sichuan, southern Guangxi, northern Guangdong, Fujian, Hainan, and southern Jiangxi. The lost habitats were located in the Qinling Mountains of Shaanxi and the junction between Henan, Hebei, and Shandong. During the LGM, new habitat areas amounted to 18.92×10^5 km^2, whereas 0.37×10^5 km^2 of old habitat was lost. Therefore, the habitat of sect. *Tuberculata* underwent large changes during the cold climatic conditions of the LGM, especially in southwestern China. Predictions based on the MaxEnt model indicate that sect. *Tuberculata* has emerged in new habitats in Guangxi, Guizhou, eastern Sichuan, Chongqing, Hunan, southern Hubei, western Chongqing, and northern Jiangxi in China. Therefore, the karst region in southwest China may be a natural refuge for sect. *Tuberculata*. The lost habitat areas were mainly located in Henan and the border between Anhui and Hubei. During the MH, new, retained, and lost habitats amounted to 20.31×10^5, 1.16×10^5, and 1.75×10^5 km^2, respectively. The lost habitat areas were located in Henan, the Qilian Mountains in central Shaanxi, and the junction between Shandong, Henan, Jiangsu, and Anhui. The difference between the MH and LGM in terms of added habitat areas was small, at only 1.39×10^5 km^2.

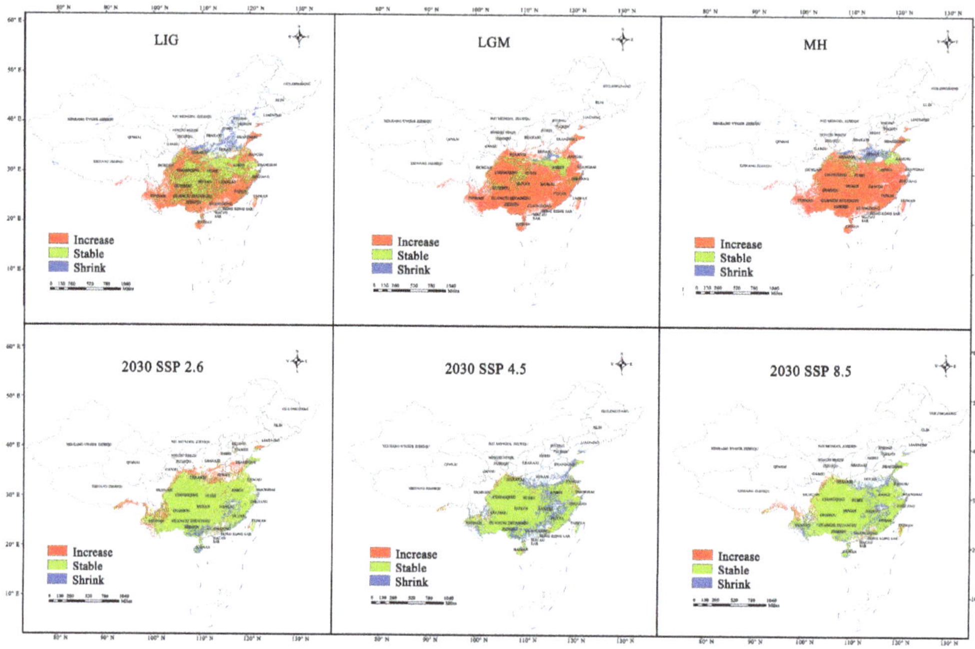

Figure 7. Changes in the spatial pattern of the habitats of sect. *Tuberculata* in the historical period and in 2030. (LIG, Last Interglacial; LGM, Last Glacial Maximum; MH, Mid-Holocene; SSP2.6, Low-concentration greenhouse gas emissions scenario; SSP4.5, Medium-concentration greenhouse gas emissions scenario; SSP8.5, High-concentration greenhouse gas emissions).

Table 3. Changes in the range of sect. *Tuberculata* in each period.

Climate Scenarios	Increase ($\times 10^5$ km^2)	Stable ($\times 10^5$ km^2)	Shrink ($\times 10^5$ km^2)
Last interglacial	12	9.98	1.86
Last Glacial Maximum	18.92	3.14	0.37
Mid Holocene	20.31	1.75	1.16
2030sSSP2.6	2.12	20.92	1.14
2030sSSP4.5	0.4	18.05	4.01
2030sSSP8.5	0.78	19.78	2.28
2050sSSP2.6	2.64	21.33	0.73
2050sSSP4.5	3.44	21.65	0.41
2050sSSP8.5	4	21.71	0.35
2070sSSP2.6	1.33	20.22	1.84
2070sSSP4.5	2.15	19.78	2.28
2070sSSP8.5	0.31	15.6	6.46
2090sSSP2.6	2.47	21.84	0.23
2090sSSP4.5	3.2	20	2.06
2090sSSP8.5	2.97	18.3	3.76

Between the 12 climate scenarios, SSP4.5 and SSP4.5 for the 2050s and 2090s, respectively, were predicted to have the largest additions in habitat area compared with the current period, at 3.44×10^5 and 3.2×10^5 km^2, respectively (Figure 8). The newly added habitats will be located in the junction between Shandong, Henan, Jiangsu, and Anhui, southern Gansu, and the Qinling Mountains in central Shaanxi. Habitat addition was smallest in the SSP4.5 and SSP8.5 scenarios for the 2030s and 2070s, respectively (at 0.4×10^5 and 0.31×10^5 km^2, respectively), whereas habitat loss was largest in the SSP8.5 scenario for the 2070s, at 6.46×10^5 km^2; in this scenario, the primary areas of habitat loss were predicted to be in the Wuyi Mountains (Jiangxi and Fujian), Dabie Mountains, and Qinling Mountains in eastern and northern Hubei, and parts of the Qinling Mountains in southern Shaanxi, southern Guangxi, and central Guangdong. Habitat loss was smallest in the SSP8.5 and SSP2.6 scenarios for the 2050s and 2090s, respectively (0.35×10^5 and 0.23×10^5 km^2, respectively).

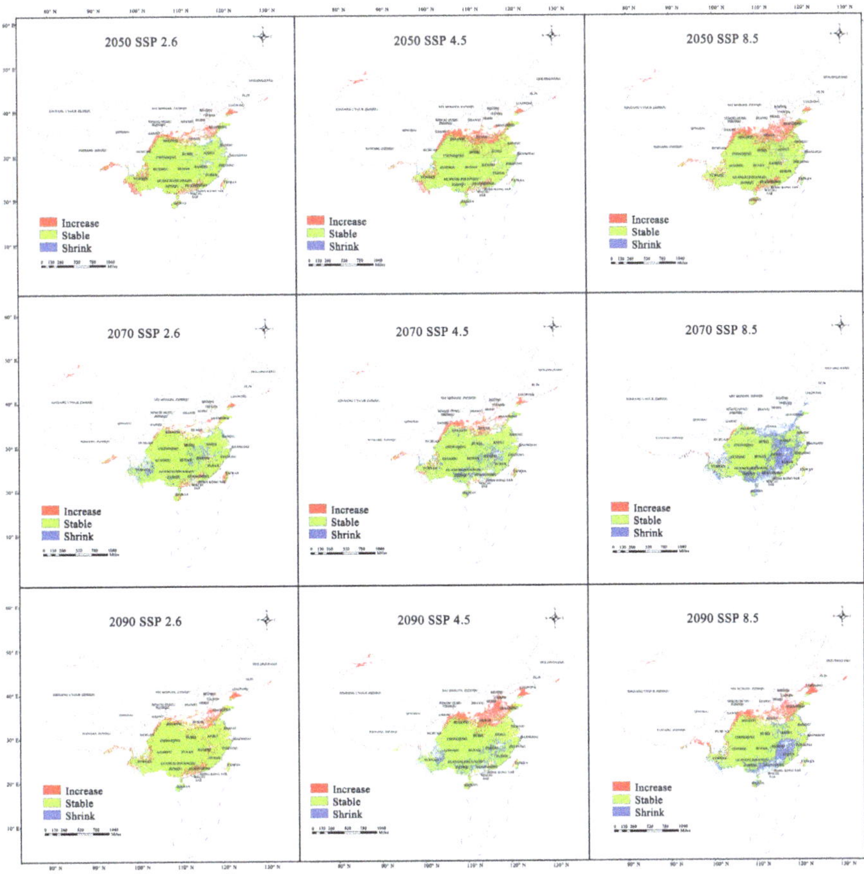

Figure 8. Changes in the spatial pattern of the habitats of sect. *Tuberculata* in 2050, 2070, and 2090. (SSP2.6, Low-concentration greenhouse gas emissions; SSP4.5, Medium-concentration greenhouse gas emissions; SSP8.5, High-concentration greenhouse gas emissions).

In summary, these predictions show that the range of sect. *Tuberculata* will become increasingly concentrated and less fragmented. The areas of habitat loss were mainly present in the Dabie Mountains and Qinling Mountains in eastern and northern Hubei, and parts of the Wuyi Mountains in Jiangxi and Fujian, which were centered on Guizhou and radiated outward to parts of the neighboring provinces. Conservation measures should be prioritized in these areas, as they are likely to be sensitive areas for habitat loss in the future.

2.6. Shifts in the Centroid of the Sect. Tuberculata Range

Here, the shifts in the centroid of the range of sect. *Tuberculata* from one period to another were simulated by calculating its geometric center in each period [24]. As presented in Table 4 and shown in Figure 9, the centroid of the current range of sect. *Tuberculata* is located in the western part of Yiyang City in Hunan Province. During the LIG, LGM, and MH, the centroid was located 252.75, 360.18, and 623.48 km, respectively, away from its current position to the north. These shifts indirectly indicate that the climate has undergone significant changes between the LIG and MH. Overall, the centroid progressively shifted northward during these periods, but shifted southwest during the transition from the MH to the current period (Table 4 and Figure 9).

Table 4. Shifts in the centroid of the range of sect. *Tuberculata* in each period.

Index		Longitude (°)	Latitude (°)	Distance (km)
LIG		111.68	30.28	252.75
LGM		112.05	31.20	360.18
MH		113.17	33.43	623.48
Current		110.89	28.12	0.00
2030s	SSP2.6	110.61	28.60	60.06
	SSP4.5	110.44	28.02	44.76
	SSP8.5	110.38	28.02	50.58
2050s	SSP2.6	110.68	28.26	25.40
	SSP4.5	110.80	28.77	73.02
	SSP8.5	110.92	28.77	72.98
2070s	SSP2.6	110.88	28.39	30.79
	SSP4.5	110.67	28.76	75.29
	SSP8.5	110.01	28.19	84.53
2090s	SSP2.6	111.02	28.50	44.62
	SSP4.5	111.39	29.38	149.10
	SSP8.5	110.72	29.50	155.18

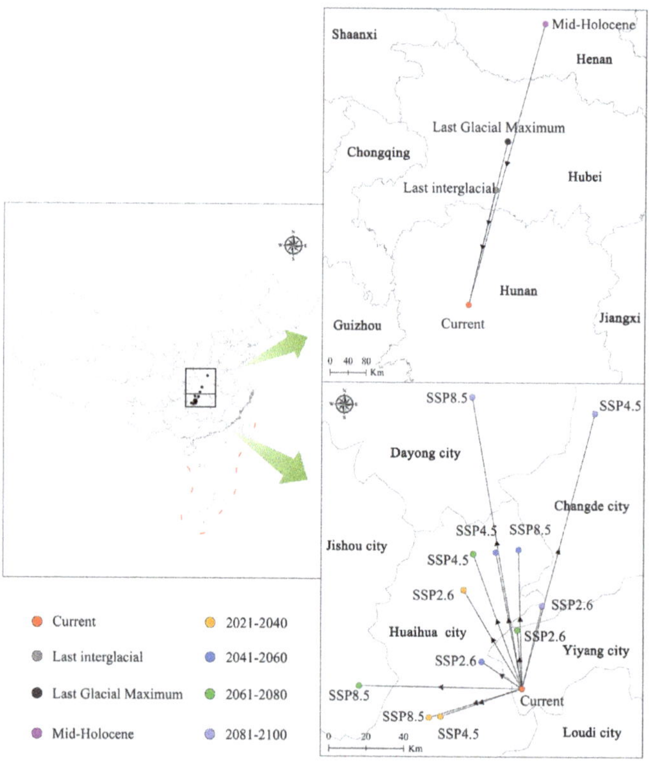

Figure 9. Location of the shift of the centroid of suitable areas for sect. *Tuberculata* under different climate scenarios in different periods.

In the SSP2.6 scenario, the centroid was predicted to shift northwest in 2030, and subsequently southward from 2030 to 2050. It will then shift toward the northeast direction from 2050 to 2070 and continue to shift in this direction from 2070 to 2090. In the SSP4.5 scenario, the shifts in the centroid will be southwestward from the current period to 2030, northeastward from 2030 to 2050, westward from 2050 to 2070, and northeastward from 2070 to 2090. In the SSP8.5 scenario, the shifts in the centroid will be southwestward from the current period to 2030, northeastward from 2030 to 2050, southwestward from 2050 to 2070, and northeastward from 2070 to 2090. The largest shift in the centroid (relative to the current position) was observed in the SSP8.5 scenario for the 2090s, at 155.18 km. The shortest shift in the centroid was observed in the SSP2.6 scenario for the 2050s, at 25.40 km.

Overall, the centroid is expected to shift northwest in the future; from its current position, it will shift towards the Wuyi Mountains in northwestern Hunan. In particular, climate change will cause a shift in the range of sect. *Tuberculata* toward higher latitudes.

3. Discussion

3.1. The Most Important Environmental Factors for the Distribution of Sect. Tuberculata

The MaxEnt model is an important research tool for biodiversity, conservation, and evolution studies [25–28]. In the study, the MaxEnt model was used to predict the potential ranges of sect. *Tuberculata* plants in the past, present, and future. The six dominant environmental factors for sect. *Tuberculata* were obtained using the jackknife method, contribution rates, and univariate response curves, and were determined as bio2, S_bs, bio7, bio9, uvb2, and uvb3. The most important factor is bio2, namely, the mean diurnal temperature range. Therefore, temperature is the most important environmental factor for the range of sect. *Tuberculata*, followed by the intensity of UV radiation. In particular, the climate restricts the geographic range of these plants on the regional scale, and the most important climatic factors are hydrothermal. In a previous study on the climatic factors affecting *C. rubimuricata* in the Guizhou Maolan National Nature Reserve [29], the temperature was found to be the most important factor of all, which is consistent with our findings. Guo et al. studied the germination and physiological characteristics of *C. rubituberculata* in karst regions and found that temperature was the most important factor for seed germination, followed by illumination [30]. Hence, temperature factors are more important than illumination factors for it. Its range was mainly located in southwestern China and centered on Guizhou, from which it dispersed to adjacent provinces. Southwestern China has a wet subtropical and mountainous climate and is a classic mountain-and-valley habitat for East Asian flora, with a complex terrain typical of karst regions. Owing to this unique geology and climate, the flora of this region are usually xerophytes and lithophytes [31,32]. To achieve respiration and photosynthesis, a plant must be provided with a suitable range of temperatures, and the optimal range varies from one species to another [33]. Excessively low temperatures can lead to frost damage, which, in turn, inhibits growth [34]. Sect. *Tuberculata* is a typical karst flora whose distribution can span a wide range of environmental conditions. In phytomorphological studies, its leaves have characteristics closely related to cold and drought resistance, such as low stomatal density and trichomes on the leaves. These features allow it to have high photosynthetic and heat dissipation efficiencies, which helps them survive in harsh environments [18]. The next most important factor is UV intensity, which indirectly affects photosynthesis, thus influencing the plant's distribution [35].

3.2. Changes in the Potential Habitable Range of Sect. Tuberculata Plants and Centroid Analysis

Studies regarding the potential distribution pattern of sect. *Tuberculata* plants in the presence of climate change play an important role in informing species conservation and ecosystem preservation efforts. The distribution of each species reacts differently to climate change. In this study, we simulated the possible geographic ranges of sect. *Tuberculata* at different periods using the MaxEnt model and the ArcGIS system. This is the first study of its distribution; we found that it is mainly distributed between 22.54°–30.96° N latitude and 103.83°–111.76° E longitude. This range is primarily located in southwestern China, which is consistent with previous findings. This also shows that the optimized MaxEnt model can accurately and effectively predict its range [36]. Climate plays an important role in determining the distribution of a species; hence, the distribution of some species is a reflection of the climate [5]. Based on the results, we found that sect. *Tuberculata* is mainly located in the karst highlands of Guizhou, which is consistent with the results of previous studies [14]. Between the LIG and current period, its habitat initially decreased and subsequently increased to its maximum in the current period. Therefore, its habitat has expanded to some extent during the current period. The unique climate of the karst region creates a humid and shady environment, which is an ideal habitat for sect. *Tuberculata*. Although its potential habitats were relatively contiguous during the LIG, LGM, and MH periods, they have become fragmented during the current period. We speculate that this could be caused by increased CO_2 emissions and human activities, such as logging. In the four future periods, the habitable area for sect. *Tuberculata* will decrease by varying degrees (depending on the emissions mitigation scenario) compared with that in the current period. Therefore, it may not adapt well to the climatic conditions of the future. In the predictions, we found that the loss of habitats in the 2070s and 2090s was relatively low in the SSP2.6 scenario but large in the SSP8.5 scenario. We speculate that this is because the high-emissions SSP8.5 scenario leads to worsened environmental conditions and higher temperatures, which exacerbates fragmentation in the habitat of sect. *Tuberculata*. Therefore, we show that global warming poses a tremendous challenge to it.

According to centroid analysis, the current centroid of the sect. *Tuberculata* habitat was located in the western part of Yiyang city in Hunan. The centroid shifted northward during the LIG-MH, and subsequently shifted southwestward during the current period. Global warming was found to shift the centroid of its potential habitat toward higher latitudes, which is consistent with previous findings [37,38]. In the past, sect. *Tuberculata* gradually adapted to the wet and humid mountainous environment owing to changes in the terrain; in the future, it may evolve to migrate towards lower altitudes. However, the potential range of a plant is not only affected by climate and terrain, but also by human activities and other abiotic factors, which can also its their growth and development [39]. Therefore, future studies on the range of sect. *Tuberculata* should simultaneously consider the plant's intrinsic physiology and ecosystem, and human factors, in their models to improve the predictive accuracy.

4. Materials and Methods

4.1. Data Collection and Processing

The occurrence records of sect. *Tuberculata* plants were obtained from online platforms, such as the Chinese Virtual Herbarium (https://www.cvh.Ac.cn/, accessed on 10 February 2024), Global Biodiversity Information Facility (https://www.gbif.org/, accessed on 10 February 2024), and Plants of the World Online (https://powo.science.kew.org/, accessed on 10 February 2024). Moreover, 86 occurrence records were obtained from field surveys in 23 locations. The occurrence records were sorted by species name, longitude (X), and latitude (Y), and stored as .csv files. By keeping only one point out of all points within a range of 10 km, 71 data points were obtained. To avoid the occurrence points close to each other, the "trim duplicate occurrences" tool in ENMTools was used to exclude redundant data points in each 10 × 10 km grid. Finally, 65 valid occurrence points were obtained (Figure 10), which were stored in a .csv file and subsequently used in the MaxEnt model [40].

4.2. Acquisition and Selection of Environmental Data

To study the potential habitat range of sect. *Tuberculata* plants, range prediction was conducted using 60 environmental factors in eight periods, namely, the LIG, LGM, MH, current period, 2020–2040 (2030), 2040–2060 (2050), 2060–2080 (2070), and 2080–2100 (2090). The environmental factors include 19 bio-climatic factors, three terrain factors, 32 soil factors, and six ultraviolet (UV) factors. The 19 climatic factors were obtained from Worldclim (http://www.worldclim.org/, accessed on 25 February 2024) [41], which includes data for the current period (1970–2000) and projections for future periods

(2021–2100). The SSP2.6, SSP4.5, and SSP8.5 (aggressive, moderate, and weak emission mitigation) climate scenarios from Community Climate System Model version 4 (CCSM4) were used as future climate projections. Version 2 of the Beijing Climate Center Climate System Model (BCC-CSM2-MR_2.5) was used as the atmospheric circulation model, as it performs well in the prediction of air temperatures and precipitation in China [42–44]. The terrain data were obtained from Worldclim, and the slope and aspect data were obtained through ArcGIS. The soil factors were obtained from the China dataset of the Harmonized World Soil Database (v1.1) (http://vdb3.soil.csdb.cn, accessed on 25 February 2024). The UV data were obtained from the glUV dataset (http://www.ufz.de/gluv/, accessed on 25 February 2024) [19]. In order to prevent overfitting due to multicollinearity between environmental factors, we performed a Spearman correlation test the environmental data based on the model contribution in R (4.1.1) [45]. All environmental factors with correlations higher than 0.8 and low model contributions were excluded; thus, the remaining 20 statistically and biologically significant environmental factors were used for range modeling (Table 5).

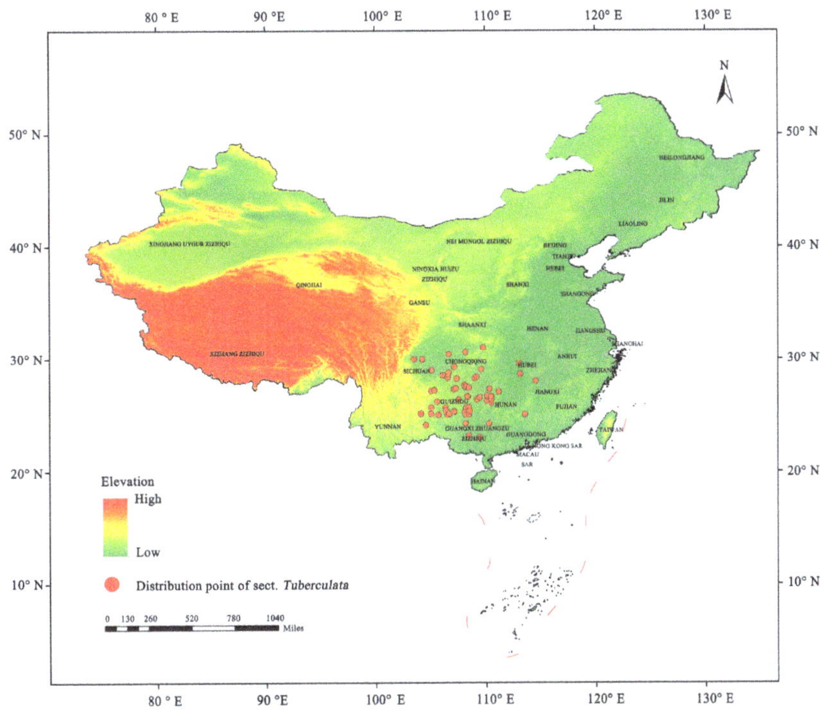

Figure 10. Screened occurrence points of the sect. *Tuberculata* plants.

Table 5. Relative importance of each environmental factor in the MaxEnt model.

Variable	Description	Percent Contribution (%)	Permutation Importance (%)
bio2	Mean diurnal range (Mean of monthly (max temp–min temp))	16.9	11.2
s_bs	Basic saturation (lower level)	10.4	8.4
bio7	Temperature annual range	10.1	30.3
bio9	Mean temperature of driest quarter	8.9	5.4
uvb2	Annual UV-B seasonality (standard deviation)	7.1	0.6
uvb3	Mean UV-B of highest month	3.4	0

Table 5. *Cont.*

Variable	Description	Percent Contribution (%)	Permutation Importance (%)
bio15	Precipitation seasonality	2.7	0.7
elevation	Altitude	2.3	16.3
t_sand	Sand content (upper layer)	2.3	1.4
bio12	Annual precipitation	2.3	10
slope	Slope	1.6	1.1
bio14	Precipitation of driest month	1.1	8
s_teb	Exchangeable salt base (lower layer)	0.3	1.3
t_gravel	Percent of gravel (upper layer)	0.3	0.3
aspect	Aspect	0.2	0.3
bio16	Precipitation of wettest quarter	0.1	1.1
uvb4	Mean UV-B of lowest month	0.1	0.4
s_gravel	Percent of gravel volume (lower layer)	0	0
s_esp	Exchangeable sodium salt (lower layer)	0	0.1
t_caso4	Sulfate content (upper layer)	0	0

4.3. Building and Optimizing the Model

Optimization of MaxEnt model parameters was carried out using "kuenm" in R v3.6.3. [21]. Accordingly, parameter combinations were created by combining regularization multiplier (RM) values from 0.5 to 4 (in increments of 0.5) with six feature classes (FCs), namely, L, LQ, H, LQH, LQHP, and LQHPT (L: linear, Q: quadratic, H: hinge, P: product, T: threshold) [46,47]. The kuenm package was then used to test the parameter combinations, and the fit and complexity of each candidate model was assessed using the delta of their Akaike information criterion (AIC) values (ΔAIC) [48,49]. The parameter combination resulting in the lowest ΔAIC was used to construct the MaxEnt model, and the resulting model was evaluated using the area under the curve (AUC) of its receiver operating characteristic (ROC) curve. The AUC scores were interpreted as follows: <0.6: failed, 0.6–0.7: poor, 0.7–0.8: fair, 0.8–0.9: good, and >0.9: excellent [40].

4.4. Model Assessment

Twenty environmental variables and 65 occurrence points were included in the MaxEnt v3.4.4 model [48–50]. The model parameters were configured as follows: (1) In the model, 75% of the distribution points were used as training data, while the remaining 25% were used as test data, with other settings kept at their default values. (2) The environmental response curve and predicted range map were plotted. (3) The jackknife method was used to evaluate the importance and contribution of each environmental factor, and the logistic output format was used. All other parameters remained in their default settings. Accuracy validation was conducted by evaluating the ROC AUC value (0 to 1) derived from the MaxEnt model's output; the larger the AUC, the higher the credibility of the model's predictions [21]. The AUC values were interpreted as follows: <0.7: poor, 0.7–0.8: fair, 0.8–0.9: good, and >0.9: excellent [42].

4.5. Classification of Habitat Areas

Based on the classification criterion used for sect. *Tuberculata* plants in China [50] and the output of the MaxEnt model, the potential habitats for sect. *Tuberculata* plants were classified into four types using the natural breaks method. In particular, the potential area of distribution was segmented into four areas based on the average logistic values and actual area of distribution of the plants, namely, non-suitable (0–0.2), low suitability (0.2–0.4), medium suitability (0.4–0.6), and high suitability (0.6–1) areas. Breaks were then input manually, and the total habitat area was obtained by combining the medium- and high-suitability areas.

4.6. Centroid Analysis

The direction and distance of species habitat shifts can be inferred from the changes in the habitat's centroid [49]. Here, the kuenm R package was used to compute changes in the spatial pattern of the total habitat area in each period, as well as the habitat's centroid [51]. Finally, the SDM was used to track the coordinates of the centroid, thus determining the centroid of sect. *Tuberculata* plants in each period. The calculated coordinates of the centroid were then used to determine the distance of habitat shift [52].

5. Conclusions

In the study, a MaxEnt model optimized by the kuenm R package was used to predict the distribution and potential habitat areas of sect. *Tuberculata* plants in three historical periods, the present, and 12 future climate change scenarios. The size of its potential habitat area in the current climate was 1.05×10^5 km^2, and the highly suitable habitats were located in Guizhou, central-southern Sichuan, the Wuling Mountains in Chongqing, the Panjiang Basin, and Southwestern Hunan. In the centroid analysis, the centroid of its potential habitat will shift from lower to higher latitudes in the northwest direction. We expect that our findings will help inform future field surveys of sect. *Tuberculata* plants, as well as their conservation and rehabilitation. Additionally, our results will act as auxiliary evidence for its geographic range in phylogenetic studies.

Author Contributions: Conceptualization, X.X. and Z.L.; methodology, X.X. and Z.R.; software, Z.R. and X.X.; validation, Z.L. and C.Y.; investigation, Z.R. and X.X., resources, J.C.; data curation, Z.R.; writing—original draft preparation, X.X. and Z.R.; writing—review and editing, Z.L., Z.R. and X.X.; visualization, Z.L.; supervision, X.X.; project administration, Z.L.; funding acquisition, Z.L. All authors have read and agreed to the published version of the manuscript.

Funding: This work was supported by the National Natural Science Foundation of China (32400179) and the Guizhou Provincial Basic Research Program (Natural Science) 2022 (072).

Data Availability Statement: Data are contained within the article.

Conflicts of Interest: The authors declare no conflicts of interest.

References

1. Klein, T.; Anderegg, W.R.L. A vast increase in heat exposure in the 21st century is driven, by global warming and urban population growth. *Sustain. Cities Soc.* **2021**, *73*, 103098. [CrossRef]
2. Lehikoinen, A.; Virkkala, R. Norh by north-west: Climate change and diredions of density shifts in birds. *Glob. Chang. Biol.* **2015**, *22*, 1121–1129. [CrossRef] [PubMed]
3. Wu, J.G.; Lü, J.J.; Ai, L. The impacts of climate change on the biodiversity: Vulnerability and adaptation. *Ecol. Environ. Sci.* **2009**, *18*, 693–703. Available online: https://www.jeesci.com/CN/Y2009/V18/I2/693 (accessed on 7 November 2024).
4. Parmesan, C. Ecological and evolutionary responses to recent climate change. *Ann. Rev. Ecol. Evol. Syst.* **2006**, *37*, 637–669. [CrossRef]
5. Li, L.; Chen, J.K. Influence of climate change on wild plants and the conservation strategies. *Biodivers. Sci.* **2014**, *22*, 549–563. [CrossRef]
6. Ji, L.T.; Zheng, T.Y.; Chen, Q.; Zhong, J.J.; Kang, B. Responses of potential suitable area of *Paris verticillata* to climate change and its dominant climate factors. *Chin. J. Appl. Ecol.* **2020**, *31*, 89–96. [CrossRef]
7. Fuller, D.O.; Ahumada, M.L.; Quiñones, M.L.; Herrera, S.; Beier, J.C. Near-present and future distribution of *Anopheles albimanus* in Mesoamerica and the Caribbean Basin modeled with climate and topographic data. *Int. J. Health Geogr.* **2012**, *11*, 13. [CrossRef]
8. Barbosa, F.G.; Schneck, F. Characteristics of the topcited papers in species distribution predictive models. *Ecol. Model.* **2015**, *313*, 77–83. [CrossRef]
9. Chang, H.T. A taxonomy of the genus *Camellia*. *Acta Sci. Nat. Univ. Sunyatseni* **1981**, *4*, 18–52.
10. Chang, H.T.; Ren, S.X. A classification on the section *Tuberculata* of *Camellia*. *Acta Sci. Nat. Univ. Sunyatseni* **1991**, *30*, 86–91.
11. Chang, H.T. Camellia. In *Flora Reipublicae Popularis Sinicae*; Science Press: Beijing, China, 1998; pp. 37–48.
12. An, M.T. Present Status of the Natural Resource of Camellias in Guizhou Province. *Guizhou For. Sci. Technol.* **2005**, *2*, 25–29.
13. Min, T.L.; Zhang, W.J. The evolution and distribution of genus *Camellia*. *Acta Bot. Yunnan* **1996**, *18*, 1–13.
14. Min, T.L. A systematic synopsis of genus *Camellia*. *Acta Bot. Yunnan* **1999**, *21*, 149–159.
15. Ran, Z.H.; Li, Z.; Xiao, X.; An, M.T.; Yan, C. Complete chloroplast genomes of 13 species of sect. *Tuberculata* Chang (*Camellia* L.): Genomic features, comparative analysis, and phylogenetic relationships. *BMC Genom.* **2024**, *25*, 108. [CrossRef]

16. Xiao, X.; Lu, J.T.; Yang, G.Y.; Li, Z. The complete chloroplast genome of *Camellia leyeensis* (theaceae). *Mitochondrial DNA Part B* **2022**, *7*, 735–737. [CrossRef]
17. Xiao, X.; Lu, J.T.; Yang, G.Y.; Li, Z. Analysison chloroplast genome characteristics and codon usage biases of *Camellia rubituberculata*. *Seed* **2022**, *41*, 19–26. [CrossRef]
18. Lu, H.F.; Jiang, B.; Shen, Z.G.; Shen, J.B.; Peng, Q.F.; Cheng, C.G. Comparative leaf anatomy, FTIR discrimination and biogeographical analysis of *Camellia* section *Tuberculata* (Theaceae) with a discussion of its taxonomic treatments. *Plant Syst. Evol.* **2008**, *274*, 223–235. [CrossRef]
19. Jiang, B.; Peng, Q.F.; Shen, Z.G.; Moller, M.; Pi, E.X.; Lu, H.F. Taxonomic treatments of *Camellia* (Theaceae) species with secretory structures based on integrated leaf characters. *Plant Syst. Evol.* **2010**, *290*, 1–20. [CrossRef]
20. Guo, M.X.; Yang, N.K.; Liu, H.Y.; Tang, S.H.; Fan, Z.W.; Zou, T.C. Spatial distribution pattern and quantitative dynamics of the endemic plant *Camellia rubituberculata* in Guizhou Province. *Guihaia* **2019**, *39*, 1359–1369. [CrossRef]
21. Liu, Y.L.; Du, X.Q.; Fang, X.P. The analysis of and spatial distribution pattern of *Camellia rubimuricata* population. *Seed* **2010**, *29*, 53–55. [CrossRef]
22. Hosmer, D.W.; Leneshow, S. Applied logistic regression. *Biometrics* **1991**, *4*, 1632–1633. [CrossRef]
23. Du, Z.Y.; He, Y.M.; Wang, H.T.; Wang, C.; Duan, Y.Z. Potential geographical distribution and habitat shift of the genus Ammopiptanthus in China under current and future climate change based on the MaxEnt model. *J. Arid. Environ.* **2021**, *184*, 104328. [CrossRef]
24. Zhang, Y.B.; Liu, Y.L.; Qin, H.; Meng, Q.X. Prediction on spatial migration of suitable distribution of *Elaeagnus mollis* under climate change conditions in Shanxi Province, China. *Chin. J. Appl. Ecol.* **2019**, *30*, 496–502. [CrossRef]
25. Guisan, A.; Zimmermann, N.E. Predictive habitat distribution models in ecology. *Eeological Model.* **2000**, *135*, 147–186. [CrossRef]
26. Zimmermann, N.E.; Edwards, T.C.; Grham, C.H.; Peaman, P.B.; Svenning, J.C. New trends in species distribution modeling. *Ecography* **2010**, *33*, 985–989. [CrossRef]
27. Elith, J.; Leathwick, J.R. Species distribution models: Ecological explanation and prediction across space and time. *Annu. Rev. Ecol. Evol. Syst.* **2009**, *40*, 677–697. [CrossRef]
28. Li, G.; Du, S.; Wen, Z. Mapping the climatic suitable habitat of oriental arborvitae (*Platycladus orientalis*) for introduction and cultivation at a global scale. *Sci. Rep.* **2016**, *6*, 30009. [CrossRef]
29. Liu, Y.L.; Du, X.Q. Correlation between climatic factor and ring width of *Camellia rubimuricata*. *Guizhou Agric. Sci.* **2010**, *38*, 38–41
30. Guo, M.X.; Lv, X.M.; Hong, J.; Liu, H.Y.; Zou, T.C. Studies on seed germination regulated by temperature and light and physiological characteristics change of *Camellia rubituberculata*. *Seed* **2019**, *38*, 36–42. [CrossRef]
31. Li, Q.; Wang, Y.; Li, L.; Zhang, H.; Wang, B. Ecological restoration scheme of lake basins on the karst plateau based on natural solution: Take nine lakes on the Yunnan Plateau as example. *Carsologica Sin.* **2023**, *42*, 391–401. [CrossRef]
32. Liu, Y.T.; Wang, L.; Li, X.H.; Guo, L. Analysis on spatio-temporal variability of fractional vegetation cover and influencing factors from 2000 to 2020 in Southwestern China. *Plateau Meteorol.* **2024**, *43*, 264–276. [CrossRef]
33. Ma, W.H.; Liu, Z.L.; Wang, Z.H.; Wang, W.; Liang, C.Z.; Tang, Y.H.; He, J.S.; Fang, J.Y. Climate change allers interanual varation of gassland aboveground productivity: Evidence from a 22-year measurement seres in the lmer Mongolian grassland. *J. Plant Res.* **2010**, *123*, 509–517. [CrossRef] [PubMed]
34. Ladwig, L.M.; Ratajczak, Z.R.; Ochelree, T.W.; Hafich, K.A.; Churhil, A.C.; Frey, S.J.K.; Fuss, C.B.; Kazanski, C.E.; Munoz, J.D.; Petrie, M.D.; et al. Beyond aretie and alpine: The inuence of winer cimate on temperate ecosystems. *Ecology* **2016**, *97*, 372–382. [CrossRef]
35. Liu, Y.N.; Ao, M.; Li, B.; Guan, Y.X. Effect of ultraviolet-B (UV-B) radiation on plant growth and development and its application value. *Soils Crops* **2020**, *9*, 191–202. Available online: http://sc.iga.ac.cn/article/2020/2 (accessed on 7 November 2024).
36. Zhang, Q.; Shen, X.B.; Jiang, X.L.; Fan, T.T.; Liang, X.C.; Yan, W.D. MaxEnt modeling for predicting suitable habitat for endangered tree *Keteleeria davidiana* (Pinaceae) in China. *Forests* **2023**, *14*, 394. [CrossRef]
37. Sekercioglu, C.H.; Schneider, S.H.; Fay, J.P.; Loarie, S.R. Climate change, elevational range shifts, and bird extinctions. *Conserv. Biol.* **2007**, *22*, 140–150. [CrossRef] [PubMed]
38. Fang, J.Y.; Zhu, J.L.; Shi, Y. The responses of ecosystems to global warming. *Chin. Sci. Bull.* **2018**, *63*, 136–140. [CrossRef]
39. Fick, S.E.; Hijmans, R.J. World Clim 2: New 1-km spatial resolution climate surfaces for global land areas. *Int. J. Climatol.* **2017**, *37*, 4302–4315. [CrossRef]
40. Su, B.; Huang, J.; Mondal, S.K.; Zhai, J.Q.; Wang, Y.J.; Wen, S.S.; Gao, M.N.; Lu, Y.R.; Jiang, S.; Jiang, T.; et al. Insight from CMIP6 SSP-RCP scenarios for future drought characteristics in China. *Atmos. Res.* **2021**, *250*, 105375. [CrossRef]
41. Sang, Y.H.; Ren, H.L.; Shi, X.L.; Xu, X.F.; Chen, H.H. Improvement of Soil moisture simulation in eurasia by the Beijing climate center climate system model from CMIP5 to CMIP6. *Geosci. Model Dev.* **2019**, *12*, 1573–1600. [CrossRef]
42. Beckmann, M.; Václavík, T.; Manceur, A.M.; Šprtová, L.; Henrik, V.W.; Erik, W.; Anna, F.C. glUV: A global UV-B radiation data set for macroecological studies. *Methods Ecol. Evol.* **2014**, *5*, 372–383. [CrossRef]
43. Yang, X.Q.; Kushwaha, S.P.S.; Saran, S.; Xu, J.C.; Roy, P.S. MaxEnt modeling for predicting the potential distribution of medicinal plant, *Justicia adhatoda* L. in Lesser Himalayan foothills. *Ecol. Eng.* **2013**, *51*, 83–87. [CrossRef]
44. Cobos, M.E.; Peterson, A.T.; Barve, N.; Luis, O.O. kuenm: An R package for detailed development of ecological niche models using Maxent. *PeerJ* **2019**, *7*, e6281. [CrossRef] [PubMed]

5. Zhao, X.; Lei, M.; Wei, C.; Guo, X.X. Assessing the suitable regions and the key factors for three Cd-accumulating plants (*Sedum alfredii*, *Phytolacca americana* and *Hylotelephium spectabile*) in China using MaxEnt model. *Sci. Total Environ.* **2022**, *852*, 158202. [CrossRef]
6. Zhu, G.P.; Qiao, H.J. Effect of the MaxEnt model's complexity on the prediction of species potential distributions. *Biodivers. Sci.* **2016**, *24*, 1189–1196. [CrossRef]
7. Muscarella, R.; Galante, P.J.; Soley-Guardia, M.; Boria, R.A.; Kass, J.M.; María, U.; Anderson, R.P. ENMeval: An R package for conducting spatially independent evaluations and estimating optimal model complexity for MaxEnt ecological niche models. *Method. Ecol. Evolut.* **2015**, *5*, 1198–1205. [CrossRef]
8. Zhao, G.H.; Cui, X.Y.; Sun, J.J.; Li, T.T.; Wang, Q.; Ye, X.Z.; Fan, B.G. Analysis of the distribution pattern of Chinese *Ziziphus jujuba* under climate change based on optimized biomod2 and MaxEnt models. *Ecol. Indic.* **2021**, *132*, 108256. [CrossRef]
9. Wang, X.G.; Duan, Y.X.; Jin, L.L.; Wang, C.Y.; Peng, M.C.; Li, Y.; Wang, X.H.; Ma, Y.F. Prediction of historical, present and future distribution of *Quercus* sect. *Heterobalanus* based on the optimized MaxEnt model in China. *Acta Ecol. Sin.* **2023**, *43*, 6590–6604. Available online: https://kns.cnki.net/kcms/detail/11.2031.Q.20230413.1604.002.html (accessed on 7 November 2024).
0. Bateman, B.L.; Pidgeon, A.M.; Radeloff, V.C.; Vanderwal, J.; Thogmartin, W.E.; Vavrus, S.J.; Heglund, P.J. The pace of past climate change vs. potential bird distributions and land use in the United States. *Glob. Chang. Biol.* **2016**, *22*, 1130–1144. [CrossRef]
1. Cong, M.Y.; Xu, Y.Y.; Tang, L.Y.; Yang, W.J.; Jian, M.F. Predicting the dynamic distribution of sphagnum bogs in China under climate change since the last interglacial period. *PLoS ONE* **2020**, *15*, e0230969. [CrossRef]
2. Walther, G.R.; Berger, S.; Sykes, M.T. An ecological 'footprint' of climate change. *Proc. R. Soc. B Biol. Sci.* **2005**, *272*, 1427–1432. [CrossRef] [PubMed]

plants

MDPI

Article

Examining the Sensitivity of Satellite-Derived Vegetation Indices to Plant Drought Stress in Grasslands in Poland

Maciej Bartold [1], Konrad Wróblewski [1,*], Marcin Kluczek [1], Katarzyna Dąbrowska-Zielińska [1] and Piotr Goliński [2]

[1] Remote Sensing Centre, Institute of Geodesy and Cartography, Modzelewskiego 27, 02-679 Warsaw, Poland; maciej.bartold@igik.edu.pl (M.B.); marcin.kluczek@igik.edu.pl (M.K.); katarzyna.dabrowska-zielinska@igik.edu.pl (K.D.-Z.)

[2] Department of Grassland and Natural Landscape Sciences, Poznań University of Life Sciences, Dojazd 11, 60-632 Poznań, Poland; piotr.golinski1@up.poznan.pl

* Correspondence: konrad.wroblewski@igik.edu.pl

Abstract: In this study, the emphasis is on assessing how satellite-derived vegetation indices respond to drought stress characterized by meteorological observations. This study aimed to understand the dynamics of grassland vegetation and assess the impact of drought in the Wielkopolskie (PL41) and Podlaskie (PL84) regions of Poland. Spatial and temporal characteristics of grassland dynamics regarding drought occurrences from 2020 to 2023 were examined. Pearson correlation coefficients with standard errors were used to analyze vegetation indices, including NDVI, NDII, NDWI, and NDDI, in response to drought, characterized by the meteorological parameter the Hydrothermal Coefficient of Selyaninov (HTC), along with ground-based soil moisture measurements (SM). Among the vegetation indices studied, NDDI showed the strongest correlations with HTC at $r = -0.75$, $R^2 = 0.56$, RMSE = 1.58, and SM at $r = -0.82$, $R^2 = 0.67$, and RMSE = 16.33. The results indicated drought severity in 2023 within grassland fields in Wielkopolskie. Spatial–temporal analysis of NDDI revealed that approximately 50% of fields were at risk of drought during the initial decades of the growing season in 2023. Drought conditions intensified, notably in western Poland, while grasslands in northeastern Poland showed resilience to drought. These findings provide valuable insights for individual farmers through web and mobile applications, assisting in the development of strategies to mitigate the adverse effects of drought on grasslands and thereby reduce associated losses.

Keywords: drought stress; grasslands; hydrothermal coefficient of selyaninov; plant response; satellite imagery; vegetation indices

1. Introduction

Plant stress is a critical focus of environmental research, particularly in the context of climate change, ecosystem ecology, and conservation efforts [1,2]. Understanding how plants respond to various stressors such as drought, heat, cold, salinity, and pollution is essential for predicting how ecosystems will respond to changing environmental conditions [3]. In the face of climate change, extreme weather events such as heatwaves, droughts, and floods are becoming more frequent and severe. These events have had a significant impact on plant health and productivity. By studying the responses of plants to these stressors, researchers can better predict which species are most vulnerable and which may be more resilient in a changing climate. This knowledge is crucial for developing effective conservation strategies and mitigating the impacts of climate change on plant biodiversity. In grassland ecosystem ecology, plant stress research is particularly important because grasslands are highly sensitive to environmental changes [4,5].

Grassland ecosystems provide essential services such as carbon sequestration, water filtration, and habitat for a diverse range of species [6,7]. Understanding how different grassland species respond to stress can help researchers predict how these ecosystems will change in the future and develop strategies to conserve their biodiversity and ecosystem functions [8]. Additionally, plant stress research is essential for agriculture, as it helps farmers select crop varieties that are more resilient to environmental stressors, ultimately increasing crop yields and food security [9]. By studying plant stress responses, researchers can also develop new techniques for sustainable agriculture, such as drought-resistant crops and more efficient irrigation systems [10]. Therefore, plant stress plays a crucial role in environmental research, helping scientists understand how plants respond to changing environmental conditions and developing strategies to conserve biodiversity, protect ecosystems, and ensure food security in a changing climate [11].

Remote sensing has revolutionized the way we detect and monitor plant stress by providing valuable information on the spectral responses of plants to various stressors [12]. It involves the collection and interpretation of data from a distance, often using satellites, aircraft, or drones equipped with sensors capable of detecting different wavelengths of light. One of the key benefits of remote sensing for detecting plant stress is its ability to capture large-scale, spatially explicit information about plant health and vigor [13]. By measuring the spectral reflectance of plants across different wavelengths of light, remote sensing can provide valuable insights into the physiological and biochemical changes that occur in plants under stress. For example, stressed plants often exhibit changes in chlorophyll content, leaf water content, and canopy structure, which can alter their spectral reflectance properties. Remote sensing techniques such as multispectral and hyperspectral imaging can detect these changes by measuring the amount of light reflected by plants at specific wavelengths [14].

Multispectral imaging systems typically capture light in a few discrete bands, such as the visible and near-infrared spectra (NIR), allowing us to calculate vegetation indices such as the commonly used Normalized Difference Vegetation Index (NDVI). These indices provide valuable information about plant health and vigor, with low values indicating stressed or unhealthy vegetation [15]. This allows researchers to detect subtle changes in plant physiology and biochemistry associated with stress [16]. It is important to note, however, that this relationship does not hold for all vegetation indices. Other indices, such as the Enhanced Vegetation Index (EVI) or the Water Band Index (WBI), may respond differently to plant stress. In some cases, higher values could indicate stress or other physiological changes. Therefore, remote sensing science and techniques play a crucial role in detecting and monitoring plant stress by providing valuable information on the spectral responses of plants to various stressors [17]. By combining remote sensing data with ground-based observations and models, we can understand how plants respond to environmental stress and develop strategies to mitigate the impacts of stress on agricultural productivity, ecosystem health, and biodiversity conservation [18,19].

A widely used satellite data source comes from the Sentinel-2 satellites, which are highly popular due to their open-access policy, providing freely available data to the public [20,21]. The Sentinel-2 satellites provide high-resolution imagery at 10 m up to 60 m with a wide swath at 290 km, allowing for detailed observation and monitoring of changes in land cover [22], water bodies [23], vegetation, and ecosystems [24]. Secondly, the Sentinel-2 satellites capture imagery in 13 spectral bands, ranging from the visible to the shortwave infrared (SWIR), enabling researchers to analyze various vegetation indices and detect changes in plant health, biomass, and stress levels [25]. Additionally, the satellites revisit the same area on the Earth's surface every 5 days with both satellites in operation, providing a frequent revisit time for regular monitoring of changes in vegetation, land cover, and environmental conditions. Lastly, the global coverage provided by the Sentinel-2 satellites allows researchers to monitor changes in vegetation, land cover, and environmental conditions across the entire planet.

A notable limitation of using Sentinel-2 for monitoring vegetation, including drought stress, is its optical sensor's susceptibility to cloud cover, which can obscure the satellite's view of the Earth's surface [26]. This issue is particularly challenging in regions such as

Poland, known for frequent cloud cover, where it can significantly hinder the continuity and reliability of the data [27,28]. Therefore, in our study, we used near-cloud-free images.

Our study aims to explore which commonly used vegetation indices retrieved from Sentinel-2 satellite imagery best capture changes in grassland vegetation caused by drought stress. We evaluated in situ measurements conducted during the growing season, as well as meteorological datasets, to identify conditions indicating drought. Then the meteorological and satellite imagery-based retrievals were examined to figure out the plant stress due to drought conditions. The manuscript proceeds with further analyses focusing on (1) the contribution of different vegetation indices from Sentinel-2 to determining plant drought stress, (2) the impact of utilizing high-resolution temporal satellite observations under frequent cloud coverage on the detection of drought severity in grasslands in Poland, and (3) the potential applications of our study's findings for plant science and mapping drought stress using earth observation data.

The innovative aspect of this research lies in its evaluation of various Sentinel-2 vegetation indices specifically tailored to detect plant stress, determining which most accurately reflect changes in grassland vegetation due to drought, and incorporating high-resolution temporal analysis under conditions of frequent cloud coverage, with implications for enhancing drought stress mapping using remote sensing data.

2. Materials and Methods

The following steps were taken to examine the sensitivity of satellite-derived vegetation indices to plant drought stress: (1) conducting field measurements in grasslands from 2020 to 2023; (2) creating a database of meteorological parameters to investigate drought conditions; (3) developing a database of vegetation indices primarily for identifying drought using Sentinel-2 imagery; (4) investigating and detecting drought through cross-verification of in situ and meteorological data with satellite information; and (5) culminating in the creation of illustrations that present spatiotemporal variations of drought intensity in Polish grasslands. We have included brief descriptions for context and employed color coding to visually distinguish between each step (Figure 1).

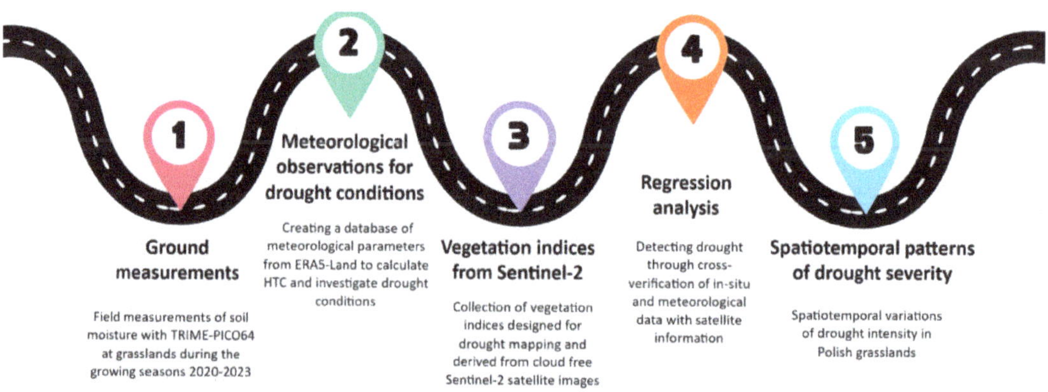

Figure 1. The process of examining the sensitivity of satellite-derived vegetation indices to plant drought stress at grasslands in Poland from 2020 to 2023.

2.1. Study Areas

The study areas, which are highlighted with a red frame in Figure 2, are spread across different regions of the country, each characterized by unique natural conditions that include variations in terrain, water bodies, and vegetation structures. These areas vary from lowlands to hilly terrains, influencing the types of soil, their structure, and their properties. Furthermore, the sites range from regions with intense human activity to more pristine areas, adding to their diversity and making them valuable for research. In the Podlaskie region of northeastern Poland, coded as PL84 under NUTS 2 (NUTS2 is a classification level in the NUTS (Nomenclature of Territorial Units for Statistics) system of the European Union, used for regional analysis and allocation of structural funds), the landscape is mostly lowland but features a variety of terrain. The presence of rivers like the Biebrza and Bug enhances the region's ecological diversity by supporting a rich mosaic of aquatic and wetland ecosystems. These river valleys are often humid, fostering unique ecological conditions that are conducive to the growth and development of the extensive grasslands [29].

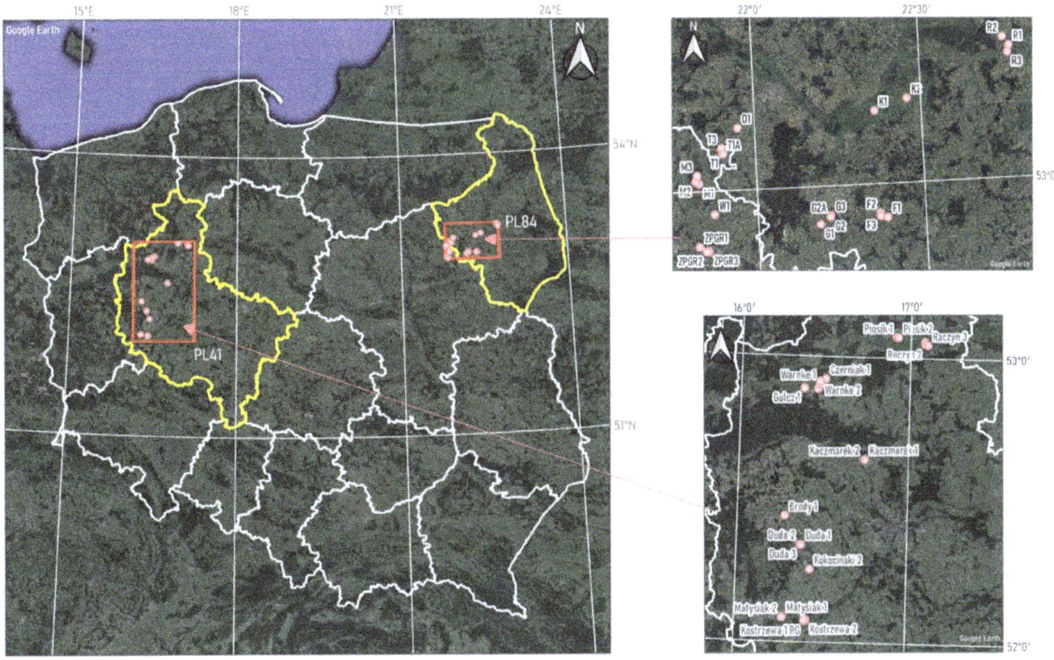

Figure 2. Location of field measurements conducted at grasslands highlighted by red dots in Wielkopolskie (PL41) and Podlaskie (PL84) provinces in Poland.

Conversely, the Wielkopolskie region in western Poland, coded PL41 under NUTS 2, features a more balanced topography, predominantly composed of lowlands interspersed with some hilly areas. The Wielkopolskie Lowland, lying at the heart of this region, exemplifies a typical lowland terrain characterized by flat or gently rolling landscapes ideal for intensive farming [30]. This region is traversed by numerous rivers and streams, including the significant Warta River, which is a key hydrographic feature of the landscape.

Climatically, the regions span various climatic zones of Poland, leading to differences in temperature, precipitation, and vegetative periods. These climatic conditions directly influence soil development and its agricultural utility. Podlaskie experiences a moderately warm continental climate with a shorter growing season, with an average annual temperature of 6.5 °C and annual rainfall of 550 mm [31]. In contrast, Wielkopolskie is characterized by a warmer transitional climate with milder winters and warmer summers, where the average annual temperature is 7.5 °C and the total rainfall is 500 mm [32]. Both regions have experienced increases in temperatures and more frequent extreme weather events, such as droughts and heavy rainfall, in recent years [33]. In Wielkopolskie, key challenges include severe drought [34] and soil erosion [35].

The distribution and characteristics of grasslands across Poland present significant challenges for nature conservation [36]. Typically, these grassland patches are relatively small in size, increasing the risk of local species extinction. Additionally, they are often situated at considerable distances from one another, impeding the movement of populations between patches and hindering their ability to re-establish [30]. Analyses have shown that the proportion of permanent grasslands in the agricultural land structure in 2020 is significantly higher in the Podlaskie Voivodeship compared to Wielkopolskie, with proportions of 38.2% and 13.9%, respectively [30].

2.2. Field Measurements

Ground measurements were conducted in the grasslands recognized in the Podlaskie and Wielkopolskie voivodeships (Figure 1). A total of 46 field sites were selected for ground measurements, with 23 sites in each of the two voivodeships. Field sites were chosen to capture the variability and spatial distribution of the vegetation. In the Wielkopolskie region, the study area consisted of three fields smaller than one hectare, eighteen fields ranging from one to ten hectares, and two fields larger than ten hectares. Conversely, in the Podlaskie region, there were four fields smaller than one hectare and nineteen fields between one and ten hectares.

Ground measurements were synchronized with Sentinel-2 overpasses and were carried out every three to four weeks during the growing season from April to September. Table 1 lists the field campaign dates from 2020 to 2023. The frequency of ground measurements depended on cloud cover observations, the distances between specific fields, and the accessibility of convenient locations for easy access. During the ground campaigns, two biophysical parameters characterizing soil conditions and vegetation state were measured at the field sites: soil moisture (SM; measured to a depth of 15 cm) using the TRIME-PICO64 (Ettlingen, Germany) and grass height using the Electronic Bluetooth® Plate Meter EC-20, supported by photo documentation.

Table 1. Ground measurement dates were conducted at Wielkopolskie and Podlaskie voivodeships across the years 2020–2023.

Year	Wielkopolskie	Podlaskie
2020	16.06, 22–23.06, 30.06, 15.07, 18.07, 06.08, 12–13.08, 23.08, 09–10.09, 17.09, 24.09	06–08.07, 28–29.07, 18–20.08, 30.09, 01.10
2021	24.04, 26.04, 08.05, 10.05, 21.05, 04–05.06, 17.06, 21.06, 02–03.07, 17.07, 29.07, 31.07, 02.08, 21–22.08, 04.09, 06.09, 19.09, 23.09	10–11.05, 09–10.06, 29–30.06, 26–27.07, 23–24.08
2022	22.04, 30.04, 07.05, 20.05, 22–23.05, 27.05, 18.06, 25.06, 08.07, 20.07, 24.07, 06–07.08, 25.08, 28–29.09	10–11.05, 22–23.06, 27–28.07
2023	29.04, 02.05, 12.05, 21.05, 17.06, 24.06, 09.07, 16.07, 27.07, 05.08, 13.08, 26–27.08, 10.09, 23–24.09	19–20.04, 24–25.05, 05–06.07

2.3. Satellite Data Acquisition

The Sentinel-2 program, part of the Copernicus program, offers several advantages for environmental research and monitoring (https://sentinel.esa.int/web/sentinel/missions/sentinel-2 (accessed on 13 July 2024). The grassland areas in the Podlaskie region include three Sentinel-2 granules: 34UED, 34UFD, and 34UFE. Similarly, the grass fields in the Wielkopolskie region include three Sentinel-2 granules: 33UWT, 33UWU, and 33UXU. The satellite data are accessible from orbits 22, 122, 79, 36, and 136, allowing for image acquisition every 5–6 days. Sentinel-2A and Sentinel-2B satellite images at processing level 2A were automatically retrieved using Google Earth Engine (GEE), a cloud-based platform offering geospatial data, tools, and computational power for analyzing and visualizing satellite imagery and other geospatial data. Users can access and analyze the data using various programming languages, such as Python and JavaScript. GEE also provides a range of tools for data processing and analysis, including machine learning algorithms for image classification and time-series analysis [37]. For this study, we utilized the JavaScript API within the Earth Engine Code Editor. The climate in the two regions is moderately continental, characterized by a short growing season, prolonged snow and ice cover, and significant cloudiness. Consequently, we filtered the satellite data to include only those images with less than 10% cloud cover. Table 2 provides a comprehensive list of the selected Sentinel-2 images for the years 2020–2023 over the study areas.

Table 2. Number of S2 imaging scenes across years 2020–2023 used for the study.

Year	Wielkopolskie	Podlaskie
2020	108	60
2021	97	48
2022	107	45
2023	100	59

Next, satellite meteorological datasets from ERA-5 Land, developed by the European Centre for Medium-Range Weather Forecasts (ECMWF), were utilized. ERA-5 Land is renowned as a top-tier tool for analyzing worldwide weather conditions and represents an enhanced iteration thanks to the incorporation of currently used technologies and refined algorithms [38]. The ERA5-Land dataset offers a spatial resolution of 0.1 degrees, equivalent to approximately 9 km × 9 km, providing detailed meteorological insights across diverse Earth regions. Spanning from 1950 to the present, this dataset facilitates the examination of climate trends over decades. Encompassing a wide range of meteorological variables, including air temperature, humidity, wind speed, atmospheric pressure, precipitation, and more, ERA5-Land provides a comprehensive toolkit for analysis. These parameters are accessible at different atmospheric levels. Demonstrating its versatility, ERA5-Land has proven invaluable in various scientific domains such as climate studies [38], land cover and vegetation seasonality [39], and weather forecasting [40]. In our study, we exploited ERA5-Land data, specifically daily averaged air temperature and daily total precipitation, to find out the sensitivity of remote sensing-based drought indices at grasslands.

2.4. Meteorological Drought Assessment

To investigate the meteorological conditions for grass development and detect drought, ERA-5 Land datasets comprising daily air temperature (Figure 3) and daily total precipitation (Figure 4) at grasslands during the growing season from 2020 to 2023 were utilized. It was observed that the distributions of average daily temperatures from 2020 to 2023 for grasslands in two distant voivodeships did not differ significantly. However, the thermal conditions for grass growth in both regions exhibited frequent temperature fluctuations, especially in 2022 and 2023, when temperature observations often surpassed the critical threshold of 20–25 degrees Celsius for average daily temperature. This threshold is considered the limit for thermal stress in plants under Polish climate conditions [41]. Additionally, concerning daily rainfall totals, distinct differences were observed in the temporal distribution of grasslands in both regions. In the Podlaskie Voivodeship, there were frequent instances of large daily rainfall totals exceeding 50 mm in 2020. Conversely, during a similar period in the Wielkopolskie Voivodeship, relatively consistent rainfall events were recorded, with daily totals not exceeding 40 mm. In subsequent years, daily rainfall totals generally ranged from 10 to 20 mm, depending on the observation period.

Next, the Hydrothermal Coefficient (HTC), also known as Selyaninov's coefficient [42], was calculated to estimate meteorological drought conditions in grasslands. This widely used metric for drought monitoring in central and eastern European countries [43,44] characterizes the atmospheric moisture conditions that contribute to drought. The HTC is closely linked to the productivity of various grassland types [45]. Considering the widespread use of HTC in research, it was deemed an appropriate method for assessing the occurrence and intensity of drought.

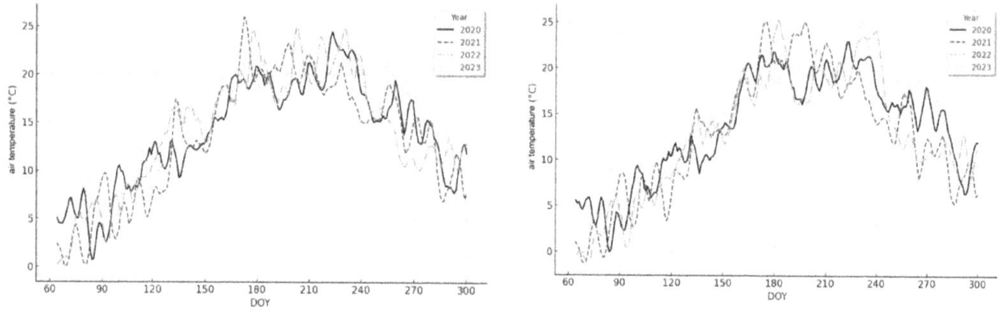

(a) (b)

Figure 3. Average daily air temperature from March to October in 2020–2023 in the Wielkopolskie (**a**) and Podlaskie (**b**) voivodeships.

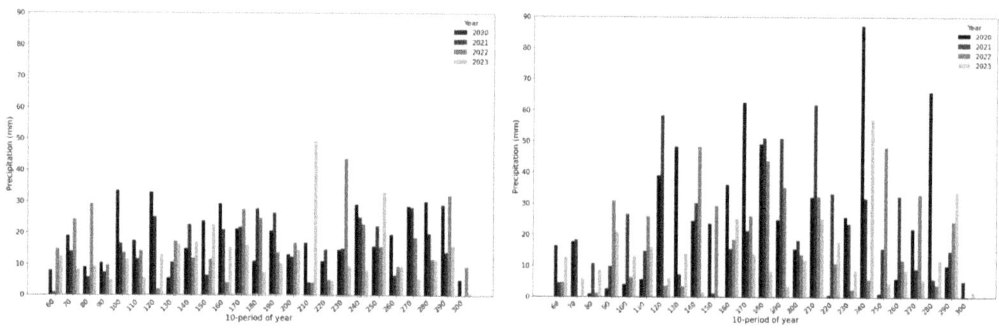

(a) (b)

Figure 4. Daily total precipitation from March to October in 2020–2023 in the Wielkopolskie (**a**) and Podlaskie (**b**) voivodeships.

The HTC combines air temperature and precipitation parameters over a specific period. HTC is calculated as the ratio of accumulated air temperature to accumulated precipitation over a given period. It is defined by the following Equation (1) [42]:

$$HTC = \frac{10\sum_{i=1}^{n} P_i}{\sum_{i=1}^{n} T_i} \quad (1)$$

where

n—length of the preceding period in days;
P_i—precipitation amount on the ith day (mm);
T_i—daily average of the air temperature on the ith day (°C).

This study adopted the widely accepted classification into nine HTC classes [46]. The HTC values are as follows: extremely dry (HTC < 0.4), very dry (HTC 0.4–0.8), dry (HTC 0.8–1.1), quite dry (HTC 1.1–1.4), optimum (HTC 1.4–1.7), quite humid (HTC 1.7–2.1), humid (HTC 2.1–2.6), very humid (HTC 2.6–3.0), and extremely humid (HTC > 3.0). In order to investigate the complexity and dynamics of meteorological drought for each analyzed moment within the growing season from 2015 to 2023, the median of the HTC index over the preceding 30 days (HTC30) was determined. The 30-day median HTC index is described as valuable because it smooths out short-term fluctuations and outliers, offering a more consistent and reliable measure of heat transfer performance [44].

2.5. Vegetation Indices Calculations

To assess how sensitive satellite-derived vegetation indices (VIs) are to plant stress, we selected and calculated VIs specifically designed for drought mapping using Sentinel-2 images of the study area. Initially, we identified the most commonly used spectral vegetation indices, which capture various aspects of plant growth, drought detection, and canopy water content. These VIs are detailed in Table 3.

Table 3. Spectral indices used in study.

Short Name	Full Name	Formula	Authors
NDVI	Normalized Difference Vegetation Index	$NDVI = \frac{R_{842} - R_{665}}{R_{842} + R_{665}}$	Rouse et al., 1973 [47]
NDII	Normalized Difference Infrared Index	$NDII = \frac{R_{842} - R_{1610}}{R_{842} + R_{1610}}$	Hardisky et al., 1983 [48]
NDWI	Normalized Difference Water Index	$NDWI = \frac{R_{860} - R_{1240}}{R_{860} + R_{1240}}$	Gao, 1996 [49]
NDDI	Normalized Difference Drought Index	$NDDI = \frac{NDVI - NDWI}{NDVI + NDWI}$	Gu et al., 2007 [50]

The Normalized Difference Vegetation Index (NDVI) is a widely used measure of plant health and density, derived from the difference between the maximum absorption of red light and the maximum reflectance of near-infrared light. NDVI values range from −1 to 1, with higher values representing healthier and more abundant vegetation. This index is instrumental in monitoring vegetation and detecting changes in ecosystems and biodiversity. However, its accuracy can be influenced by atmospheric conditions, the soil background, and the structure of the plant canopy [51]. Moreover, when vegetation is dense and covers 100% of the ground, NDVI gives poor estimates of vegetation productivity because it is saturated and does not reflect the increase in biomass [52,53]. The Normalized Difference Infrared Index (NDII) specifies vegetation moisture and leaf water content. NDII values can range from −1 to 1, with higher values indicating greater water content in vegetation and lower values indicating less moisture. NDII is particularly useful for monitoring grassland productivity [54], detecting mowing frequency [55], and assessing plant hydric stress [56]. The Normalized Difference Water Index (NDWI), primarily developed by Gao in 1996 [49], is utilized to monitor changes in moisture levels and water content in grassland plants [57]. The last one, the Normalized Difference Drought Index (NDDI), was developed by Gu in 2007 [50] for assessing drought in grasslands. It is based on the relationships between two previously established indices: NDVI and NDWI. The value of the NDDI index increases with the severity of the drought. Assuming minimum drought criteria for NDVI < 0.5 and NDWI < 0.3, a drought condition occurs at NDDI > 0.25. The NDDI has been validated for evaluating drought conditions in grasslands in India [58] and Mongolia [59].

2.6. Estimating Drought Response to VIs

Considering that the NDVI and NDWI indices are included in the NDDI calculation and given that NDII is regarded as a less efficient indicator for monitoring drought conditions [57], we focused on investigating the sensitivity of the latest NDDI indicator values in relation to meteorological records. Additionally, ground measurements were used for cross-validation to assess the usefulness of the index that best characterizes drought severity. Furthermore, to verify drought severity, which is described within the value ranges of NDII and NDDI, we examined the significance of the relationships between them. We conducted a verification of the usefulness of vegetation indices through regression analysis in two stages. First, we assessed the extent to which changes in HTC values are reflected in the values of S-2 NDDI used as a drought indicator, and we examined their consistency with in situ soil moisture measurements taken at grassland sites from 2020 to 2023. The second stage focused on investigating drought severity by examining the relationship between NDDI and NDII. Finally, temporal variations of drought severity with S-2 NDDI at individual grassland parcels in the Wielkopolskie region during the growing season across the years 2020–2023 were studied. The performance statistics of the correlations were evaluated using various metrics, including the Pearson correlation coefficient (r), the coefficient of determination (R^2), which is the square of the correlation coefficient, as well as the mean bias error (MBE), mean absolute error (MAE), and root mean square error (RMSE).

3. Results

The heatmaps (Figure 5) visualize temporal patterns of HTC30 during the growing seasons across years from 2015 to 2023 in Wielkopolskie and Podlaskie. The HTC30 is segmented into ten-day periods throughout the growing season, which spans from day 80 to day 300 of the year. The heatmaps illustrate a transition in meteorological conditions from wetter to drier over the years. Until the end of March (DOY 60–90), the data show predominantly "wet" and "extreme wet" conditions (blue shades), while during the growing season lasting until the end of September (DOY 90–270), there is a mix of all categories, indicating diverse conditions. In the most recent years, there has been a clear shift towards "dry" and "extremely dry" conditions (red shades). Specifically in Wielkopolskie, we noted frequent extreme dry and very dry conditions, i.e., 119 out of 216, which constitute 55% of all observations. This pattern suggests a long-term trend towards increased dryness, with significant seasonal variability within each year, reflecting the changing climate and its potential impact on the studied area. On the other hand, Podlaskie has experienced moderately dry and moderately wet conditions in recent years. The heatmaps provide a visual representation of changing conditions over time and highlight a shift towards drier conditions in recent years, which is critical for understanding long-term climatic trends and planning for future environmental and resource challenges.

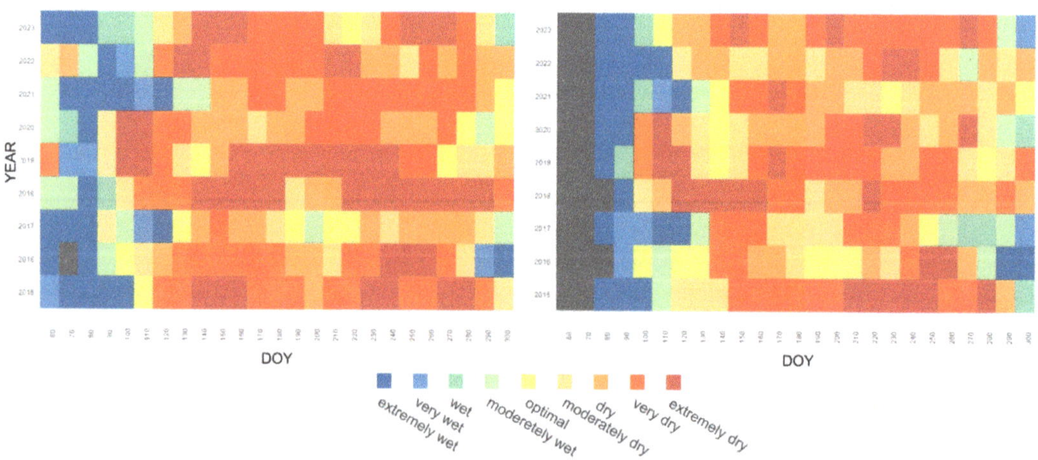

Figure 5. Temporal patterns of HTC30 at Wielkopolskie (**left**) and Podlaskie (**right**) regions during the growing season from March (DoY 60) until the end of October (DoY 300) in 2015–2023.

Figure 6 shows the relationship between NDDI and HTC, characterized by a significant negative correlation coefficient r = −0.75 and R^2 = 0.56. The errors are relatively low, i.e., MBE = 0.04, MAE = 1.28, and RMSE = 1.58. The blue trend line illustrates that as NDDI increases, HTC decreases, indicating that higher NDDI values, which signify greater drought severity, are associated with lower HTC values, reflecting drier conditions. The data points are predominantly concentrated within the lower NDDI range (0 to 1.5) and the higher HTC range (1 to 4), thereby substantiating the negative correlation. This strong relationship suggests that NDDI is an effective index for identifying areas with lower hydrothermal conditions, supporting its use for monitoring drought severity in the studied region.

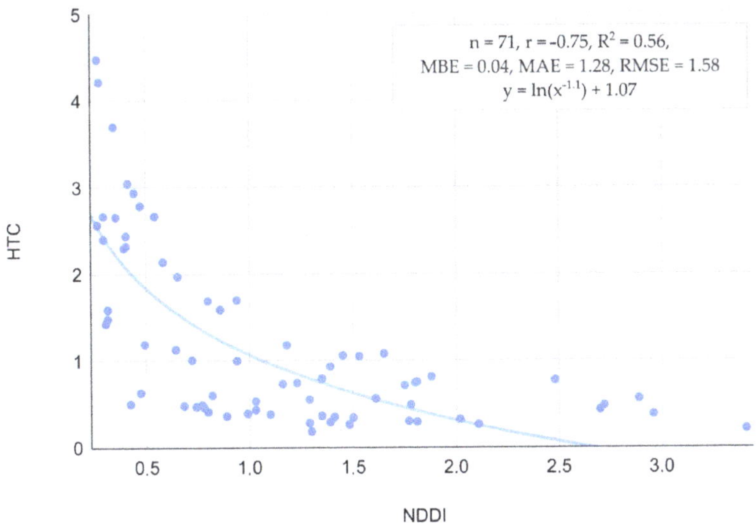

Figure 6. Relation between HTC and NDDI estimated across all satellite observations over grasslands recognized in this study.

To substantiate the hypothesis regarding the relationship between SM and NDDI, an investigation was conducted (Figure 7). The analysis revealed a strong negative correlation between the drought index and soil moisture at $r = -0.82$ and $R^2 = 0.67$, indicating that higher NDDI values are associated with lower soil moisture levels. The regression line and the scatter of the points show that soil moisture decreases as NDDI increases, reinforcing the inverse relationship.

While NDII and NDDI are considered appropriate for determining drought severity, the relationship between NDII and NDDI was investigated as well (Figure 8). A blue exponential regression line runs through the data points, highlighting the general trend between NDII and NDDI at the level $r = -0.71$, $R^2 = 0.50$, with a relatively low RMSE of 0.94. The line indicates that as NDDI values increase, NDII values tend to decrease, suggesting a negative correlation between the two indices. The plot also features annotations dividing the data into two regions: "stressed" and "severely stressed". The "stressed" region, marked with an orange ellipse, contains a cluster of data points where NDII values are positive (ranging from 0 to 0.2) and NDDI values are relatively low (up to around 0.5). This region indicates conditions of moderate stress. In contrast, the "severely stressed" region, outlined with a red ellipse, encompasses data points where NDII values are lower (ranging from −0.1 to −0.3) and NDDI values are higher (from around 0.5 to 3.0). This area represents conditions of severe stress, where higher drought conditions correspond with lower NDII values. Overall, the scatter plot provides a visual representation of the negative correlation between NDII and NDDI, emphasizing the varying levels of stress indicated by different clusters of data points. NDDI offers several advantages over NDII for determining drought conditions. NDDI is specifically designed to measure drought by combining information from both vegetation and soil moisture content. This dual focus makes it a more targeted and direct indicator of drought stress. In contrast, NDII primarily measures vegetation moisture content, which can be influenced by a variety of factors other than drought, such as plant health, soil type, or recent rainfall. Consequently, while NDII provides valuable information about vegetation health and moisture, it does not exclusively indicate drought conditions. NDDI's design to specifically detect drought stress makes it a more effective and reliable tool for identifying and assessing drought severity.

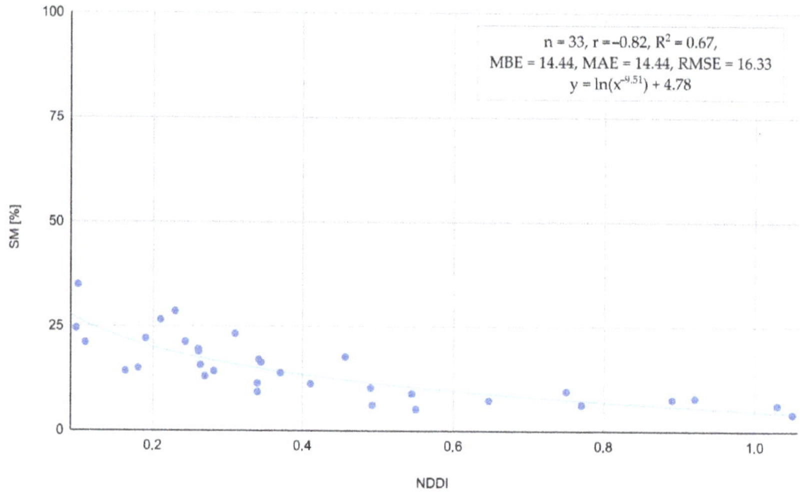

Figure 7. Relation between SM and NDDI estimated across all satellite observations over grasslands recognized in this study.

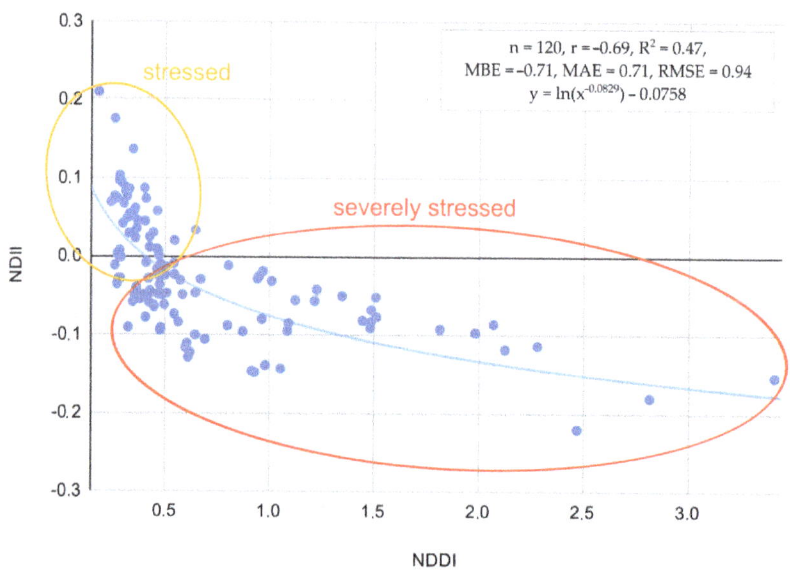

Figure 8. Relation between NDII and NDDI for classifying stressed and severely stressed plants to drought.

Figure 9 comprises four scatter plots, each representing the occurrence and intensity of drought conditions measured by NDDI across the years 2020–2023 at individual grassland fields in Wielkopolskie. Each plot showcases how NDDI values fluctuate over time, with each data point corresponding to observations from various fields or stations, indicated by distinct colored markers. In the 2020 plot, NDDI values generally remain low, with only a few instances where values exceed 1, indicating mild drought conditions. However, in 2021, there is a noticeable increase in NDDI values, especially around DOY 100–150, where several points exceed 3, signifying more severe drought conditions. The 2022 plot shows a broader distribution of higher NDDI values, particularly between DOY 100 and 200, with some points reaching up to 5, indicating very severe drought conditions during this period. The 2023 plot continues this trend, with numerous observations exceeding 2 and several reaching up to 5, particularly between DOY 100 and 200, suggesting persistent and severe drought conditions. Overall, the data indicates a trend of increasing drought severity over the four-year period, with 2022 and 2023 experiencing the most significant drought conditions. This pattern highlights the importance of continuous monitoring and analysis of NDDI values to manage and mitigate the impacts of drought on agricultural fields and other affected areas.

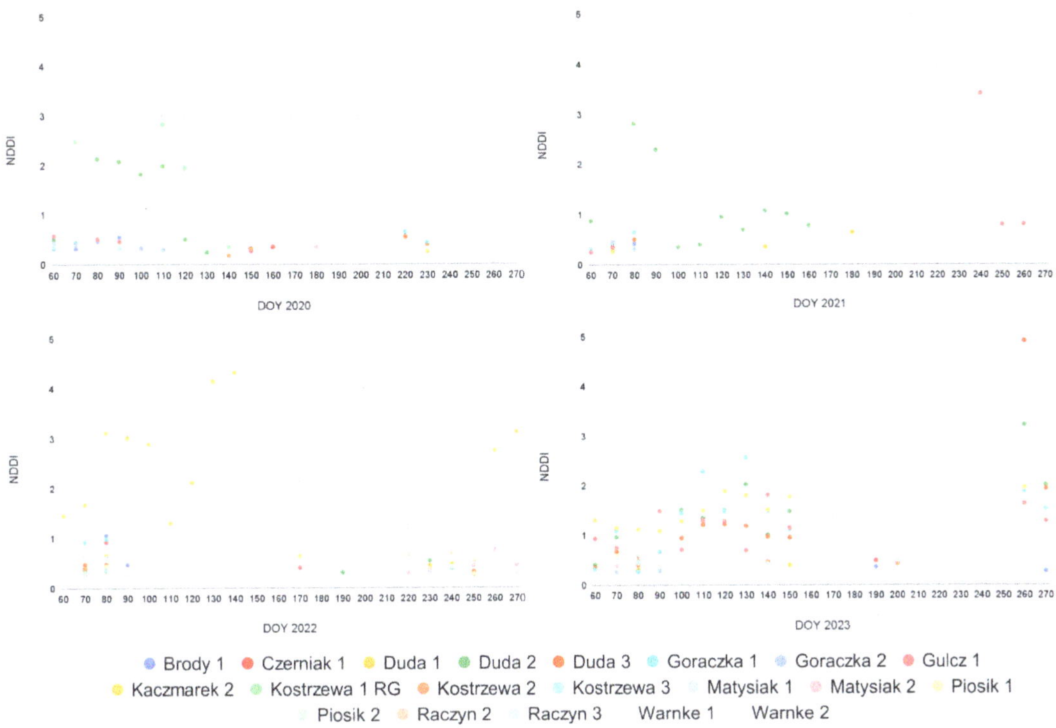

Figure 9. Temporal variations of S-2 NDDI at individual grassland parcels in the Wielkopolskie region during the growing season across the years 2020–2023.

4. Discussion

This study demonstrated that satellite-derived indices for detecting plant stress offer new opportunities for continuous temporal and spatial monitoring of grassland vegetation health. The findings support the hypothesis that terrestrial ecosystems remain under stress due to hydrothermal conditions affecting the state and development of grass plants. The severity of the drought in 2023 was attributed to a significant increase in HTC, accompanied by higher daily air temperatures and a notable decrease in daily rainfall. Our study confirmed that the utilization of hydrothermal coefficients (HTC) enables us to explore spatio-temporal patterns of meteorological drought, thereby facilitating the detection of drought stress in grass plants. While HTC offers a comprehensive approach to quantifying the combined influence of temperature and moisture on vegetation, there are some limitations associated with their application. Their calculation entails complexities, relying on sophisticated models and algorithms, which can be computationally demanding [60] and may not uniformly capture the impact of temperature and moisture variations across different vegetation types and geographical regions, potentially introducing biases in interpretation [61]. Furthermore, the accuracy of HTC estimates hinges on the availability and quality of input data, including meteorological observations and remote sensing imagery [44].

In this study, well-validated Sentinel-2 surface reflectance products were used to determine vegetation indices (VIs) and analyze drought dynamics for 2020–2023. Shepherd [62] confirmed the quality of compositions from Sentinel-2 satellite images, demonstrating the use of improved cloud-free and composited daily, weekly, or monthly mosaics for regular land monitoring. Due to unfavorable weather conditions in Poland, where clouds are present on average 150 days per year [63], S-2 imaging scenes with less than 10% cloud coverage were taken into consideration. As shown in Table 2, the number of acquired images for the two distinct areas varied significantly across different years. For the Wielkopolskie region, nearly twice as many cloud-free or minimally cloudy satellite images were obtained. This spatial relationship confirms observations of spatial variability in the number of cloudy days in Poland, as documented by Sypniewska [63]. Moreover, a consistent and repeatable amount of annual Sentinel-2 scenes was collected for the years 2020–2023, regardless of geographic location. Similar variations in the availability of cloud-free or minimally cloud-covered satellite optical data have been noted in previous grassland studies conducted under Polish climate conditions [64,65]. This variability is an important factor to consider in future studies. Therefore, we recommend that future research investigating temporal patterns in remote sensing data should account for this factor and explicitly evaluate its impact on drought stress in grass plants.

Among the tested vegetation indices and relationships presented in Figures 6 and 7, the NDDI has shown the strongest correlation with HTC and SM. The NDDI is highly applicable for detecting drought severity in grasslands due to its sensitivity to both vegetation and soil moisture conditions. It integrates NDVI and NDWI, leveraging the red and near-infrared spectra critical for NDVI and the shortwave infrared spectrum essential for NDWI. NDVI captures the photosynthetic activity of plants by distinguishing between healthy vegetation and stressed or sparse vegetation using red and NIR reflectance. NDWI, on the other hand, is sensitive to leaf water content, utilizing the SWIR spectrum to detect moisture levels in vegetation. By combining these indices, NDDI provides a comprehensive measure of both plant health and water stress, enabling more accurate and timely detection of drought conditions in grasslands. This dual sensitivity allows for better monitoring and management of drought impacts, ultimately aiding in the preservation of these critical ecosystems, as noted in the previous studies of Artikanur [66] and Patil [67]. Meteorological conditions significantly influence the values of NDDI, providing an added value in this study for detecting plant stress (Figure 9). The integration of meteorological data, such as temperature and precipitation, with NDDI allows for a more accurate assessment of drought severity and its impact on vegetation. This enhanced detection capability is crucial for understanding the extent of stress in plants, particularly in Polish grasslands under various climate zones, where the interplay between climatic factors and plant health is complex and dynamic [68–70].

The analysis of VIs revealed that plant drought stress can be detected and monitored using the NDDI provided at a 10-m spatial resolution from S-2 imagery. While many studies have analyzed drought in Poland using commonly used indices such as the Drought Information Satellite System (DISS) at a coarse 1000 m spatial resolution [71] and the Standardized Precipitation Evapotranspiration Index (SPEI) based on meteorological observations at a 0.25-degree spatial resolution [34], our approach offers local farmers the opportunity to investigate drought conditions with much greater precision at individual field levels. Farmers in the Wielkopolska region, which is strongly affected by soil erosion [35], should take particular interest in these types of analyses. Research examining drought frequency between 2001 and 2023, conducted using the DISS index methodology with satellite data from the Terra MODIS satellite [71], has highlighted that agricultural fields and grasslands in Wielkopolska are significantly more vulnerable to drought compared to the Podlaskie region (Figure 10). Therefore, the critical focus remains on linking adverse conditions for grass vegetation growth and providing precise spatial information at a resolution of 10 × 10 m, along with frequent temporal updates. These requirements are met by the satellite observations provided within the Earth Observation Copernicus S-2.

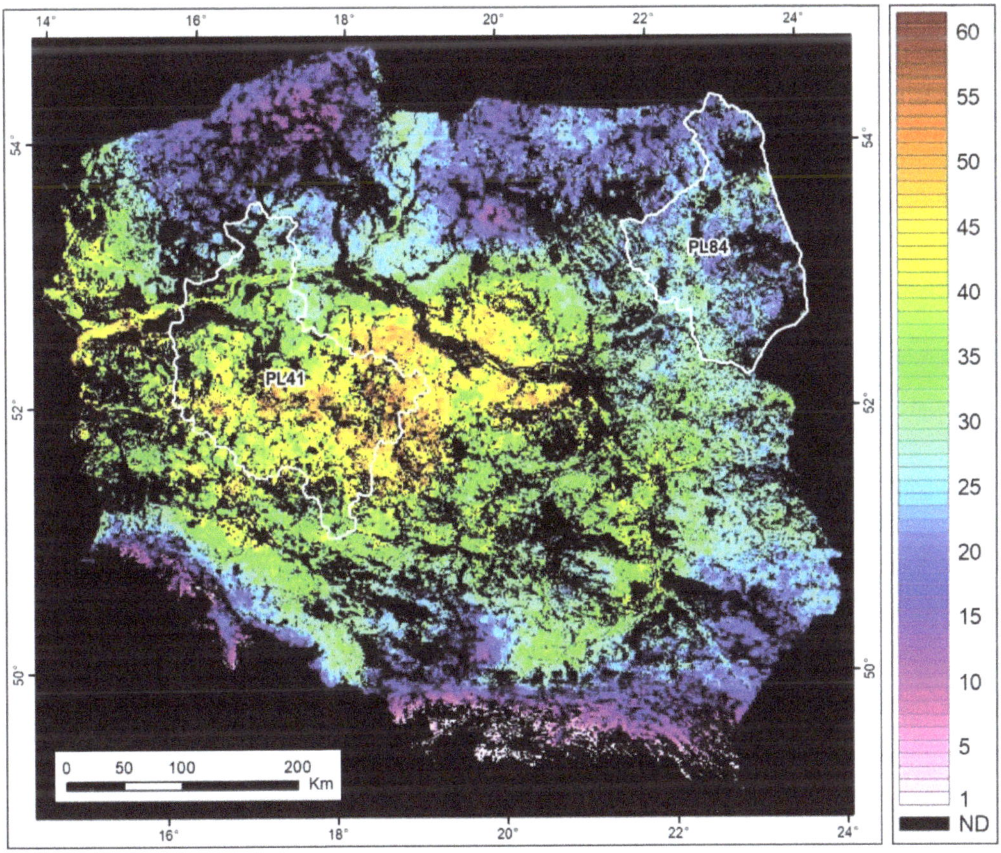

Figure 10. Frequency (%) of drought occurrences derived from satellite Terra MODIS observations across 2001–2023. White lines highlight our study areas.

In the context of tracking drought severity with high temporal and spatial accuracy for individual farmers in Wielkopolskie, Figure 11 unveils new insights for utilizing S-2 NDDI observations. The figure illustrates the proportion of grassland fields endangered by drought in ten-day periods across four years. In 2023, drought risk peaked early in the year, notably between DOY 70 and 90, with over 50% of fields affected, a significantly higher proportion compared to the same period in other years. Another substantial increase in 2023 occurred around DOY 150 and 230, once again surpassing the other years. In contrast, 2021 exhibited consistently low drought risk throughout the year, while 2020 and 2022 showed moderate risk spikes at various periods. These data indicate that 2023 experienced more severe and frequent drought conditions in grassland fields, especially in the early and mid-year periods, underscoring a notable escalation in drought risk compared to previous years.

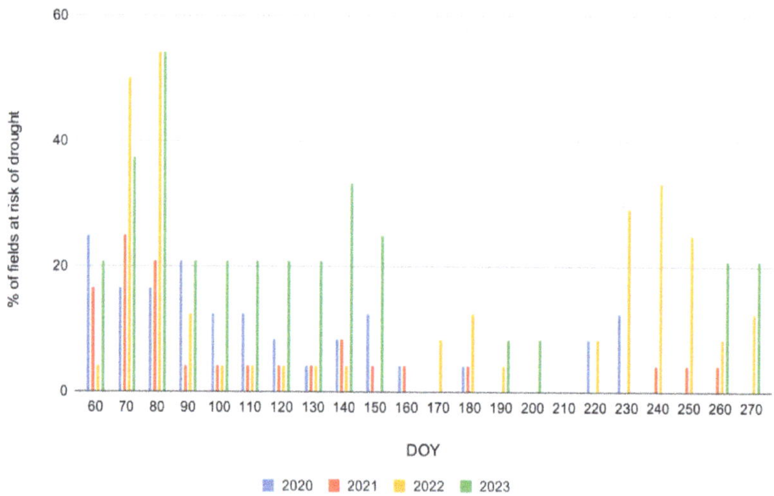

Figure 11. Frequency (%) of individual parcels in Wielkopolskie with recognized drought severity estimated from satellite Sentinel-2 NDDI observations across 2020–2023.

Examining drought sensitivity in agriculture has been the subject of numerous studies conducted by various researchers [72–74]. It has been proven that satellite-derived information provides valuable, repeatable, and near-real-time data. However, utilizing the Normalized Difference Drought Index (NDDI) with Sentinel-2 data for mapping drought severity comes with several uncertainties and limitations. One primary challenge is the variability in atmospheric conditions, such as cloud cover and aerosol presence, which can affect the accuracy of satellite imagery and its availability (Table 2). Additionally, the spatial resolution of Sentinel-2, while high, may still miss small-scale variations in drought conditions within heterogeneous landscapes. The temporal resolution also poses a limitation, as a ten-day period from five-day revisits might not capture rapid changes in soil moisture and vegetation stress. Furthermore, the NDDI relies on the accuracy of both NDVI and NDWI, which can be influenced by factors such as soil background, vegetation type, and phenological stage [51,57]. Calibration and validation of NDDI data against ground-based measurements are essential to ensure reliability but can be resource-intensive and geographically limited [67]. Thus, during field campaigns (Table 1), we collected in situ soil moisture data to confirm the reliability of the NDDI. These factors collectively introduce uncertainties in the assessment and mapping of drought severity, potentially affecting the precision of drought monitoring and decision-making processes [75].

5. Conclusions

The examination of vegetation indices to assess drought stress across various grassland regions in Poland has provided valuable insights into the intricate interplay between environmental factors and meteorological observations. This study emphasizes the critical role of effective grassland management in promoting agricultural sustainability amidst challenges posed by climate change, which often manifests in drought occurrences. The findings underscore the necessity for adaptive management strategies that prioritize resilience within agricultural systems. Distinct growth patterns in grasslands were observed, particularly concerning drought, as indicated by NDDI observations. While Wielkopolskie experienced increased drought occurrences in the first decades of the growing seasons from 2022 to 2023, Podlaskie exhibited stable conditions with minimal variations. These findings underscore the importance of considering local environmental factors when analyzing vegetation dynamics.

Additionally, the examination of meteorological conditions revealed significant regional differences, particularly in temperature and precipitation. The year 2023 presented considerable challenges for vegetation development, characterized by unfavorable conditions across all study areas. These observations underscore the vulnerability of agricultural systems to climatic variability and the importance of adaptive management strategies. The use of the NDDI index for assessing drought response provided insights into distinguishing between wet and dry conditions. Significant variations in drought occurrences and the proportion of fields at risk were identified within Poland, influenced by climatic fluctuations and grassland management practices. The use of the NDDI index proved invaluable for assessing vegetation water content, offering a potential tool for monitoring and managing grassland resources.

The implications of these findings extend beyond academic research, holding practical significance for agricultural stakeholders. By enhancing our understanding of the relationship between environmental conditions and biomass production, farmers and land managers can implement more effective cultivation practices and grassland management strategies. This knowledge is particularly crucial given changing climatic conditions, where adaptive approaches are essential for ensuring the resilience and sustainability of agricultural systems. In conclusion, this study provides valuable insights into the complex dynamics of vegetation growth and response to drought across diverse agricultural landscapes. By elucidating the influence of environmental factors on agricultural outcomes, it lays the groundwork for informed decision-making and sustainable resource management in response to evolving climatic challenges.

Author Contributions: Conceptualization, M.B. and K.W.; methodology, M.B., K.W., M.K. and K.D.-Z.; software, M.K.; validation, K.W., M.B, K.D.-Z. and P.G.; formal analysis, M.B., K.W. and M.K.; investigation, M.B., K.W., M.K., K.D.-Z. and P.G.; resources, M.B., K.W. and M.K.; data curation, M.B., K.W. and M.K.; writing—original draft preparation, M.B., K.W., K.D.-Z. and P.G.; writing—review and editing, M.B., K.W., K.D.-Z. and P.G.; visualization, M.B. and K.W.; supervision, K.D.-Z. and P.G.; project administration, K.D.-Z. and P.G.; project funding acquisition, K.D.-Z. and P.G.; article funding acquisition, M.B. All authors have read and agreed to the published version of the manuscript.

Funding: This study has been conducted within the framework of the project GrasSAT—"Tools for information to farmers on grasslands yields under stressed conditions to support management practices" NOR/POLNOR/GrasSAT/0031/2019-0 financed by The National Centre for Research and Development (NCBIR)—Norwegian Funds.

Data Availability Statement: The datasets used and analyzed during the current study are available from the corresponding author upon reasonable request.

Acknowledgments: Field measurements were conducted in Podlaskie (PL84) by the Remote Sensing Centre, Institute of Geodesy and Cartography, 27 Modzelewskiego St., 02-679 Warsaw, Poland; and in Wielkopolskie (PL41) by the Department of Grassland and Natural Landscape Sciences, Poznań University of Life Sciences, 11 Dojazd St., 60-632 Poznań, Poland.

Conflicts of Interest: The authors declare no conflicts of interest.

References

1. Rhodes, D.; Nadolska-Orczyk, A. Plant Stress Physiology. In *Encyclopedia of Life Sciences;* John Wiley & Sons, Ltd.: Chichester, UK, 2001; pp. 1–7.
2. Shabala, S. *Plant Stress Physiology;* CAB International: Oxford, UK; Oxford, MS, USA, 2012.
3. De Santis, M.A.; Soccio, M.; Laus, M.N.; Flagella, Z. Influence of Drought and Salt Stress on Durum Wheat Grain Quality and Composition: A Review. *Plants* **2021**, *10*, 2599. [CrossRef]
4. Dodd, J.L.; Lauenroth, W.K. Analysis of the Response of a Grassland Ecosystem to Stress. In *Perspectives in Grassland Ecology. Ecological Studies, Vol 32;* French, N.R., Ed.; Springer: New York, NY, USA, 1979. [CrossRef]
5. Peltzer, D.A.; Wilson, S.D. *Competition and Environmental Stress in Temperate Grasslands;* CABI: Wallingford, UK, 2001; pp. 193–212. [CrossRef]
6. Bengtsson, J.; Bullock, J.M.; Egoh, B.; Everson, T.; O'Connor, T.; O'Farrell, P.J.; Smith, H.G.; Lindborg, R. Grasslands—More important for ecosystem services than you might think. *Ecosphere* **2019**, *10*, e02582. [CrossRef]

7. Amsten, K.; Cromsigt, J.P.G.M.; Kuijper, D.P.J.; Loberg, J.M.; Jung, J.; Strömgren, M.; Niklasson, M.; Churski, M. Pyric herbivor in a temperate European wood-pasture system. *J. Appl. Ecol.* **2024**, *61*, 1081–1094. [CrossRef]
8. Naudts, K.; Van den Berge, J.; Farfan, E.; Rose, P.; AbdElgawad, H.; Ceulemans, R.; Janssens, I.A.; Asard, H.; Nijs, I. Future climat alleviates stress impact on grassland productivity through altered antioxidant capacity. *Environ. Exp. Bot.* **2014**, *99*, 150–15 [CrossRef]
9. Li, S.; Lu, S.; Li, X.; Hou, X.; Zhao, X.; Xu, X.; Zhao, N. Effects of Spring Drought and Nitrogen Addition on Productivity an Community Composition of Degraded Grasslands. *Plants* **2023**, *12*, 2836. [CrossRef]
10. Zhang, Y.; Wang, Z.; Wang, Q.; Yang, Y.; Bo, Y.; Xu, W.; Li, J. Comparative Assessment of Grassland Dynamic and Its Response t Drought Based on Multi-Index in the Mongolian Plateau. *Plants* **2022**, *11*, 310. [CrossRef] [PubMed]
11. Tester, M.; Bacic, A. Abiotic stress tolerance in grasses. From model plants to crop plants. *Plant Physiol.* **2005**, *137*, 791–79 [CrossRef] [PubMed]
12. Ramoelo, A.; Dzikiti, S.; van Deventer, H.; Maherry, A.; Cho, M.A.; Gush, M. Potential to monitor plant stress using remot sensing tools. *J. Arid. Environ.* **2015**, *113*, 134–144. [CrossRef]
13. Awais, M.; Li, W.; Cheema, M.J.M.; Zaman, Q.U.; Shaheen, A.; Aslam, B.; Zhu, W.; Ajmal, M.; Faheem, M.; Hussain, S. UAV-base remote sensing in plant stress imagine using high-resolution thermal sensor for digital agriculture practices: A meta-review. *Int J. Environ. Sci. Technol.* **2022**, *20*, 1135–1152. [CrossRef]
14. Zhu, X.; Chen, X.; Ma, L.; Liu, W. UAV and Satellite Synergies for Mapping Grassland Aboveground Biomass in Hulunbui Meadow Steppe. *Plants* **2024**, *13*, 1006. [CrossRef] [PubMed]
15. Bannari, A.; Morin, D.; Bonn, F.; Huete, A.R. A review of vegetation indices. *Remote Sens. Rev.* **1995**, *13*, 95–120. [CrossRef]
16. Cohen, W.B. Response of vegetation indices to changes in three measures of leaf water stress. *Photogramm. Eng. Remote Sens.* **1991** *57*, 185–202.
17. Baret, F.; Houlès, V.; Guérif, M. Quantification of plant stress using remote sensing observations and crop models: The case o nitrogen management. *J. Exp. Bot.* **2007**, *58*, 869–880. [CrossRef] [PubMed]
18. Jones, H.G.; Schofield, P. Thermal and other remote sensing of plant stress. *Gen. Appl. Plant Physiol.* **2008**, *34*, 19–32.
19. Barton, C. Advances in remote sensing of plant stress. *Plant Soil* **2012**, *354*, 41–44. [CrossRef]
20. Adiri, Z.; Lhissou, R.; El Harti, A.; Jellouli, A.; Chakouri, M. Recent advances in the use of public domain satellite imagery for mineral exploration: A review of Landsat-8 and Sentinel-2 applications. *Ore Geol. Rev.* **2020**, *117*, 103332–103365. [CrossRef]
21. Misra, G.; Cawkwell, F.; Wingler, A. Status of Phenological Research Using Sentinel-2 Data: A Review. *Remote Sens.* **2020**, *12*, 2760 [CrossRef]
22. Phiri, D.; Simwanda, M.; Salekin, S.; Nyirenda, V.R.; Murayama, Y.; Ranagalage, M. Sentinel-2 Data for Land Cover/Use Mapping A Review. *Remote Sens.* **2020**, *12*, 2291. [CrossRef]
23. Du, Y.; Zhang, Y.; Ling, F.; Wang, Q.; Li, W.; Li, X. Water Bodies' Mapping from Sentinel-2 Imagery with Modified Normalized Difference Water Index at 10-m Spatial Resolution Produced by Sharpening the SWIR Band. *Remote Sens.* **2016**, *8*, 354. [CrossRef]
24. Segarra, J.; Buchaillot, M.L.; Araus, J.L.; Kefauver, S.C. Remote Sensing for Precision Agriculture: Sentinel-2 Improved Features and Applications. *Agronomy* **2020**, *10*, 641. [CrossRef]
25. Spoto, F.; Sy, O.; Laberinti, P.; Martimort, P.; Fernandez, V.; Colin, O.; Hoersch, B.; Meygret, A. Overview of Sentinel-2. In Proceedings of the IEEE International Geoscience and Remote Sensing Symposium, Munich, Germany, 22–27 July 2012; pp. 1707–1710. [CrossRef]
26. Tiede, D.; Sudmanns, M.; Augustin, H.; Baraldi, A. Investigating ESA Sentinel-2 products' systematic cloud cover overestimation in very high altitude areas. *Remote Sens. Environ.* **2021**, *252*, 112163. [CrossRef]
27. Konik, M.; Kowalewski, M.; Bradtke, K.; Darecki, M. The operational method of filling information gaps in satellite imagery using numerical models. *Int. J. Appl. Earth Obs. Geoinf.* **2019**, *75*, 68–82. [CrossRef]
28. Paszkuta, M. Impact of cloud cover on local remote sensing—Piaśnica River case study. *Oceanol. Hydrobiol. Stud.* **2022**, *51*, 283–297. [CrossRef]
29. Dembicz, I.; Kozub, Ł.; Bobrowska, I.; Dengler, J. Grasslands of the mineral islands in the Biebrza National Park, Poland. *Palaearct. Grassl.* **2020**, *47*, 43–51. [CrossRef]
30. Gabryszuk, M.; Barszczewski, J.; Wróbel, B. Characteristics of grasslands and their use in Poland. *J. Water Land Dev.* **2021**, *51*, 243–249. [CrossRef]
31. Górniak, A. *Podlasie Voivodeship Climate during Global Warming*; University of Bialystok Press: Białystok, Poland, 2021; ISBN 978-83-7431-694-1. (In Polish)
32. Miler, A. Climate of the Wielkopolski National Park in the Period of 1848–2017. *For. Lett.* **2018**, *111*, 6–11.
33. Przybylak, R.; Oliński, P.; Koprowski, M.; Filipiak, J.; Pospieszyńska, A.; Chorążyczewski, W.; Puchałka, R.; Dąbrowski, H.P. Droughts in the area of Poland in recent centuries in the light of multi-proxy data. *Clim. Past* **2020**, *16*, 627–661. [CrossRef]
34. Pińskwar, I.; Choryński, A.; Kundzewicz, Z.W. Severe Drought in the Spring of 2020 in Poland—More of the Same? *Agronomy* **2020**, *10*, 1646. [CrossRef]
35. Wojtasik, M.; Wiśniewski, P.; Loranc, L. Problems of soil erosion on example some communes of Kujawy-Pomerania and Wielkopolska provinces. *Sci. Rev. Eng. Environ. Sci.* **2008**, *3*, 41–49.
36. Szymura, T.H.; Szymura, M. Spatial structure of grassland patches in Poland: Implications for nature conservation. *Acta Soc. Bot. Pol.* **2019**, *88*, 3615. [CrossRef]

37. Gorelick, N.; Hancher, M.; Dixon, M.; Ilyushchenko, S.; Thau, D.; Moore, R. Google Earth Engine: Planetary-scale geospatial analysis for everyone. *Remote Sens. Environ.* **2017**, *202*, 18–27. [CrossRef]
38. Muñoz-Sabater, J.; Dutra, E.; Agustí-Panareda, A.; Albergel, C.; Arduini, G.; Balsamo, G.; Boussetta, S.; Choulga, M.; Harrigan, S.; Hersbach, H.; et al. ERA5-Land: A state-of-the-art global reanalysis dataset for land applications. *Earth Syst. Sci. Data* **2021**, *13*, 4349–4383. [CrossRef]
39. Nogueira, M.; Boussetta, S.; Balsamo, G.; Albergel, C.; Trigo, I.F.; Johannsen, F.; Miralles, D.G.; Dutra, E. Upgrading land-cover and vegetation seasonality in the ECMWF coupled system: Verification with FLUXNET sites, METEOSAT satellite land surface temperatures, and ERA5 atmospheric reanalysis. *J. Geophys. Res. Atmos.* **2021**, *126*, e2020JD034163. [CrossRef] [PubMed]
40. Hersbach, H.; Bell, B.; Berrisford, P.; Hirahara, S.; Horányi, A.; Muñoz-Sabater, J.; Thépaut, J.N. The ERA5 global reanalysis. *Q. J. R. Meteorol. Soc.* **2020**, *146*, 1999–2049. [CrossRef]
41. Lewak, S.; Kopcewicz, J. Introduction, Polish Scientific Publishers PWN. *Plant Physiol.* **2013**, *2*, 612–667. (In Polish)
42. Selyaninov, G.T. About climate agricultural estimation. *Proc. Agric. Meteorol.* **1928**, *20*, 165–177.
43. Cherenkova, E.; Kononova, N.; Muratova, N. *Summer Drought* 2010 in the European Russia. *Geogr. Environ. Sustain.* **2013**, *6*, 55–66. [CrossRef]
44. Dabrowska-Zielinska, K.; Bochenek, Z.; Malinska, A.; Bartold, M.; Gurdak, R.; Lagiewska, M.; Paradowski, K. Drought Assessment Applying Joined Meteorological and Satellite Data. In Proceedings of the 2021 IEEE International Geoscience and Remote Sensing Symposium IGARSS, Brussels, Belgium, 11–16 July 2021; pp. 6591–6594. [CrossRef]
45. Rybashlykova, L.; Sivceva, S.; Mahovikova, T. Relationship between hydrothermal coefficient (HTC) and productivity of pastures in the arid zone of Northwestern Caspian Sea. *J. Agrometeorol.* **2023**, *25*, 454–457. [CrossRef]
46. Vlăduţ, A.; Nikolova, N.; Licurici, M. Aridity assessment within southern Romania and northern Bulgaria. *Croat. Geogr. Bull.* **2017**, *79*, 5–26. [CrossRef]
47. Rouse, J.W.; Haas, R.H.; Schell, J.A.; Deering, D.W. Monitoring Vegetation Systems in the Great Plains with E.R.T.S. In Proceedings of the 3rd Earth Resources Technology Satellite Symposium, Washington, DC, USA, 10–14 December 1973; pp. 309–317.
48. Hardisky, M.; Klemas, V.; Smart, M. The influence of soil salinity, growth form, and leaf moisture on the spectral radiance of Spartina Alterniflora Canopies. *Photogramm. Eng. Remote Sens.* **1983**, *49*, 77–83.
49. Gao, B.-C. NDWI—A normalized difference water index for remote sensing of vegetation liquid water from space. *Remote Sens. Environ.* **1996**, *58*, 257–266. [CrossRef]
50. Gu, Y.; Brown, J.F.; Verdin, J.P.; Wardlow, B. A five-year analysis of MODIS NDVI and NDWI for grassland drought assessment over the central Great Plains of the United States. *Geophys. Res. Lett.* **2007**, *34*, L06407. [CrossRef]
51. Jones, J.R.; Fleming, C.S.; Pavuluri, K.; Alley, M.M.; Reiter, M.S.; Thomason, W.E. Influence of soil, crop residue, and sensor orientations on NDVI readings. *Precis. Agric.* **2015**, *16*, 690–704. [CrossRef]
52. Dabrowska-Zielinska, K.; Budzynska, M.; Tomaszewska, M.; Bartold, M.; Gatkowska, M. The study of multifrequency microwave satellite images for vegetation biomass and humidity of the area under Ramsar convention. In Proceedings of the 2015 IEEE International Geoscience and Remote Sensing Symposium (IGARSS), Milan, Italy, 26–31 July 2015; Volume 2015, pp. 5198–5200. [CrossRef]
53. Yan, J.; Zhang, G.; Ling, H.; Han, F. Comparison of Time-Integrated NDVI and Annual Maximum NDVI for Assessing Grassland Dynamics. *Ecol. Indic.* **2022**, *136*, 108611. [CrossRef]
54. Dusseux, P.; Guyet, T.; Pattier, P.; Barbier, V.; Nicolas, H. Monitoring of grassland productivity using Sentinel-2 remote sensing data. *Int. J. Appl. Earth Obs. Geoinf.* **2022**, *111*, 102843. [CrossRef]
55. Andreatta, D.; Gianelle, D.; Scotton, M.; Vescovo, L.; Dalponte, M. Detection of Grassland Mowing Frequency Using Time Series of Vegetation Indices from Sentinel-2 Imagery. *GISci. Remote Sens.* **2022**, *59*, 481–500. [CrossRef]
56. Davidson, A.; Wang, S.; Wilmshurst, J. Remote sensing of grassland–shrubland vegetation water content in the shortwave domain. *Int. J. Appl. Earth Obs. Geoinf.* **2006**, *8*, 225–236. [CrossRef]
57. Ding, C.; Liu, X.N.; Huang, F.; Li, Y.; Zou, X.Y. Onset of drying and dormancy in relation to water dynamics of semi-arid grasslands from MODIS NDWI. *Agric. For. Meteorol.* **2017**, *234*, 22–30. [CrossRef]
58. Biswas, T.; Ramsey, R.D.; Bissonette, J.A.; Symanzik, J. Integration of two spectral indices to monitor loss of moist grasslands within the Jaldapara Wildlife Sanctuary, India. *Int. J. Remote Sens.* **2014**, *35*, 1038–1063. [CrossRef]
59. Chang, S.; Wu, B.; Yan, N.; Davdai, B.; Nasanbat, E. Suitability Assessment of Satellite-Derived Drought Indices for Mongolian Grassland. *Remote Sens.* **2017**, *9*, 650. [CrossRef]
60. Svoboda, M.; Fuchs, B. *Handbook of Drought Indicators and Indices*; World Meteorological Organization (WMO): Geneva, Switzerland, 2016.
61. Alahacoon, N.; Edirisinghe, M. A Comprehensive Assessment of Remote Sensing and Traditional Based Drought Monitoring Indices at Global and Regional Scale. *Geomat. Nat. Hazards Risk* **2022**, *13*, 762–799. [CrossRef]
62. Shepherd, J.D.; Schindler, J.; Dymond, J.R. Automated Mosaicking of Sentinel-2 Satellite Imagery. *Remote Sens.* **2020**, *12*, 3680. [CrossRef]
63. Sypniewska, L.; Szyga-Pluta, K. Zmienność czasowa i zróżnicowanie przestrzenne zachmurzenia w Polsce w latach 2001–2016. *Bad. Fizjogr.* **2018**, *IX*, 193–213. (In Polish)
64. Dabrowska-Zielinska, K.; Budzynska, M.; Kowalik, W.; Malek, I.; Gatkowska, M.; Bartold, M.; Turlej, K. Biophysical parameters assessed from microwave and optical data. *Int. J. Electron. Telecom.* **2012**, *58*, 99–104. [CrossRef]

65. Budzynska, M.; Dabrowska-Zielinska, K.; Tomaszewska, M.; Bartold, M.; Gatkowska, M. Studies on carbon exchange in grassland areas. *Grassl. Sci. Pol.* **2015**, *18*, 47–58. (In Polish)
66. Artikanur, S.D.; Widiatmaka; Setiawan, Y.; Marimin. Normalized Difference Drought Index (NDDI) computation for mapping drought severity in Bojonegoro Regency, East Java, Indonesia. *IOP Conf. Ser. Earth Environ. Sci.* **2022**, *1109*, 012027. [CrossRef]
67. Patil, P.; Jagtap, M.; Khatri, N.; Madan, H.; Vadduri, A.; Patodia, T. Exploration and advancement of NDDI leveraging NDVI and NDWI in Indian semi-arid regions: A remote sensing-based study. *Case Stud. Chem. Environ. Eng.* **2024**, *9*, 100573. [CrossRef]
68. Jankowska-Huflejt, H. The need of protecting permanent grasslands as a premise for the development of organic meadow farms. *J. Res. Appl. Agric. Eng.* **2016**, *61*, 186–192.
69. Dabrowska-Zielinska, K.; Budzynska, M.; Gatkowska, M.; Kowalik, W.; Bartold, M.; Kiryla, W. Importance of grasslands monitoring applying optical and radar satellite data in perspective of changing climate. In Proceedings of the 2017 IEEE International Geoscience and Remote Sensing Symposium (IGARSS), Fort Worth, TX, USA, 23 July 2017; pp. 5782–5785. [CrossRef]
70. Medyńska-Gulij, B.; Szoszkiewicz, K.; Cybulski, P. Permanent areas and changes in forests, grasslands, and wetlands in the North European Plain since the eighteenth century—A case study of the Kościan Plain in Poland. *Sci. Rep.* **2024**, *14*, 10305. [CrossRef]
71. Dabrowska-Zielinska, K.; Malinska, A.; Bochenek, Z.; Bartold, M.; Gurdak, R.; Paradowski, K.; Lagiewska, M. Drought Model DISS Based on the Fusion of Satellite and Meteorological Data under Variable Climatic Conditions. *Remote Sens.* **2020**, *12*, 2944. [CrossRef]
72. Iizuka, K.; Kato, T.; Silsigia, S.; Soufiningrum, A.Y.; Kozan, O. Estimating and Examining the Sensitivity of Different Vegetation Indices to Fractions of Vegetation Cover at Different Scaling Grids for Early Stage *Acacia* Plantation Forests Using a Fixed-Wing UAS. *Remote Sens.* **2019**, *11*, 1816. [CrossRef]
73. Luna, D.A.; Pottier, J.; Picon-Cochard, C. Variability and drivers of grassland sensitivity to drought at different timescales using satellite image time series. *Agric. For. Meteorol.* **2023**, *331*, 109325. [CrossRef]
74. Fang, J.; Shi, G.; Wei, S.; Ma, J.; Zhang, X.; Wang, J.; Chen, L.; Liu, Y.; Zhao, X.; Lu, Z. Drought Sensitivity of Spring Wheat Cultivars Shapes Rhizosphere Microbial Community Patterns in Response to Drought. *Plants* **2023**, *12*, 3650. [CrossRef] [PubMed]
75. Wardlow, B.; Anderson, M.; Hain, C.; Crow, W.; Otkin, J.; Tadesse, T.; AghaKouchak, A. *Advancements in Satellite Remote Sensing for Drought Monitoring*; CRC Press: Boca Raton, FL, USA, 2017; ISBN 9781315265551. [CrossRef]

plants

MDPI

Article

The Influence of Climate Change on the Distribution of *Hibiscus mutabilis* in China: MaxEnt Model-Based Prediction

Lu Zhang [1], Beibei Jiang [1], Yu Meng [2], Yin Jia [1], Qian Xu [1] and Yuanzhi Pan [3,*]

[1] College of Landscape Architecture, Sichuan Agricultural University, Chengdu 611130, China; sicauzl@outlook.com (L.Z.)
[2] College of Landscape Architecture and Tourism, Hebei Agricultural University, Baoding 071000, China
[3] College of Forestry, Sichuan Agricultural University, Chengdu 611130, China
* Correspondence: scpyzls@163.com

Abstract: Our study utilized 374 geographical distribution records of *H. mutabilis* and 19 bioclimatic factors, employing the MaxEnt model and the Geographic Information System (ArcGIS). The key environmental variables influencing the suitable distribution areas of *H. mutabilis* were analyzed through the comprehensive contribution rate, permutation importance, and Pearson correlation coefficient. Based on this analysis, the contemporary and future suitable distribution areas and their extents were predicted. The results indicate that the key limiting factor affecting the suitable distribution areas of *H. mutabilis* is the precipitation of the driest month (bio14), with secondary factors being annual precipitation (bio12), annual mean temperature (bio1), and annual temperature range (bio7). Under contemporary climate conditions, the total suitable area for *H. mutabilis* is approximately 2,076,600 km^2, primarily concentrated in the tropical and subtropical regions of southeastern China. Under low-to-medium-emission scenarios (SSP1-2.6, SSP2-4.5), the total suitable area of *H. mutabilis* shows a trend of first decreasing and then increasing compared to the current scenario. In contrast, under high-emission scenarios (SSP5-8.5), it exhibits a trend of first increasing and then decreasing. The spatial pattern changes indicate that the retention rate of suitable areas for *H. mutabilis* ranges from 95.28% to 99.28%, with the distribution centers primarily located in Hunan and Guizhou provinces, showing an overall migration trend towards the west and north. These findings suggest that *H. mutabilis* possesses a certain level of adaptability to climate change. However, it is crucial to consider regional drought and sudden drought events in practical cultivation and introduction processes. The results of our study provide a scientific basis for the rational cultivation management, conservation, and utilization of germplasm resources of *H. mutabilis*.

Keywords: climate change; *Hibiscus mutabilis*; MaxEnt model; potential suitable habitats

1. Introduction

Climate, as a primary controlling factor in ecosystems, has profound impacts on species' geographical distributions and ecological processes [1,2]. Numerous reports indicate that climate change can lead to restricted growth or death of certain plants [3], further causing habitat fragmentation [4] and even species extinction [5]. Against the backdrop of global climate change, research on species' suitable habitats has become increasingly urgent. Establishing species distribution models to simulate and predict potential geographical distributions of species has become a crucial method for understanding species–environment relationships, conserving biodiversity, and rationally utilizing species resources.

The application of species distribution models (SDMs) has become a pivotal tool in the study of ecology and conservation biology [6,7]. Among these models, the MaxEnt model, developed by Phillips, is a spatial distribution model at the geographic scale based on the maximum entropy theory [8]. It is highly favored for its efficiency and accuracy, as it integrates limited species presence data with environmental information to effectively understand species distribution patterns and explore their relationships with environmental variables. The MaxEnt model is now widely used to study the impact of climate change on species distribution [9,10], such as *Osmanthus fragrans* [11], *Cotoneaster multi florus* [12], and *Primula filchnerae* [13]. The findings from these studies provide valuable guidance for the conservation of genetic diversity and the management of germplasm resources [14,15], and have further heightened public concern about climate change [16].

Hibiscus mutabilis (*H. mutabilis*), belonging to the genus *Hibiscus* in the family Malvaceae, is native to China and is now widely distributed in tropical and subtropical regions of Asia, Africa, North America, and South America [17]. As a common ornamental plant, it is extensively planted and utilized for its outstanding landscape aesthetics and ecological functions [18,19]. Additionally, the leaves of *H. mutabilis* possess significant medicinal value, typically offering anti-inflammatory, detoxifying, expectorant, antipyretic, and anesthetic effects [20]. Current research on *H. mutabilis*, both in China and internationally, primarily focuses on its medicinal value [21,22], stress tolerance [23,24], and pest and disease management [25,26]. However, there have been no systematic studies on its potential suitable habitats, influencing factors, or possible responses to future climate change.

Our study focuses on *H. mutabilis*, utilizing the MaxEnt model and Geographic Information System (ArcGIS) technology to simulate and predict its suitable habitats in both contemporary and future periods. By investigating the primary bioclimatic factors influencing its geographical distribution and analyzing the spatial variation patterns of its suitable habitats, this research aims to provide theoretical support for the rational cultivation and management of *H. mutabilis*. Additionally, it seeks to offer scientific evidence for the conservation and utilization of its germplasm resources.

2. Materials and Methods

2.1. Data Collection

2.1.1. Occurrence Data

The distribution data for *H. mutabilis* were primarily sourced from the Global Biodiversity Information Facility (GBIF, http://www.gbif.org, accessed on 20 May 2024), the Chinese Virtual Herbarium (CVH), and the National Specimen Information Infrastructure (NSII). The collected distribution data were processed and filtered as follows: (1) Duplicate records were removed, and for records with specific locations but without coordinate information, latitude and longitude coordinates were obtained using Google Earth. (2) To avoid spatial autocorrelation, the collected distribution data were imported into the ArcGIS system for buffer analysis. For distribution points less than 10 km apart, only one point was randomly retained. Ultimately, we obtained 985 valid distribution records of *H. mutabilis* worldwide and 374 valid distribution records in China (Figures 1 and 2). Data on the distribution of *H. mutabilis* in China (Figure 2) were used for modeling according to the methodology of Du [27] and Zhang [28].

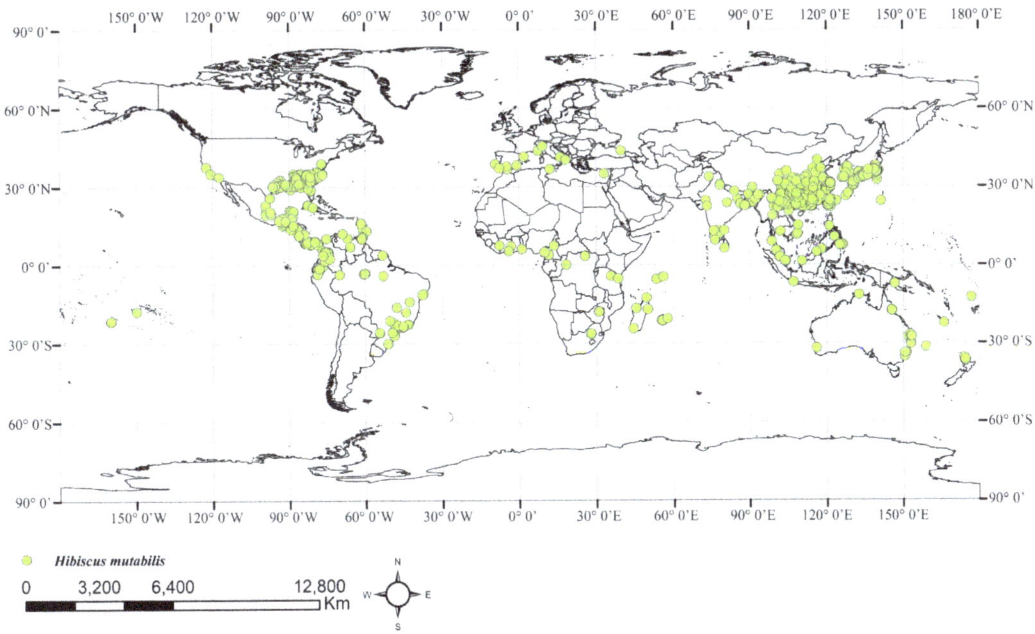

Figure 1. Current distribution of *H. mutabilis* in the world.

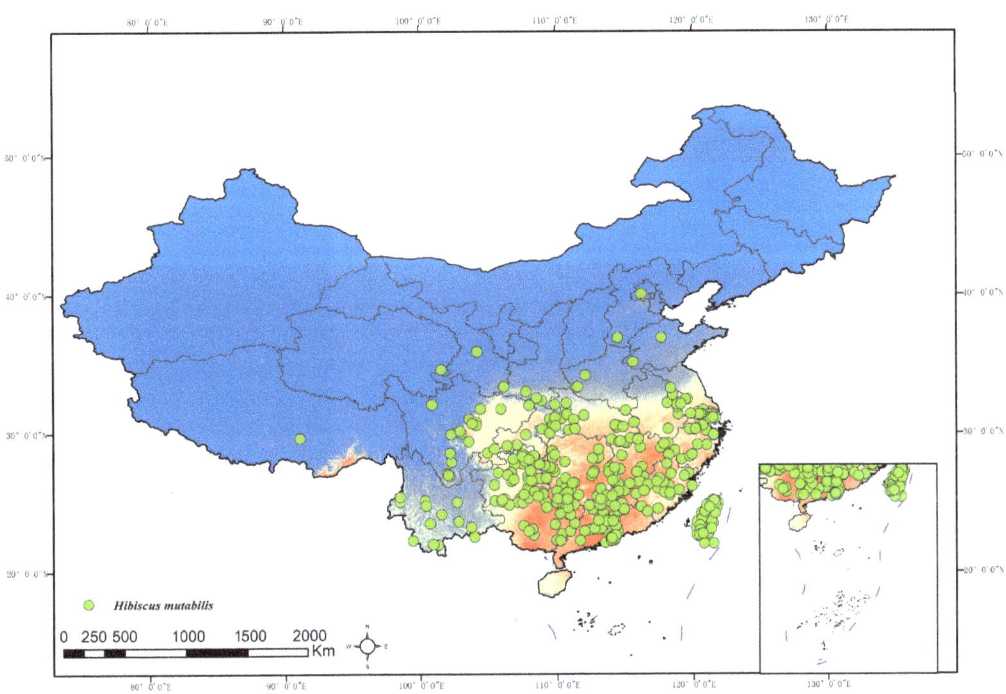

Figure 2. Current distribution of *H. mutabilis* in China.

2.1.2. Predictor Variables

The contemporary and future climatic data used in this study were sourced from the WorldClim database (http://www.worldclim.org, accessed on 20 May 2024). The climate data for the contemporary period (1970–2000) and three future periods (2050s, 2070s, and 2090s) were obtained using the BCC-CSM2-MR climate model. The selected climate scenarios include low forcing (SSP1-2.6), medium forcing (SSP2-4.5), and high forcing (SSP5-8.5) [29].

2.2. Data Processing and Selection

2.2.1. Bioclimatic Variables Screening

To increase the accuracy of the simulation and prediction results, we first conducted a preliminary simulation of the distribution points of *H. mutabilis* using MaxEnt v.3.4.4 combined with 19 bioclimatic factors, then excluded bioclimatic factors with a contribution rate of 0 [30]. Subsequently, we used ArcGIS 10.8 to extract the values of the remaining bioclimatic factors for the valid distribution points of *H. mutabilis*. Pearson correlation analysis was performed on the extracted values using SPSS Statistics 22.0. When the absolute value of the correlation coefficient between two variables $|r|$ was less than 0.8, the correlated variables were retained. When $|r|$ exceeded 0.8, the variables were selected based on their contribution rates to the model and their permutation importance [31]. The final set of bioclimatic factors was then used to study the contemporary and future suitable habitat distribution of *H. mutabilis* in China.

2.2.2. Species Distribution Model Parameter Setting

The potential distribution of *H. mutabilis* was mapped using MaxEnt version 3.4.4. In the model, 75% of the distribution points were used as training data, while the remaining 25% were used as test data, with other settings kept at their default values [32,33]. The MaxEnt model includes a jackknife test to analyze the contribution and importance of environmental variables. It uses the area under the receiver operating characteristic curve (AUC) to estimate model accuracy. The AUC value ranges from 0 to 1, with a model AUC value exceeding 0.8 considered to indicate excellent accuracy [34].

2.2.3. Prediction of Potential Suitable Habitats

The asc format files from the MaxEnt model results were imported into ArcGIS software(10.8 version). Following the classification methods of Li and Deng et al. [35,36], the suitable habitats were divided into four categories: non-suitable ($p \leq 0.2$), low suitability ($0.2 < p \leq 0.4$), medium suitability ($0.4 < p \leq 0.6$), and high suitability ($p > 0.6$). Using ArcGIS, the data were converted to raster format and reclassified. A map of China was used as the base map to visualize the potential distribution of *H. mutabilis* for the present, 2050s, 2070s, and 2090s. The area of potential suitable habitats for each period was calculated using the spatial analysis module in ArcGIS.

2.2.4. Changes in the Area and Shifts in the Distribution Center of Suitable Habitats for *H. mutabilis*

Using the current suitable habitat area as a reference, the MaxEnt model was employed to cross-validate the suitable habitats under past and future scenarios. This approach was used to predict and calculate the changes in suitable habitat area for *H. mutabilis* across different scenarios. Subsequently, by tracking the changes in the centroid of the habitat layers, the distribution centers and migration routes of suitable habitats for *H. mutabilis* from the present to nine future scenarios were determined.

3. Results

3.1. Model Accuracy Evaluation

In the current scenario, the ROC curve results indicate an AUC value of 0.920 (Figure 3), significantly higher than the random prediction value of 0.5. In the output results for the three different climate scenarios across three future periods, the AUC values for both the training and testing datasets are also above 0.9. This demonstrates that the MaxEnt model is stable and reliable, making it suitable for accurately predicting the suitable habitats for *H. mutabilis* in China.

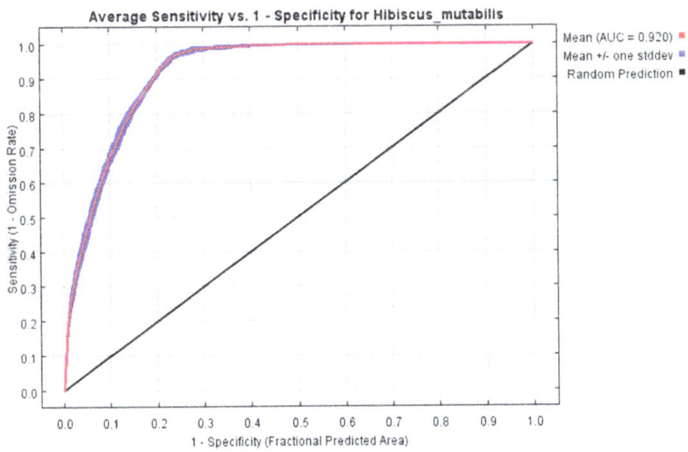

Figure 3. ROC prediction results of MaxEnt model.

3.2. Key Environmental Factors Influencing the Distribution of Suitable Habitats for H. mutabilis

We retained all 19 bioclimatic factors after the pre-simulation results showed that all of them had positive contribution rates. Subsequently, factors with absolute correlation values greater than 0.8 were removed, resulting in 11 climatic factors retained for further analysis. According to the MaxEnt model's contemporary prediction for *H. mutabilis* (Table 1), the precipitation of the driest month (bio14) had the highest contribution rate, followed by annual precipitation (bio12), with these two factors accounting for 82.2% of the total contribution. The contribution rates of mean annual temperature (bio1), temperature annual range (bio7), mean diurnal range (bio2), precipitation of the coldest quarter (bio19), and precipitation of the warmest quarter (bio18) were relatively low, being 3.4%, 2.4%, 2.2%, 2.2%, and 2%, respectively. The lowest contribution rates were observed for precipitation of the wettest quarter (bio8), isothermality (bio3), mean temperature of the warmest quarter (bio10), and precipitation seasonality (bio15), all of which were below 2%. These results indicate that precipitation-related bioclimatic factors accounted for 89.3% of the total contribution, while temperature-related factors accounted for 10.7%. This suggests that precipitation is the key environmental variable influencing the growth of *H. mutabilis* and plays a decisive role in its geographical distribution.

Table 1. Contribution of screened climatic factors to the potential distribution area of *H. mutabilis*.

No.	Bioclimatic Factor Variable	Contribution Rate/%	Permutation Importance
Bio14	Precipitation of the driest month	59.4	6.6
Bio12	Annual precipitation	22.8	13.1
Bio1	Annual mean temperature	3.4	13.1
Bio7	Variation range of annual average temperature	2.4	38.1
Bio2	Mean diurnal range	2.2	2.6
Bio19	Precipitation of the coldest quarter	2.2	0.9
Bio18	Precipitation of the warmest quarter	2	11.1
Bio8	Mean temperature of the wettest quarter	1.8	7.8
Bio3	Isothermality	1.4	2.4
Bio10	Mean temperature of the warmest quarter	1.4	2
Bio15	Coefficient of variation of precipitation	1.2	2.4

The response curves of the main climatic factors (Figure 4) indicate a positive correlation between the probability of *H. mutabilis* presence and precipitation. When the precipitation of the driest month exceeds 19.55 mm, the habitat becomes more suitable for *H. mutabilis* growth, with a presence probability greater than 0.6 [25]. When the precipitation of the driest month ranges from 64.16 to 222.20 mm, the presence probability exceeds 0.9, indicating optimal growth and distribution conditions. Regarding annual precipitation, when it exceeds 1372.96 mm, the presence probability is greater than 0.6, peaking at around 0.989 when annual precipitation reaches 2791.76 mm. After this point, the probability starts to decline, but remains suitable for survival. Additionally, temperature also plays a role in the growth of *H. mutabilis*. The presence probability is high when the mean annual temperature ranges from 16.22 °C to 25.85 °C, peaking at approximately 0.81 at 24.47 °C. When the annual temperature range is between 5 °C and 30.95 °C, the presence probability exceeds 0.6, peaking at around 0.987 when the annual temperature range is 14.79 °C.

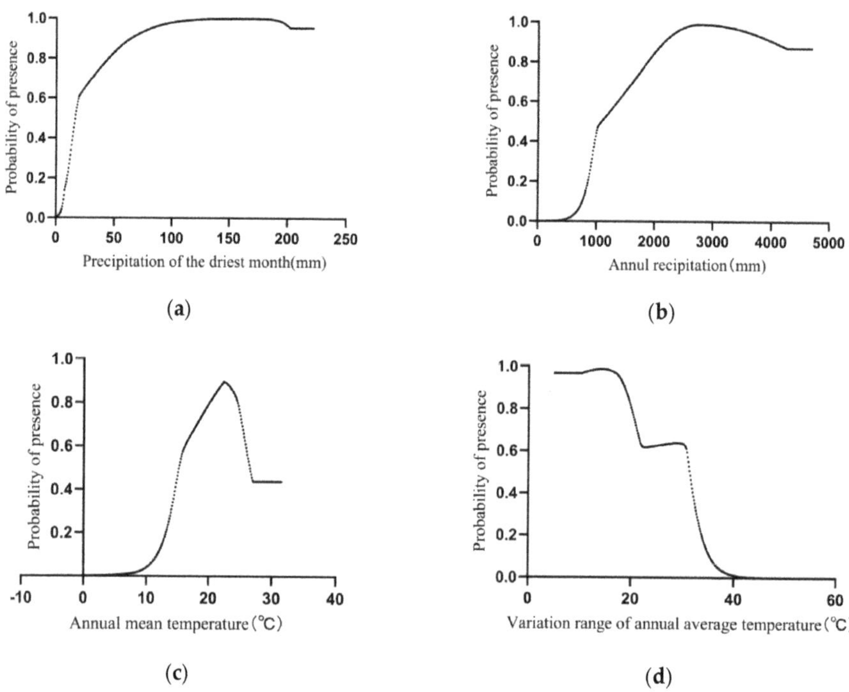

Figure 4. Response curves of main bioclimatic factors. (**a**) the relationship between precipitation of the driest month(X-axis) and the probability of the *H. mutabilis* distribution(Y-axis); (**b**) shows the relationship between annul precipitation(X-axis) and the probability of the *H. mutabilis* distribution(Y-axis); (**c**) shows the relationship between annul mean temperature(X-axis) and the probability of the *H. mutabilis* distribution(Y-axis); (**d**) shows the relationship between variation range of annul average temperature(X-axis) and the probability of the *H. mutabilis* distribution(Y-axis).

3.3. *Predicting the Suitable Habitat Range of* H. mutabilis *under Climate Change*

3.3.1. Prediction of Contemporary Potential Habitats for *H. mutabilis*

The simulation results of the MaxEnt model for the potential suitable habitat of *H. mutabilis* in contemporary China (Figure 5) indicate that the potential moderately to highly suitable habitat is mainly distributed in the southern regions, including Guangdong, Guangxi, Hainan, and Taiwan. In addition, it extends to eastern regions such as Fujian, Zhejiang, Jiangxi, and Shanghai, as well as most parts of Jiangsu and Zhejiang

Provinces. Moderate to highly suitable habitats are also observed in central regions such as Hunan, Hubei, and southern parts of Henan, as well as in southwestern regions including Chongqing, eastern and southern parts of Sichuan, Chengdu Plain, and most parts of Guizhou, except the northwest, which borders Yunnan Province. The total area of potential moderately to highly suitable habitats is approximately 158.72 × 10^4 km^2 (Table 2).

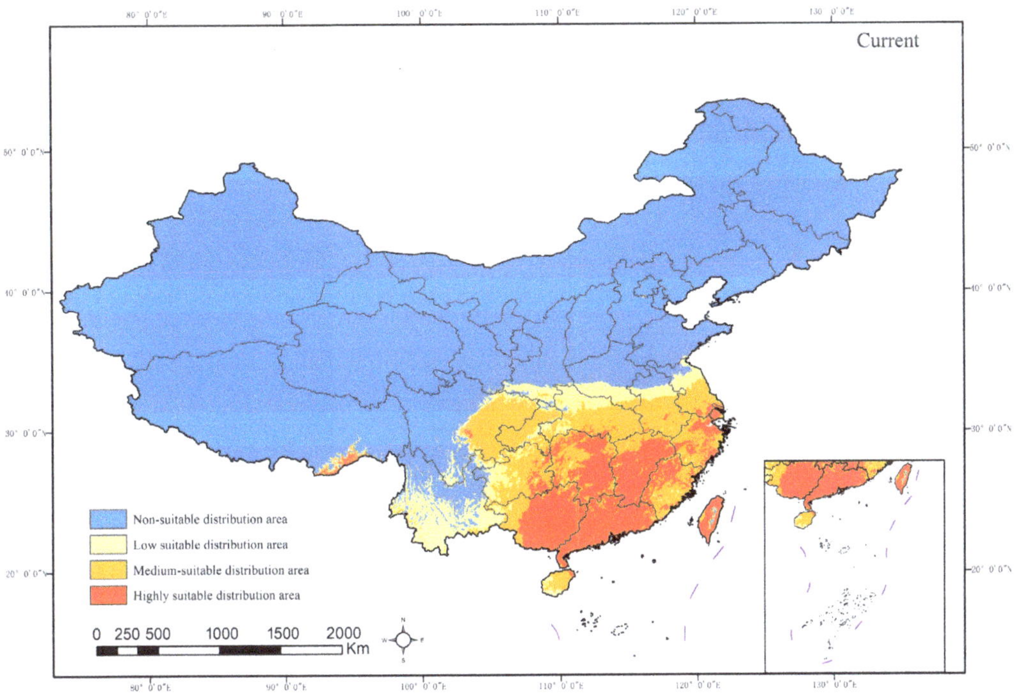

Figure 5. Current potential distribution area of *H. mutabilis*.

Table 2. Predicted area of potential habitats in different periods.

		Climate Scenarios								
Area (10^4 km^2)	**Current**	**2050s**			**2070s**			**2090s**		
		SSP1-2.6	**SSP2-4.5**	**SSP5-8.5**	**SSP1-2.6**	**SSP2-4.5**	**SSP5-8.5**	**SSP1-2.6**	**SSP2-4.5**	**SSP5-8.5**
LSDA	48.94	53.85	58.69	73.28	62.75	59.67	57.13	58.92	53.36	48.42
MSDA	87.02	87.07	78.93	88.27	69.76	78.49	80.27	82.57	76.06	88.61
HSDA	71.70	58.64	64.06	85.99	66.12	67.60	75.25	70.97	76.83	71.59
TSDA	207.66	199.56	201.68	247.55	198.63	205.75	212.65	212.46	206.25	208.63
HSDA-Current		−13.06	−7.64	14.30	−5.58	−4.10	3.55	−0.73	5.13	0.11
HSDA/TSDA (%)	0.35	0.29	0.32	0.35	0.33	0.33	0.35	0.33	0.37	0.34
TSDA-Current		−8.10	−5.98	39.89	−9.03	−1.91	4.98	4.80	−1.41	0.96

Note: LSDA represents low-suitable distribution area; MSDA represents medium-suitable distribution area; HSDA represents highly suitable distribution area; TSDA represents total suitable distribution area; HSDA-Current represents the change in the highly suitable area compared to current times; TSDA-Current represents the change in the total suitable area compared to current times; SDA/TSDA (%) represents the proportion of highly suitable distribution area to total suitable distribution area.

3.3.2. Future Potential Habitat Prediction for *H. mutabilis*

In this study, we employed the MaxEnt model to predict the potential suitable habitats of *H. mutabilis* under three climate scenarios (SSP1-2.6, SSP2-4.5, and SSP5-8.5) for the 2050s, 2070s, and 2090s. Spatial distribution maps of potential suitable habitats (Figure 6) and the changes in habitat area for each category are presented in Table 2.

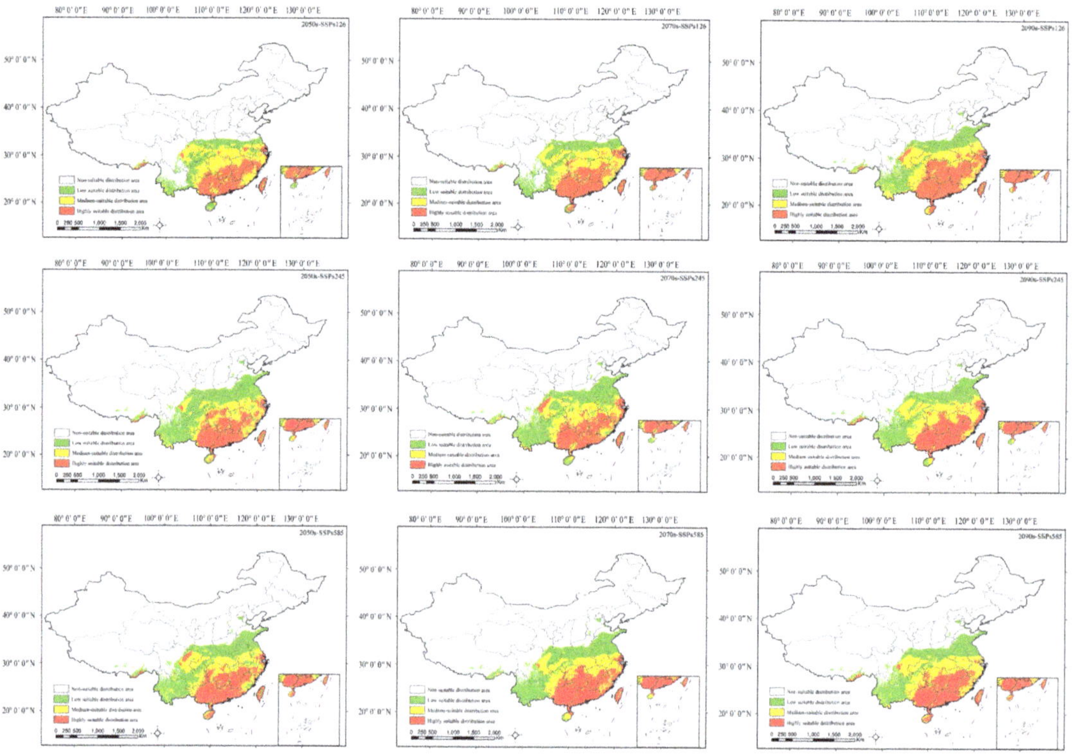

Figure 6. Optimum distribution of *H. mutabilis* in China under future climate scenarios at different periods.

From Figure 6, it is evident that in the forthcoming decades, the moderately to highly suitable habitats of *H. mutabilis* will be primarily concentrated in provinces such as Guangxi, Guangdong, Fujian, Jiangxi, Hunan, and Zhejiang, with a tendency for the highly suitable habitats to shift southward and contract. Under the SSP126 and SSP245 climate scenarios, the area of suitable habitats will initially decrease and then increase. Specifically, by the 2050s and 2070s, the area of highly suitable habitats will decrease by 13.06×10^4 km^2 and 7.64×10^4 km^2 and 7.64×10^4 km^2 and 3.06×10^4 km^2, respectively. By the 2090s, under the SSP245 scenario, the area of highly suitable habitats will increase by 5.13×10^4 km^2 compared to contemporary levels. Conversely, under the SSP5-8.5 scenario, the area of highly suitable habitats will increase by 14.30×10^4 km^2 and 3.55×10^4 km^2 in the 2050s and 2070s, respectively, exhibiting an overall increasing trend followed by a decrease. By the 2090s, the area of highly suitable habitats will decrease below contemporary levels. Across all climate scenarios, the area of moderately suitable habitats exhibits an overall decreasing trend, while the area of low suitable habitats shows a significant increasing trend compared to contemporary levels.

3.4. Spatial Pattern Changes and Distribution Center Migration Trends of H. mutabilis Habitat under Climate Change

3.4.1. Spatial Pattern Changes of *H. mutabilis* Habitat under Different Climate Scenarios

Through the spatial pattern changes of *H. mutabilis* habitat under different future climate scenarios (Figure 7, Table 3), it can be observed that the majority of contemporary suitable habitats for *H. mutabilis* remain preserved, with a retention rate ranging from 95.28% to 99.28%. These preserved habitats are mainly concentrated in regions such as Guangdong, Guangxi, Taiwan, Fujian, Zhejiang, Shanghai, Hunan, Hubei, Chongqing, Guizhou, eastern Sichuan, Chengdu Plain, southern Shaanxi and Henan, and southern Anhui and Jiangsu, among others. Under the SSP1-2.6 and SSP2-4.5 climate scenarios, the loss rates of suitable habitats exceed the increase rates during the 2050s and 2070s. The areas where habitats are lost are primarily concentrated in central Henan, northern Jiangsu, and northern Anhui. However, by the 2090s, under the SSP2-4.5 scenario, the increase rate of suitable habitats will exceed the loss rate, with new habitats mainly distributed in belt-shaped patterns across Jiangsu, Anhui, Henan, and Shaanxi, as well as in scattered patterns in Sichuan and Yunnan. Under the SSP5-8.5 climate scenario, the increase rates of suitable habitats will exceed the loss rates during the 2050s and 2090s. In particular, during the 2050s, the area of newly suitable habitats will increase by 7.46×10^4 km^2, with the highest increase rates mainly concentrated in the northern parts of Jiangsu Province, southern parts of Sichuan Province, and northern parts of Yunnan Province. However, during the 2070s, the loss rate of suitable habitats will exceed the increase rate, with scattered areas of habitat loss in Yunnan Province, northern Jiangsu Province, and northern Anhui Province.

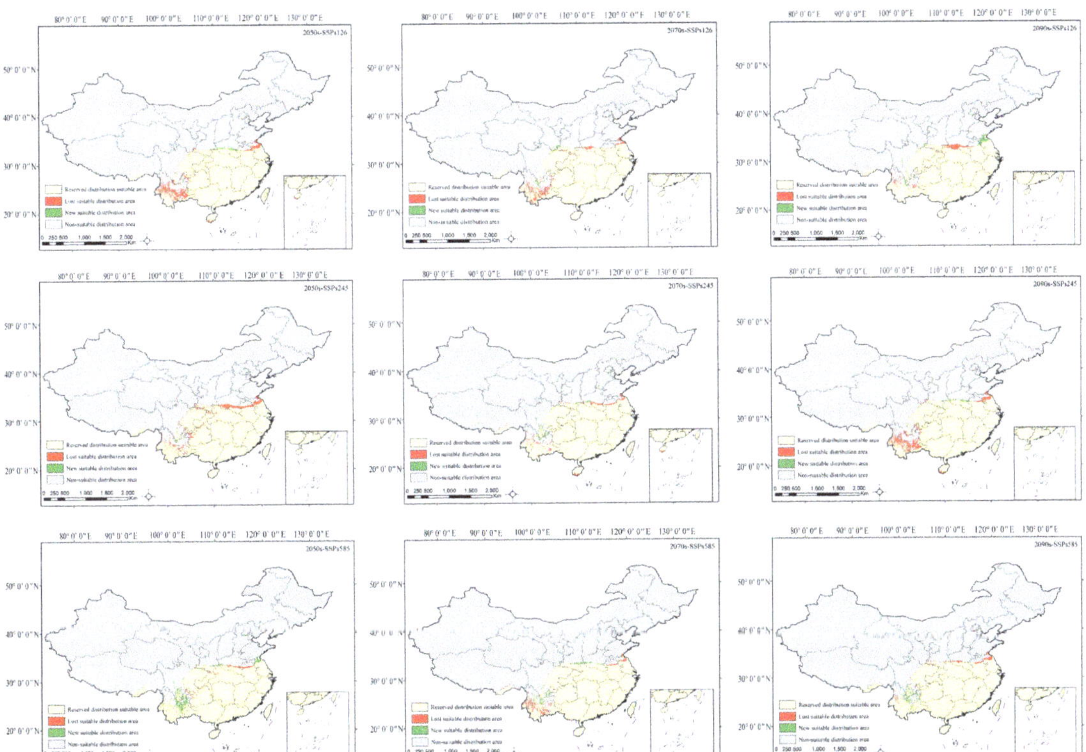

Figure 7. Spatial transformation pattern of the suitable area of *H. mutabilis* in different periods.

Table 3. Spatial transformation pattern of the suitable area of *H. mutabilis* in different periods.

Area (10^4 km^2)	Climate Scenarios								
	2050s			2070s			2090s		
	SSP1-2.6	**SSP2-4.5**	**SSP5-8.5**	**SSP1-2.6**	**SSP2-4.5**	**SSP5-8.5**	**SSP1-2.6**	**SSP2-4.5**	**SSP5-8.5**
Reserved-SDA	197.58	197.85	205.14	199.85	202.49	202.54	202.54	205.92	204.09
Lost-SDA	10.10	9.84	2.57	7.84	5.16	5.13	5.16	1.77	3.60
New-SDA	1.96	0.89	7.46	1.77	3.20	3.69	4.63	6.71	4.52
NSDA	754.20	755.27	748.68	754.38	752.93	752.47	751.52	749.42	751.60
Loss rate (%)	4.86	4.74	1.24	3.77	2.49	2.47	2.48	0.85	1.73
Increasing rate (%)	0.94	0.43	3.59	0.85	1.54	1.78	2.23	3.23	2.18
Retention rate (%)	95.14	95.28	98.79	96.24	97.51	97.53	97.53	99.16	98.28

Note: Reserved-SDA represents reserved suitable distribution area; Lost-SDA represents lost suitable distribution area; New-SDA represents new suitable distribution area; NSDA represents non-suitable distribution area.

3.4.2. Trends in *H. mutabilis* Distribution Center Migration under Different Climate Scenarios

As depicted in Figure 8, under contemporary climatic conditions, the distribution center of *H. mutabilis* is located in Huaihua City, Hunan Province (109°23′53″ E, 27°30′08″ N).

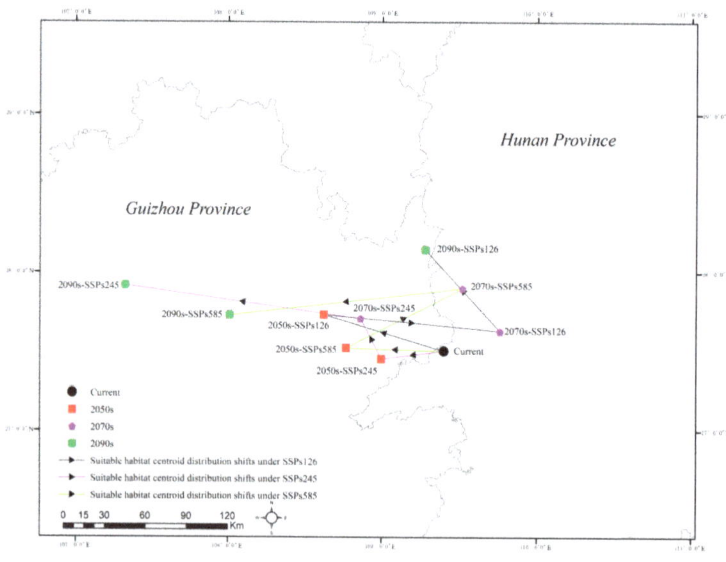

Figure 8. Centroid migration of *H. mutabilis* under future climate scenarios.

Under the SSP1-2.6 climate scenario, in the 2050s, the distribution center will shift to Jiangkou County, Tongren City, Guizhou Province (108°37′15″ E, 27°44′01″ N), moving approximately 80.95 km northwest compared to the contemporary center. By the 2070s, the distribution center will revert to Zhijiang Dong Autonomous County, Huaihua City, Hunan Province (109°45′42″ E, 27°37′40″ N), shifting about 38.81 km northeast compared to the contemporary center. In the 2090s, the distribution center will relocate to Songtao Miao Autonomous County, Tongren City, Guizhou Province (109°45′42″ E, 27°37′40″ N), approximately 72.84 km from the contemporary center.

Under the SSP2-4.5 climate scenario, in the 2050s, the distribution center will be situated in Wanshan District, Tongren City, Guizhou Province (108°59′49″ E, 27°27′10″ N), approximately 40.45 km from the contemporary center. By the 2070s, the distribution center will move to Jiangkou County, Tongren City, Guizhou Province (108°51′44″ E, 27°42′33″ N), about 57.10 km northwest of the contemporary center. In the 2090s, the distribution center will shift to Suiyang County, Zunyi City, Guizhou Province (107°19′23″ E, 27°55′10″ N), moving approximately 210.92 km northwest from the contemporary center.

Under the SSP5-8.5 climate scenario, in the 2050s, the distribution center will be located in Jiangkou County, Tongren City, Guizhou Province (108°46′05″E , 27°31′18″ N), approximately 63.36 km west of the contemporary center. By the 2070s, the distribution center will move to Fenghuang County, Xiangxi Tujia and Miao Autonomous Prefecture, Hunan Province (109°31′03″ E, 27°54′01″ N), shifting about 84.93 km northeast compared to the contemporary center. In the 2090s, the distribution center will relocate to Sinan County, Tongren City, Guizhou Province (108°00′37″ E, 27°43′47″ N), moving approximately 150.25 km northeast from the contemporary center. Overall, under different climate scenarios, the distribution center of *H. mutabilis* exhibits a tendency towards westward migration in the 2050s, followed by a general northward trend in the 2070s and 2090s.

4. Discussion

4.1. Key Climatic Factors Restricting the Distribution of H. mutabilis

This study reveals that the precipitation of the driest month is the most crucial environmental variable influencing the distribution of *H. mutabilis*, with the other three variables being annual precipitation, mean annual temperature, and temperature annual range, which are also significant factors affecting its distribution. Within a certain range, the probability of *H. mutabilis* presence is positively correlated with the driest month's precipitation. When the driest month's precipitation is less than 9.373 mm, it is not suitable for the survival of *H. mutabilis*. However, when the driest month's precipitation exceeds 64.16 mm, the probability of *H. mutabilis* presence is extremely high. Additionally, when the annual precipitation ranges from 1372.96 mm to 2791.76 mm, the probability of *H. mutabilis* presence is directly proportional to the annual precipitation. When the annual precipitation is less than 852.39 mm, it is not suitable for the survival and distribution of *H. mutabilis*. Conversely, when the annual precipitation exceeds 2791.76 mm, although the probability of *H. mutabilis* presence slightly decreases, it still remains highly suitable for its distribution. These findings indicate that the growth and distribution of *H. mutabilis* are influenced by precipitation, with drought being a limiting factor for its distribution range. Previous studies have suggested that *H. mutabilis* is intolerant to drought and tolerant to waterlogging conditions [37], exhibiting the development of adventitious roots adapting to flooded environments [24,38] without affecting flowering or ornamental characteristics. However, drought adversely affects seed germination efficiency and plant vigor [39,40], further highlighting its high water demand, which is consistent with the results of this study.

In addition to precipitation, the mean annual temperature and temperature annual range are also important climatic factors influencing the distribution of *H. mutabilis*. The MaxEnt model prediction results indicate that a mean annual temperatures exceeding 16.2 °C is highly suitable for the survival and distribution of *H. mutabilis*, with the probability of its presence peaking at 24.5 °C. Beyond this threshold, the probability of its presence noticeably declines. Conversely, mean annual temperatures ranging from −5.8 °C to 13.04 °C are not suitable for the survival and distribution of *H. mutabilis*. Studies have shown that the greatest challenge faced by *H. mutabilis* when introduced in northern regions is the inability to safely overwinter [41]. The results of this study suggest that mean annual temperature may be a key factor limiting the introduction of *H. mutabilis* to northern regions of China.

Temperature annual range also significantly influences the survival and distribution of plants. Combining response curves of climatic factors, it is observed that the probability of *H. mutabilis* presence is inversely proportional to temperature annual range. As temperature annual range increases, the probability of *H. mutabilis* presence markedly decreases. When the temperature annual range reaches 33.87 °C, the probability of *H. mutabilis* presence is less than 0.2, indicating that environments with relatively stable temperatures are more suitable for its growth.

4.2. Spatial Pattern Changes of *H. mutabilis* Habitat and Distribution Center Migration

The sixth phase of the Coupled Model Intercomparison Project (CMIP6), organized by the World Climate Research Programme (WCRP), represents the highest level of climate model simulation and prediction. With numerous participating models, scientifically designed experimental methods, and vast simulated data, CMIP6 provides more reliable estimates. Our study, based on the CMIP6 framework, effectively reveals the spatial pattern changes of *H. mutabilis* habitats and the migration of its distribution center under three new Shared Socioeconomic Pathway (SSP) scenarios. Under the SSP1-2.6 and SSP2-4.5 scenarios, the habitat area decreases in the 2050s and 2070s, but increases in the 2090s. Previous studies indicate that with the increasingly evident trend of global climate warming, future temperatures and precipitation are expected to increase compared to contemporary conditions [42]. However, the actual conditions may be more complex [43–45]. For instance, in the SSP2-4.5 climate scenario, Wen et al. [46] pointed out that, in the mid-to-late 21st century (2050s and 2070s) in the Yangtze River Basin, temperatures will increase while precipitation decreases, but by the late 21st century (2090s), precipitation will increase by 6% and 4.7%, respectively. According to the relationship between climate factors and the probability of *H. mutabilis* presence, it can be inferred that temperature and precipitation will synergistically affect and alter the spatial distribution pattern of *H. mutabilis,* with precipitation playing a more critical role. In future periods, extreme precipitation events in southeastern China are expected to increase, with the probability under SSP1-2.6 being lower than that under SSP5-8.5 [47]. Data from relevant research institutions indicate that under the SSP1-2.6, SSP2-4.5, and SSP5-8.5 scenarios, future temperatures will rise by 2 °C, 3 °C, and 5°C, respectively. Higher temperatures and increased precipitation resulting from high emission levels will, to some extent, favor the distribution of *H. mutabilis*. However, under future climate scenarios, extreme precipitation events are expected to be more complex, with droughts becoming more frequent. Therefore, attention should be paid to regional and sudden drought events during the introduction, domestication, and cultivation management of *H. mutabilis* [43].

Research suggests that species in the northern hemisphere may increasingly migrate to higher latitudes with the trend of global warming intensification [48]. According to the predictive results of this study, the contemporary potential distribution center of *H. mutabilis* is located in Zhijiang Dong Autonomous County, Hunan Province (109°23′53″ E, 27°30′08″ N). This region belongs to a subtropical monsoon humid climate zone characterized by mild climate, concentrated rainfall, and an annual average temperature ranging from 15.8 °C to 17.3 °C, with annual precipitation between 1156.4 mm and 1432.9 mm, making it highly suitable for the survival and distribution of *H. mutabilis*.

Furthermore, under different climate scenarios, the future distribution center of *H. mutabilis* shows a trend of westward and northward migration, mainly appearing in Tongren City and Zunyi City, Guizhou Province. Although this region is significantly affected by monsoons, currently, the summer temperatures in Guizhou are lower than those in Hunan, with more frequent dry spells during the summer monsoon season [49]. Additionally, the annual precipitation is lower, making the suitability conditions inferior to those of Hunan Province. However, forecasts indicate that temperatures in Guizhou Province are generally expected to increase compared to historical levels, with a tendency for increased precipitation in the future, providing favorable conditions for the migration and survival of *H. mutabilis* [50].

4.3. Conservation Strategies and Recommendations

The study on species geographical distribution responses to climate change holds significant value in providing scientific data support for species conservation [51,52]. Cindy pointed out that the shared suitable distribution areas of species in different periods often become "climate refuges" for the species due to stable climates [53]. Our study reveals that the highly suitable areas for *H. mutabilis* in both contemporary and future scenarios are located in Guangdong, Guangxi, Taiwan, Zhejiang, Shanghai, Anhui, Hunan, Hubei, Chongqing, Guizhou, eastern Sichuan, and the Chengdu Plain. Establishing conservation areas or germplasm repositories for *H. mutabilis* in these regions could effectively protect its genetic diversity. The distribution centers of *H. mutabilis* in contemporary and future climate scenarios are mainly located in Hunan and Guizhou provinces, making them key areas for *H. mutabilis* breeding. Additionally, the data indicate no records of *H. mutabilis* distribution in Hainan Province currently. However, predictive results suggest that most areas in Hainan Province will be highly suitable habitats for *H. mutabilis* under current and future climate scenarios. With an annual precipitation of 1500–2500 mm and an average annual temperature of 22.5–25.6 °C, Hainan Province possesses suitable hydrothermal conditions for the survival and distribution of *H. mutabilis*, thus warranting efforts for its introduction and cultivation.

5. Conclusions

Using ArcGIS software and the MaxEnt model, this study predicted the potential suitable habitats of *H. mutabilis* under nine different contemporary and future climate scenarios. The results indicate that the key limiting factor affecting the survival and distribution of *H. mutabilis* is the precipitation of the driest month, with precipitation playing a dominant role relative to temperature factors. From the perspective of potential distribution areas, *H. mutabilis* is suitable for growth and distribution in tropical and subtropical humid regions. In scenarios with low emissions (SSP1-2.6, SSP2-4.5), the area of highly suitable habitats decreases compared to contemporary levels, while in scenarios with high emissions (SSP5-8.5), the area of highly suitable habitats increases. This phenomenon may be attributed to the influence of changes in precipitation on future climate factors. The contemporary distribution center of *H. mutabilis* is located in Huaihua City, Hunan Province. Under different future climate scenarios, the distribution center mainly shifts to Tongren City and Zunyi City in Guizhou Province and Xiangxi Autonomous Prefecture in Hunan Province. In the 2050s, there will be a westward trend, while in the 2070s and 2090s, there will be a general trend of northward migration. Hainan Province has climatic conditions suitable for the growth and distribution of *H. mutabilis*, making it suitable for its introduction and cultivation. The results of this study provide guidance for the introduction, cultivation, genetic diversity conservation, and breeding management of *H. mutabilis*. However, factors other than bioclimatic factors, such as altitude and soil type, may also affect its survival and distribution. Future research will incorporate more factors into *H. mutabilis* habitat prediction to further provide a theoretical basis for the protection and scientific planting management of *H. mutabilis* germplasm resources.

Author Contributions: Conceptualization, L.Z.; Methodology, Q.X.; Software, L.Z. and Y.M.; Validation, L.Z.; Formal analysis, L.Z.; Investigation, Q.X.; Data curation, L.Z., Y.M. and Q.X.; Writing—original draft, L.Z.; Writing—review & editing, L.Z. and Y.P.; Visualization, L.Z.; Supervision, B.J., Y.J. and Y.P.; Project administration, Y.P.; Funding acquisition, B.J., Y.J. and Y.P. All authors have read and agreed to the published version of the manuscript.

Funding: This research was funded by China Scholarship Council, grant number 202206910043.

Data Availability Statement: Data are contained within the article.

Conflicts of Interest: The authors declare no conflict of interest.

References

1. McLaughlin, B.C.; Ackerly, D.D.; Klos, P.Z.; Natali, J.; Dawson, T.E.; Thompson, S.E. Hydrologic refugia, plants, and climate change. *Glob. Chang. Biol.* **2017**, *23*, 2941–2961. [CrossRef]
2. Erwin, D.H. Climate as a Driver of Evolutionary Change. *Curr. Biol.* **2009**, *19*, 575–583. [CrossRef] [PubMed]
3. Zandalinas, S.I.; Fritschi, F.B.; Mittler, R. Global Warming, Climate Change, and Environmental Pollution: Recipe for a Multifactorial Stress Combination Disaster. *Trends Plant Sci.* **2021**, *26*, 588–599. [CrossRef]

4. Leimu, R.; Vergeer, P.; Angeloni, F.; Ouborg, N.J. Habitat fragmentation, climate change, and inbreeding in plants. *Ann. N. Y. Acad. Sci.* **2010**, *1195*, 84–98. [CrossRef] [PubMed]
5. Parmesan, C.; Hanley, M.E. Plants and climate change: Complexities and surprises. *Ann. Bot.* **2015**, *116*, 849–864. [CrossRef]
6. Fang, Y.; Zhang, X.; Wei, H.; Wang, D.; Chen, R.; Wang, L.; Gu, W. Predicting the invasive trend of exotic plants in China based on the ensemble model under climate change: A case for three invasive plants of Asteraceae. *Sci. Total Environ.* **2021**, *756*, 143841. [CrossRef] [PubMed]
7. Wiens, J.J.; Zelinka, J. How many species will Earth lose to climate change? *Glob. Chang. Biol.* **2024**, *30*, e17125. [CrossRef]
8. Phillips, S.J.; Ik, M.D.; Schapire, R.E. A Maximum Entropy Approach to Species Distribution Modeling. In Proceedings of the Twenty-First International Conference on Machine Learning, Banff, AB, Canada, 4–8 July 2004; Association for Computing Machinery: New York, NY, USA, 2004.
9. Liu, H.; Feng, C.L.; Luo, Y.B.; Chen, B.S.; Wang, Z.S.; Gu, H.Y. Potential Challenges of Climate Change to Orchid Conservation in a Wild *Orchid* Hotspot in Southwestern China. *Bot. Rev.* **2010**, *76*, 174–192. [CrossRef]
10. Zhao, X.; Lei, M.; Wei, C.; Guo, X. Assessing the suitable regions and the key factors for three Cd-accumulating plants (*Sedum alfredii, Phytolacca americana,* and *Hylotelephium spectabile*) in China using MaxEnt model. *Sci. Total Environ.* **2022**, *852*, 158202. [CrossRef]
11. Kong, F.; Tang, L.; He, H.; Yang, F.; Tao, J.; Wang, W. Assessing the impact of climate change on the distribution of *Osmanthus fragrans* using Maxent. *Environ. Sci. Pollut. Res.* **2021**, *28*, 34655–34663. [CrossRef]
12. Huang, Q.; Liu, H.; Li, C.; Zhu, X.; Yuan, Z.; Lai, J.; Cao, M.; Huang, Z.; Yang, Y.; Zhuo, S.; et al. Predicting the geographical distribution and niche characteristics of *Cotoneaster multiflorus* based on future climate change. *Front. Plant Sci.* **2024**, *15*, 1360190. [CrossRef] [PubMed]
13. Jiang, X.; Liu, W.J.; Zhu, Y.Z.; Cao, Y.T.; Yang, X.M.; Geng, Y.; Zhang, F.J.; Sun, R.Q.; Jia, R.W.; Yan, C.L.; et al. Impacts of Climate Changes on Geographic Distribution of *Primula filchnerae*, an Endangered Herb in China. *Plants* **2023**, *12*, 3561. [CrossRef] [PubMed]
14. Wang, C.; Sheng, Q.; Zhao, R.; Zhu, Z. Differences in the Suitable Distribution Area between Northern and Southern China Landscape Plants. *Plants* **2023**, *12*, 2710. [CrossRef] [PubMed]
15. Yan, X.; Wang, S.; Duan, Y.; Han, J.; Huang, D.; Zhou, J. Current and future distribution of the deciduous shrub *Hydrangea macrophylla* in China estimated by MaxEnt. *Ecol. Evol.* **2021**, *11*, 16099–16112. [CrossRef] [PubMed]
16. Guan, L.; Yang, Y.; Jiang, P.; Mou, Q.; Gou, Y.; Zhu, X.; Xu, Y.W.; Wang, R. Potential distribution of *Blumea balsamifera* in China using MaxEnt and the ex situ conservation based on its effective components and fresh leaf yield. *Environ. Sci. Pollut. Res.* **2022**, *29*, 44003–44019. [CrossRef] [PubMed]
17. Zhang, L.; Zhang, M.Y.; Zeng, X.M.; Jia, Y.; Jiang, B.B.; Zhou, J.; Ma, J.; Xu, Q.; Shi, X.Q.; Wang, K.L.; et al. Pollen Morphology of 19 Cultivars of *Hibiscus mutabilis* in Chengdu and Its Taxonomic Significance. *J. Trop. Subtrop. Bot.* **2021**, *29*, 421–429.
18. Ding, N.; Zhong, Y.; Li, J.; Xiao, Q.; Zhang, S.; Xia, H. Visual preference of plant features in different living environments using eye tracking and EEG. *PLoS ONE* **2022**, *17*, e279596. [CrossRef] [PubMed]
19. Shang, K.; Hu, Y.H.; Vincent, G.; Labrecque, M. Biomass and phytoextraction potential of three ornamental shrub species tested over three years on a large-scale experimental site in Shanghai, China. *Int. J. Phytoremediat.* **2020**, *22*, 10–19. [CrossRef] [PubMed]
20. Liang, X.; Ma, M.; Su, W. Fingerprint analysis of *H. mutabilis* L. leaves based on ultra performance liquid chromatography with photodiode array detector combined with similarity analysis and hierarchical clustering analysis methods. *Pharmacogn. Mag.* **2013**, *9*, 238–243.
21. Feng, L.P.; Huang, L.L.; Ye, L.; Hu, P.; Xiong, J.; Yan, X.; Xia, H.L. Progress in chemical composition and pharmacological effects of medicinal plants of *Hibiscus* spp. *J. Chin. Med. Mater.* **2023**, *46*, 1812–1822.
22. Iwaoka, E.; Oku, H.; Takahashi, Y.; Ishiguro, K. Allergy-Preventive Effects of *H. mutabilis* 'Versicolor' and a Novel Allergy-Preventive Flavonoid Glycoside. *Biol. Pharm. Bull.* **2009**, *32*, 509–512. [CrossRef] [PubMed]
23. Li, X.; Chen, X.; Li, M.X.; Chen, W.K. Tolerance and enrichment characteristics of *Hibiscus mutabilis* L. in cadmium polluted soil. *Anhui J. Agric.* **2023**, *29*, 104–108.
24. Xu, Q.; Wu, M.X.; Zeng, X.M.; Jiang, B.B.; Jia, Y.; Li, M.X.; Zhang, M.Y.; Zhang, L.; Cai, L.; Huang, W.P.; et al. Response of *Hibiscus mutabilis* to Waterlogging Stress and Comprehensive Evaluation of Waterlogging Tolerance. *J. Trop. Subtrop. Bot.* **2022**, *30*, 202–212.
25. Liu, Y.; Liu, J.; Liu, W.; Xiao, L.; Zhang, K. First Report of *Alternaria alternata* Causing Leaf Spot on *H. mutabilis* in China. *Plant Dis.* **2023**, *107*, 1631. [CrossRef] [PubMed]
26. Han, S.; Yu, S.; Zhu, T.; Li, S.; Qiao, T.; Liu, Y.; Lin, T.; Yang, C. *Nigrospora oryzae* Causing Black Leaf Spot Disease of *H. mutabilis* in China. *Plant Dis.* **2021**, *105*, 2255. [CrossRef] [PubMed]
27. Du, Q.; Wei, C.H.; Liang, C.T.; Yu, J.H.; Wang, H.M.; Wang, W.J. Future climatic adaption of 12 dominant tree species in northeast china under 3 climatic scenarios by using MaxEnt modeling. *Acta Ecol. Sin.* **2022**, *42*, 9712–9725.
28. Zhang, K.; Yao, L.; Meng, J.; Tao, J. Maxent modeling for predicting the potential geographical distribution of two peony species under climate change. *Sci. Total Environ.* **2018**, *634*, 1326–1334. [CrossRef] [PubMed]
29. Xu, J.Y.; Liu, T.; Zang, G.Z.; Zheng, Y.Q. Prediction of suitable areas of *Eremochloa ophiuroides* in China under different climate scenarios based on MaxEnt model. *J. Beijing For. Univ.* **2024**, *46*, 91–102.
30. Wu, Q.; Dong, S.B.; Yang, L.; Qi, X.J.; Zhang, Y.; Yang, M.Q.; Ren, Z.H.; Liu, Q.H.; Cheng, J. Prediction of potential distribution of *Cypripedium macranthos* under climate change scenarios in China. *Acta Ecol. Sin.* **2024**, *44*, 209–223.

31. Jiang, Y.Q.; Wang, C.; Zhang, X.M.; Su, D.T.; Zhang, Z.G.; Xu, J. Effect of climate change on the potential habitat of *Hemerocallis fulva*. *J. Northeast. For.* **2024**, *52*, 59–65.
32. Phillips, S.J.; Anderson, R.P.; Schapire, R.E. Maximum entropy modeling of species geographic distributions. *Ecol. Model.* **2006**, *190*, 231–259. [CrossRef]
33. Wang, J.; Shi, T.; Wang, H.; Li, M.; Zhang, X.; Huang, L. Estimating the Amount of the Wild Artemisia annua in China Based on the MaxEnt Model and Spatio-Temporal Kriging Interpolation. *Plants* **2024**, *13*, 1050. [CrossRef] [PubMed]
34. Jiménez-Valverde, A.; Lobo, J.M. Threshold criteria for conversion of probability of species presence to either–or presence–absence. *Acta Oecol.* **2007**, *31*, 361–369. [CrossRef]
35. Li, J.; Fan, G.; He, Y. Predicting the current and future distribution of three Coptis herbs in China under climate change conditions, using the MaxEnt model and chemical analysis. *Sci. Total Environ.* **2020**, *698*, 134141. [CrossRef] [PubMed]
36. Deng, C.; Zhong, Q.; Shao, D.; Ren, Y.; Li, Q.; Wen, J.; Li, J. Potential Suitable Habitats of Chili Pepper in China under Climate Change. *Plants* **2024**, *13*, 1027. [CrossRef] [PubMed]
37. Wang, Y.Y.; Yue, C.L.; Gu, Y.X.; Li, H.P.; Wang, J.; Fang, Y.Y. Traits of bacterial communities in the rhizospheres of different woody plants in simulated vertical flow constructed wetland. *Ecol. Sci.* **2023**, *42*, 116–126.
38. Tong, L.L.; Wang, Z.Y.; Xu, X.G. Effects of flooding stress on morphological and physiological property of *Hibiscus mutabilis*. *J. Beihua Univ. (Nat. Sci.)* **2020**, *21*, 253–256.
39. Yang, Y.Z.; Zeng, X.M.; Ma, J.; Shi, X.Q.; Zhu, Z.S.; Li, F.W. Effects of Different Treatments on Seed Germination Characters of the *Hibiscus mutabilis* L. Cultivar 'Danbanhong'. *Hunan Agric. Sci.* **2020**, *8*, 16–20.
40. Liu, X.D.; Sun, Y.R.; Tang, Y.T.; Wang, M.; Xiao, B. Adaptability of Seven Woody Plants under Water Stress. *Chin. J. Trop. Agric.* **2022**, *42*, 56–63.
41. Zhang, F.; Xue, C.R.; Zhao, B. Domestication of *Hibiscus mutabilis* and Selection and Breeding of Cold Resistant Species. *Shandong For. Technol.* **2016**, *46*, 29–36.
42. Jiang, Z.H.; Zhang, X.; Wang, J. IPCC-AR4 Model Scenario Projections of 21st Century Climate Change in China Scenario Projections of Climate Change in China in the 21st Century by IPCC-AR4 Model. *Geogr. Res.* **2008**, *4*, 787–799.
43. Black, E. Global Change in Agricultural Flash Drought over the 21st Century. *Adv. Atmos. Sci.* **2024**, *41*, 209–220. [CrossRef]
44. Lu, D.Y.; Zhu, X.F.; Tang, M.X.; Guo, C.H.; Liu, T.T. Assessment of drought risk changes in China under different temperature rise scenarios. *Arid. Land Geogr.* **2024**, *47*, 369–379.
45. Guo, C.H.; Zhu, X.F.; Zhang, S.Z.; Tang, M.X.; Xu, K. Hazard Changes Assessment of Future High Temperature in China based on CMIP6. *J. Geo-Inf. Sci.* **2022**, *24*, 1391–1405.
46. Wen, S.S.; Wang, Z.C.; Zhai, J.Q.; Jiang, F.S.; Zhou, B.; N-CMIP6-Based Future Prediction of Temperature and Precipitation in the Yangtze River Basin. Yangtze River. 2024. Available online: https://link.cnki.net/urlid/42.1202.TV.20240412.1350.002 (accessed on 15 April 2024).
47. Yang, S.; Xu, L.L. Extreme Precipitation in the South China Sea and Surrounding Areas: Observation and Projection. *Chin. J. Atmos. Sci.* **2024**, *48*, 333–346.
48. Hof, C.; Levinsky, I.; Araujo, M.B.; Rahbek, C. Rethinking species' ability to cope with rapid climate change. *Glob. Chang. Biol.* **2011**, *17*, 2987–2990. [CrossRef]
49. Song, F.; Hu, Y.J.; Song, W.D.; Song, L.; Ao, Q.; Deng, S.Q. Climatic suitability and zoning of pepper varieties Jiuyeqing and Dahongpao in Suiyang County. *Agric. Technol. Serv.* **2020**, *37*, 101–103.
50. Li, X.N.; He, L.H.; Xu, L.Q. Spatial and temporal variability and prediction of extreme weather events and vegetation cover in Guizhou Province under climate change. *Pearl River*. 2024. Available online: https://link.cnki.net/urlid/44.1037.tv.20240507.1908.002 (accessed on 8 May 2024).
51. He, X.; Burgess, K.S.; Gao, L.M.; Li, D.Z. Distributional responses to climate change for alpine species of *Cyananthus* and *Primula* endemic to the Himalaya-Hengduan Mountains. *Plant Divers.* **2019**, *41*, 26–32. [CrossRef] [PubMed]
52. Teng, J.; Li, H.; Lu, S.F.; Yin, X.J.; Li, G.; Chen, Z.; Wang, Y. Responses of cold-temperature coniferous forest to climate change in southwestern China. *J. Northwest For. Univ.* **2023**, *38*, 33–44.
53. Tang, C.Q.; Matsui, T.; Ohashi, H.; Dong, Y.F.; Momohara, A.; Herrando-Moraira, S.; Qian, S.; Yang, Y.; Ohsawa, M.; Luu, H.T.; et al. Identifying long-term stable refugia for relict plant species in East Asia. *Nat. Commun.* **2018**, *9*, 4488. [CrossRef]

plants

MDPI

Article

The Spatial Shifts and Vulnerability Assessment of Ecological Niches under Climate Change Scenarios for *Betula luminifera*, a Fast-Growing Precious Tree in China

Xian-Ge Hu [1,†], Jiahui Chen [1,†], Qiaoyun Chen [1,†], Ying Yang [1], Yiheng Lin [1], Zilun Jin [1], Luqiong Sha [1], Erpei Lin [1], El-Kassaby Yousry [2] and Huahong Huang [1,*]

1 The State Key Laboratory of Subtropical Silviculture, Institute of Biotechnology, College of Forestry and Biotechnology, Zhejiang International Science and Technology Cooperation Base for Plant Germplasm Resources Conservation and Utilization, Zhejiang A&F University, Hangzhou 311300, China; xiangehu@zafu.edu.cn (X.-G.H.); chenn@stu.zafu.edu.cn (J.C.); chenqy6926@stu.zafu.edu.cn (Q.C.); yingy@stu.zafu.edu.cn (Y.Y.); yihenglin@stu.zafu.edu.cn (Y.L.); jin.zilun@foxmail.com (Z.J.); slq@stu.zafu.edu.cn (L.S.); zjulep@hotmail.com (E.L.)

2 Department of Forest and Conservation Sciences, Faculty of Forestry, The University of British Columbia, 2424 Main Mall, Vancouver, BC V6T 1Z4, Canada; y.el-kassaby@ubc.ca

* Correspondence: huanghh@zafu.edu.cn; Tel.: +86-18368155272

† These authors contributed equally to this work.

Abstract: The spatial shifts and vulnerability assessments of ecological niches for trees will offer fresh perspectives for sustainable development and preservation of forests, particularly within the framework of rapid climate change. *Betula luminifera* is a fast-growing native timber plantation species in China, but the natural resources have been severely damaged. Here, a comprehensive habitat suitability model (including ten niche-based GIS modeling algorithms) was developed that integrates three types of environmental factors, namely, climatic, soil, and ultraviolet variables, to assess the species contemporary and future distribution of suitable habitats across China. Our results suggest that the habitats of *B. luminifera* generally occur in subtropical areas (about 1.52×10^6 km^2). However, the growth of *B. luminifera* is profoundly shaped by the nuances of its local environment, the most reasonable niche spaces are only 1.15×10^6 km^2 when limiting ecological factors (soil and ultraviolet) are considered, generally considered as the core production region. Furthermore, it is anticipated that species-suitable habitats will decrease by 10 and 8% with climate change in the 2050s and 2070s, respectively. Our study provided a clear understanding of species-suitable habitat distribution and identified the reasons why other niche spaces are unsuitable in the future, which can warn against artificial cultivation and conservation planning.

Keywords: ensemble model strategy; habitat suitability; climate change; *Betula luminifera*

1. Introduction

Future agriculture and forestry will use integrated technologies such as remote sensing and geographic information system (GIS) to ensure that farms and plantations are more efficient, safe, and environmentally friendly. Indeed, the aggregate area of forest plantations in China has been on a consistent rise; nevertheless, the recent trajectory has exhibited a 19.75% decrement, dwindling from 10.1×10^4 km^2 (ninth national inventory) to 8.4×10^4 km^2 (eighth national inventory) [1–3]. Can we provide more reasonable protection and utilization strategies for perennial trees based on niche-based GIS modeling technology? At present, niche-based GIS modeling is commonly adopted to project species' potential habitats, especially at large spatiotemporal scales [4–6]. Through niche-based GIS modeling theoretically, with the available species occurrence and environmental variables, which can calculate species niche characteristics in mathematical space consist of corresponding variables by using various mathematics algorithms [7,8]. It resulted in mapping the species' potential distribution across landscapes and projecting its habitat distributions across space and time using GIS modeling technology.

Previously, several mathematics algorithms have been successfully used in niche-based GIS spatial simulations, each with their own advantages [9,10]. However, it should be stated that no single niche-based GIS modeling algorithm is appropriate for all future climate scenarios [11,12]. Hence, in order to decrease nondeterminacy due to various algorithms' choice biases, an ensemble model (EM) strategy that integrates the message from individual niche-based GIS models and allocated various statistical algorithms has been offered. The EM strategy has already been proven to dramatically promote models' precision and reliability [11,13–15]. Recently, it has been generally recommended that the species spread's restrictions be examined in habitat estimation, particularly accounting for future variable environments during the modeling processes [16,17]. Generally, the environmental factors that restrict spread contain categorical variables (e.g., different types of soil). In the scope of the currently available niche-based GIS model technology, only maximal entropy algorithms (MaxEnt) could effectively handle categorical variables [18,19].

Precision forestry for sustainability and environmental protection cannot ignore the impact of rapid climate change. In fact, climate change has brought new uncertainties to the productivity of existing and future large-scale afforestation activities. The rapidly changing climate may move species' suitable climate niches from their contemporary ranges, which is expected to cause maladapted and compromised plantation productivity [20] and broadly influence existing ecosystems and their socioeconomic activities [21,22]. Indeed, scientific evidence has proven that global warming is the chief reason for genetic diversity loss, suitable niche space fragmentation, and the changes in species niche spatial distribution [23,24]. Hence, predicting the changes in these tree plantation distribution areas will be of great practical and theoretical importance to studying the influence of climate change. On the other hand, given that environmental factors have perennially served as the primary driving force influencing the ecological niches of species, spatial shifts as well as ecological niche vulnerability assessments are necessary for plants, irrespective of the presence or absence of climate change.

Betula luminifera (H.) Winkl, Betulaceae, is a representative fast-growing and high-yielding valuable timber tree in China [25,26]. The species is well known for its high-quality timber, short juvenile phase, and fast growth. It is widely cultivated to provide commercial timber and promote the space quality of the environment based on ecological restoration and afforestation. However, the natural resources of *B. luminifera* are threatened by fragmentation from overharvesting and populations' genetic diversity reduction [1,27–29]. Therefore, in 2015, to thoroughly maintain the sustainable development of forests, including *B. luminifera*, China prohibited commercial harvesting in natural forests, leading to a shortage in the domestic timber supply [1,30,31]. As a representative fast-growing timber plantation, *B. luminifera* has great development potential for addressing China's future large timber gaps. However, rapid global warming has affected the climate variables in the original niche habitats of *B. luminifera* [17,24,32], which in turn are expected to increase the risk of maladaptation and productivity reduction.

Here, *B. luminifera*, a representative native plantation tree in China, was selected as the target to analyze habitat area (plantation) changes in contemporary and potential future climate scenarios using niche-based GIS modeling technology. For this objective, a comprehensive habitat suitability (CHS) model was developed to implement this analysis. During the procedure of model conformation, the EM strategy was performed to develop the climate suitability model and project the potential dynamic change under future climate scenarios. In addition, the maximal entropy algorithm was utilized to develop habitat limitation models to forecast spread restrictions. Our study is expected to provide a theoretical basis for *B. luminifera* planting and management, and the generated information is also expected to contribute to promoting domestic timber output.

2. Results

2.1. Model Performance and Key Environmental Variables

To guarantee the reliability and validity of the occurrence data of *B. luminifera*, three standardized processing steps (described in material and method) were implemented. Finally, a total of 204 samples of *B. luminifera* existence information (with clear longitude and latitude) were retained to create the comprehensive habitat suitability model (Figure 1A and Table S1).

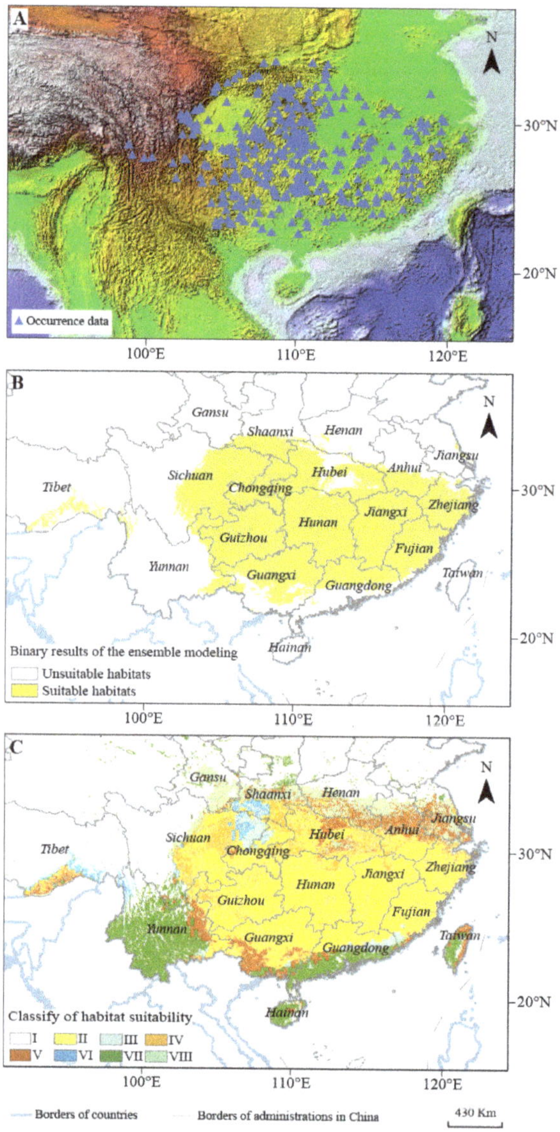

Figure 1. The geographical distribution of *B. luminifera* suitable habitats under the current climate condition. (**A**) the distribution of *B. luminifera* occurrence sample points; (**B**) the suitable habitats of *B. luminifera* developed by combining multi-climate model results. (**C**) The geographical distribution of different habitat-suitability types of *B. luminifera*.

Judging by the model's evaluation results, all models with TSS values equal to or exceeding 0.60 and AUC values surpassing 0.85 deemed satisfactory (Figure 2). For each individual model, all of the ten simulation technologies were closely associated with the stable and efficient operation of each algorithm. And each model produced higher estimation values: the average TSS and AUC results were 0.749 and 0.900, respectively. Particularly, the MARS models exhibited the highest accuracy among the four algorithms, followed by GLM, GBM, and GAM (Figure 2). Hence, along with the satisfying TSS and AUC values, this shows that our ensemble model performs well and could offer accurate results. In addition, the soil and ultraviolet distribution limitation models also produced high AUC values, with 0.873 and 0.903, respectively. Therefore, the results of the climate suitability model and the distribution limitation model could be regarded as reasonable for *B. luminifera*.

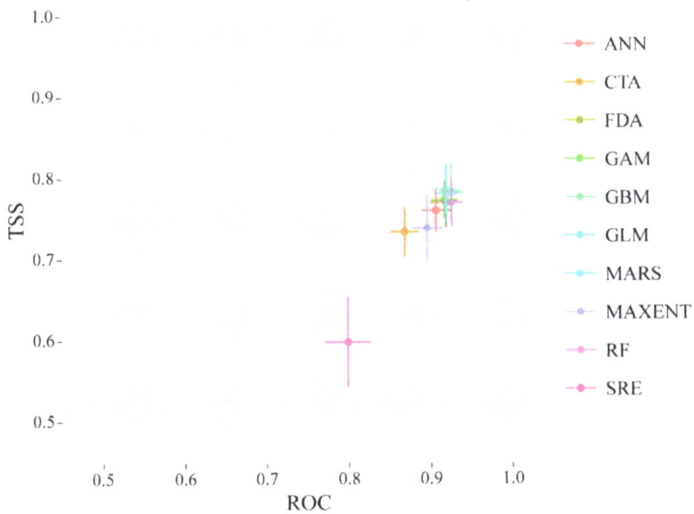

Figure 2. Mean model estimation values based on two different estimation metrics: the receiver operating characteristic (ROC) and the true skill statistic (TSS). SRE: surface range envelope, FDA: flexible discriminant analysis, GLM: generalized linear model, GAM: generalized additive model, MARS: multiple adaptive regression splines, GBM: generalized boosting model, CTA: classification tree analysis, ANN: artificial neural network, RF: random forest, and MaxEnt: maximum entropy.

Among the selected predictive bioclimatic variables (Table S2), eight climate factors related to temperature and precipitation were used to build the bioclimate suitability model of *B. luminifera* (Figures S1 and S2). The analysis of the climate factors' contribution to model formulation reveals that Bio2, Bio3, Bio7, and Bio12 were the dominant bioclimatic variables (with a contribution rate of more than 10%) affecting the distribution range of the suitable habitat of *B. luminifera*. Then, the nonparametric Kruskal–Wallis multiple-range test was used to disclose the distinct adaptability ranges for these four dominant bioclimatic variables (Figure 3). Therefore, based on the frequency presentation pattern, we obtained the thresholds for the main bioclimatic parameters: the mean diurnal temperature range (Bio2) was from 7 to 8 °C; the isothermality (Bio3) range was from 24 to 28%; the temperature annual range (Bio7) was from 26 to 31 °C; and the annual precipitation range was from 1100 to 1600 mm (Figure 3).

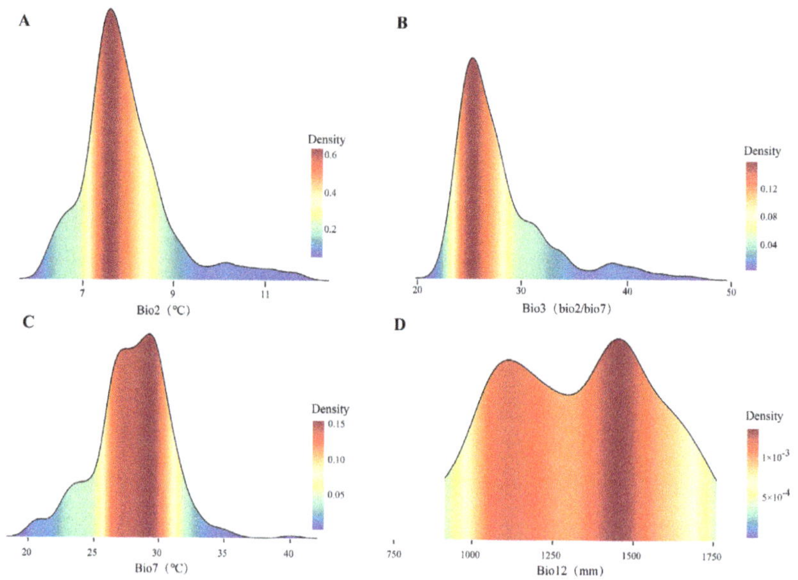

Figure 3. Kernel density plots for climatically related predictive variables in *B. luminifera* suitable habitats. (**A**) Mean diurnal temperature range (Bio2); (**B**) isothermality (Bio3); (**C**) temperature annual range (Bio7); (**D**) annual precipitation (Bio12).

2.2. The Geographical Scope of B. luminifera Suitable Habitat

The dynamic analysis showed that the core area (about 1.51×10^6 km) of *B. luminifera* suitable habitat (Table 1) is mainly distributed in the south and southeast regions of China (Figure 1B). The distribution areas mainly covered Guizhou, Chongqing, northeast of Yunnan, southeast of Sichuan, and Tibet provenances (Yunnan–Guizhou Plateau area and Hengduan mountains); south of Shaanxi and Henan provenances (belongs to the southern slope of the Qinling Mountain); Hubei, Hunan, and Jiangxi, south of Anhui and Jiangsu provinces (Yangtze valley area); Fujian and Zhejiang provinces (Wuyi mountains area); and north of Guangxi and Guangdong provinces (belongs to the Nanling Mountains) (Figure 1B).

Table 1. *B. luminifera* dynamics of habitat area under future climate scenario/year.

Future	Suitable Habitats ($\times 10^6$ km^2)		
	Unchanged	**Gain**	**Loss**
2050s	1.33	0.15	0.18
2070s	1.26	0.10	0.25

By the 2050s, suitable *B. luminifera* habitats will shift towards the north of the current niche (Figure 4). Moreover, new potential distribution areas will appear in the south of Shaanxi, Henan, Anhui, Jiangsu, and northern Yunan province; the newly expended habitats are about 0.15×10^6 km^2 (Table 1). However, part of the current habitats is expected to be lost (about 0.18×10^6 km^2), mainly located in the south of Guangxi and Guangdong provinces, which are covered in the south of current habitats. Obviously, under these scenarios/periods, habitat areas with increasing and decreasing trends are almost equal, but the geography of the suitable habitats will shift towards higher altitudes.

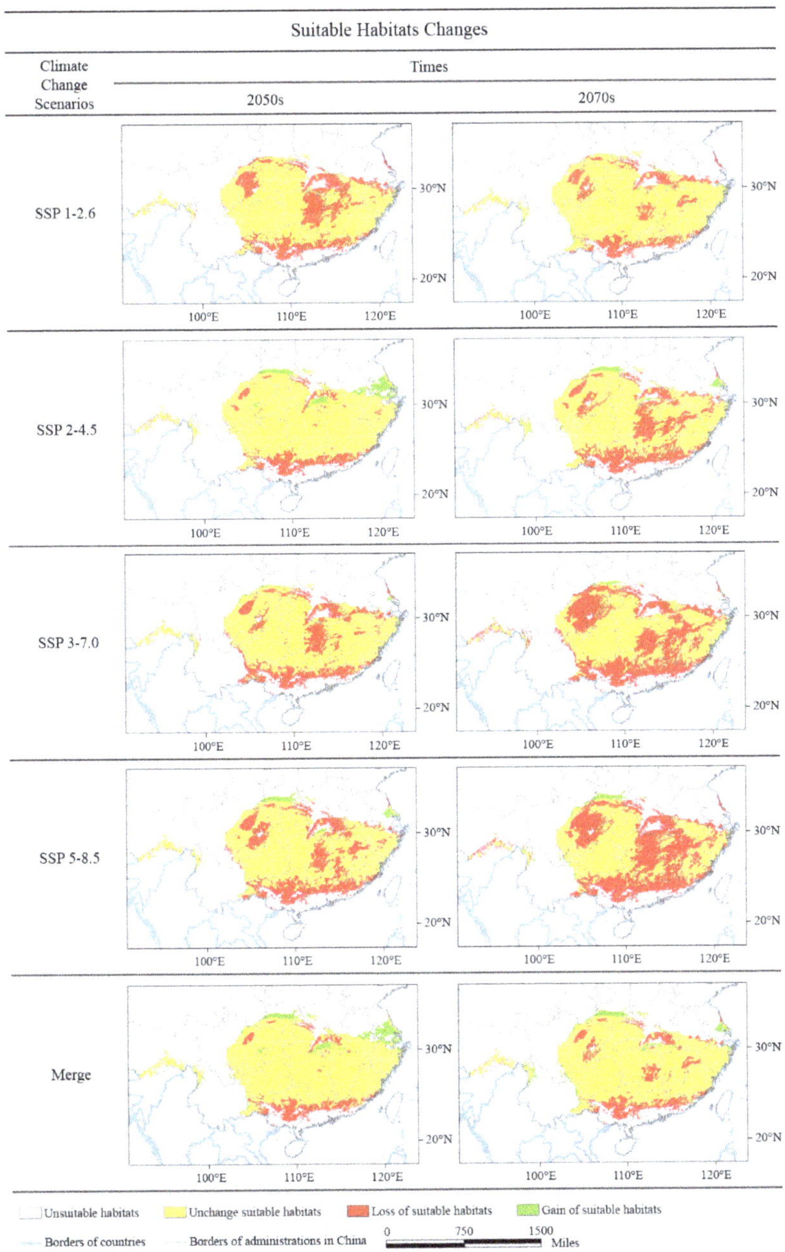

Figure 4. Projection of climatic habitat suitability for *B. luminifera* at various periods in China. Merge represents the intersection region of all possible suitable habitats at the same period.

With the extension of time, the trend of migration to higher latitudes of suitable habitat will continue in the 2070s, but the amplitude will decrease. New *B. luminifera* potential distribution areas will also appear in the north of the current niche, about 0.10×10^6 km^2 (Figure 4 and Table 1). However, it is worth noting that under these scenarios/periods, a large area of suitable habitats will be lost (0.25×10^6 km^2). Additionally, the degraded potential distribution will not only cover the south of the current habitats under baseline conditions, but it will also appear in the north of the current niche (Figure 4), which highlights the uncertainty brought about by climate change. Because the habitat areas with a decreasing trend are far beyond those with an increasing trend in the 2070s, the whole potential distribution area will considerably contract. Here, the changes in the area in potential habitat distribution under all climate change scenarios/periods are mapped (Figure 4 and Table 2), which clearly show the suitable habitat characteristics and dynamic migration trend of *B. luminifera* under future climate change scenarios.

Table 2. Statistical area for suitable habitats distribution changes for *B. luminifera* under different representative concentration pathways (SSPs: SSP1-2.6, SSP2-4.5, SSP3-7.0, and SSP5-8.5) at different times in China.

	2050s ($\times 10^6$ km)			2070s ($\times 10^6$ km)		
	Unchanged	**Gain**	**Loss**	**Unchanged**	**Gain**	**Loss**
SSP1-2.6	1.09	0.02	0.42	1.22	0.04	0.30
SSP2-4.5	1.31	0.12	0.20	1.07	0.07	0.44
SSP3-7.0	1.14	0.03	0.38	0.86	0.03	0.65
SSP5-8.5	1.08	0.07	0.43	0.81	0.05	0.71

2.3. The Spatial Distribution of Different Types of B. luminifera Habitats

According to the ecological niche model, along with comprehensive consideration of the limitations on soil and UV-B environment variables, a comprehensive habitat suitability (CHS) model was developed to estimate the *B. luminifera* suitable habitat distribution range, and this model provided the CHS value for every estimation grid. Hence, eight various habitat-suitability types of *B. luminifera* habitats were obtained in this study (Table 3).

Table 3. The codes for the eight habitat-suitability types.

Code	Climate Condition	Soil Condition	UV-B Condition
I	Unsuitable	Unsuitable	Unsuitable
II	Suitable	Suitable	Suitable
III	Suitable	Suitable	Unsuitable
IV	Suitable	Unsuitable	Suitable
V	Unsuitable	Suitable	Suitable
VI	Suitable	Unsuitable	Unsuitable
VII	Unsuitable	Suitable	Unsuitable
VIII	Unsuitable	Unsuitable	Suitable

In this study, the CHS results showed that *B. luminifera's* suitable niche space in China will decrease if restrictive ecological variables are considered. Based on the area statistics results, the area of habitat with suitable climate, soil, and UVB conditions (II type) is about 1.15×10^6 km^2; the regions with suitable climate and soil environments but unsuitable UVB conditions (III type) are about 0.08×10^6 km^2; and regions including suitable climate and UVB environments, unsuitable soil conditions (IV type) are about 0.25×10^6 km^2 (Table 4). The optimal habitats (II type) were mainly located in Guizhou and Hunan provinces (Yunnan–Guizhou Plateau), Fujian and Zhejiang provinces (Wuyi mountains area) (Figure 1C). The most significant and highly suitable regions appeared in the Yunnan–Guizhou Plateau area and Wuyi mountains area, generally considered to be core *B. luminifera* production region.

In the 2050s, the model forecast indicates that *B. luminifera's* suitable habitats will shift towards the north and will slightly decrease (Figure 5). Large losses and degraded niche space will mainly appear in the southern of Nanling mountains regions (in the south of the current niche), containing Guangxi and Guangdong, and south of the Tibet provinces. In addition, some new potential distribution areas will appear in the north and northeast regions of China, mainly in northern Zhejiang and Hubei, Anhui, and Jiangxi provinces. The core production area of *B. luminifera* (suitable habitat type of II) will contract to 1.07 × 10^6 km^2 (a decreased of 6.95%).

Table 4. Habitat area with different suitability types of *B. luminifera* at various periods in China. I: unsuitable; II: suitable habitat; III: habitats with an unsuitable UV-B condition; IV: habitats with an unsuitable soil condition; V: habitats with unsuitable climate conditions; VI: habitats with an unsuitable soil and UV-B condition; VII: habitats with an unsuitable climate and UV-B condition; VIII: habitats with unsuitable climate and soil conditions.

Type	**Habitats under Future Climate Scenario/Year (×10^6 km^2)**		
	Current	**2050s**	**2070s**
II	1.15	1.07	0.99
III	0.08	0.06	0.06
IV	0.25	0.29	0.25
V	0.24	0.31	0.40
VI	0.05	0.06	0.06
VII	0.50	0.51	0.52
VIII	0.27	0.22	0.25

In the 2070s, *B. luminifera*-suitable habitat areas will continue to move towards the higher latitude and decrease (Figure 5), since the loss of suitable niche space also occurs in the south of current areas in Guangxi and Guangdong provinces, but the location shifts further north than that of the 2050s. In addition, new suitable niche spaces will mainly appear in the south of Hubei, Anhui, and Jiangsu provinces. The core suitable habitat area of *B. luminifera* (suitable habitat type of II) will contract to 0.99 × 10^6 km^2 (a decrease of 13.91%). It is noteworthy that climate change will continuously impact *B. luminifera's* suitable habitat distribution, leading to severe habitat fragmentation with the progression of time. Here, the characteristics and dynamic migration trends of these eight habitat-suitability types of *B. luminifera* are mapped in Figure 5, which clearly shows the changes in the area for each type under future climate change scenarios in China.

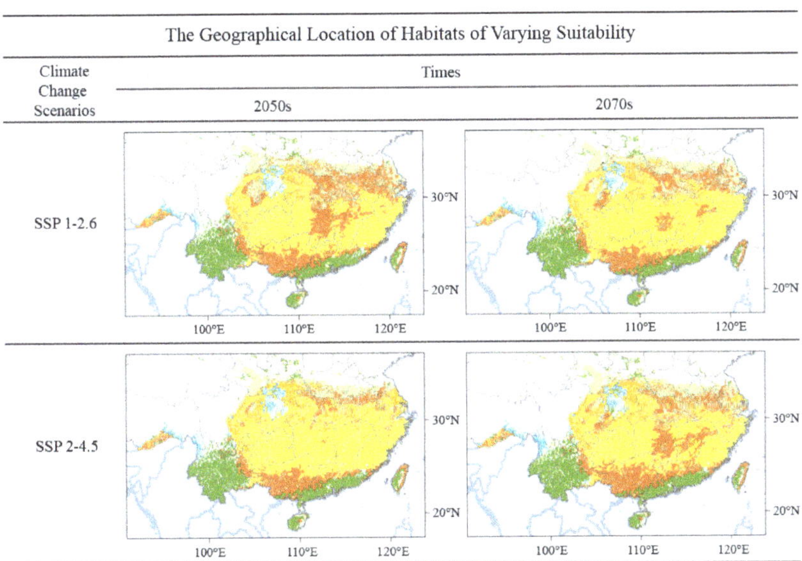

Figure 5. *Cont.*

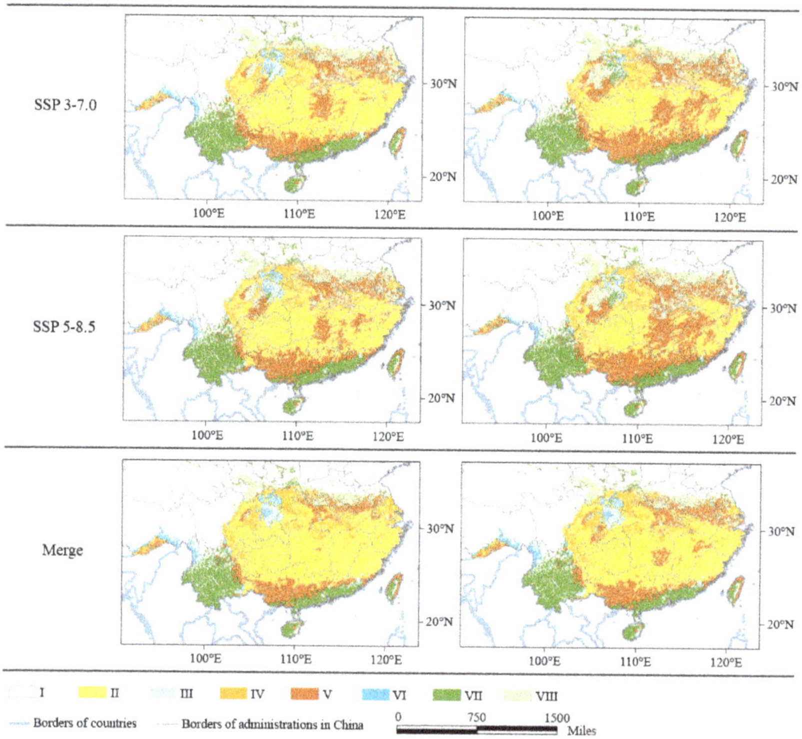

Figure 5. The geographical location of different types of suitable habitats for *B. luminifera* at various future climate change scenarios in China. Merge showing the intersection region of the suitable habitats at different scenarios. I: unsuitable; II: suitable habitat; III: habitats with an unsuitable UV-B condition; IV: habitats with an unsuitable soil condition; V: habitats with unsuitable climate conditions; VI: habitats with unsuitable soil and UV-B conditions; VII: habitats with unsuitable climate and UV-B conditions; VIII: habitats with unsuitable climate and soil conditions.

2.4. Changes in Potential Distribution Areas under Climate Change

To quantify the influence of global climate change on the geographical range of these eight habitat-suitability types, the variance in annual average temperature was counted under various climate scenarios and stages. According to the above data, variations in the suitable habitat areas under various climate change scenarios were calculated (Tables 5 and S3), and the dynamic forecast of the changing trends of the area variation for each suitable type was mapped (Figure 6).

The results indicated that, with various predicted temperature increments, the area variation trend of these eight habitat-suitability types can be divided into three groups: with projected temperature increases, the potential distribution area shows a significant decreasing trend, including II, III, IV and VI (class one); a relatively steady trend, including I, VII, and VIII (class two); and an increase significantly trend, including V (class three) (Figure 6). Notably, the habitat-suitability type area changes in suitable habitat (II), habitats with unsuitable UV-B conditions (III), habitats with unsuitable climate conditions (V), and habitats with unsuitable soil and UV-B conditions (VII) were significantly associated with the different projected temperature increases.

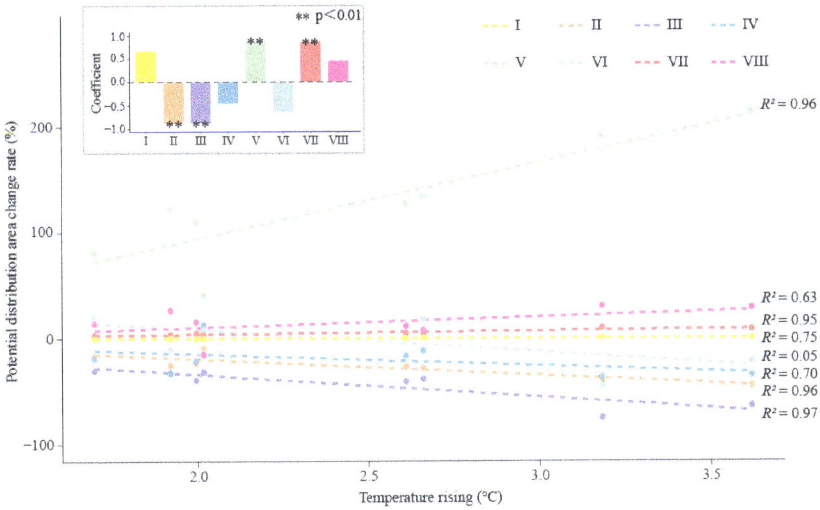

Figure 6. Area of habitats of differing suitability for *B. luminifera* at various periods in China. I: unsuitable; II: suitable habitat; III: habitats with an unsuitable UV-B condition; IV: habitats with unsuitable soil conditions; V: habitats with unsuitable climate conditions; VI: habitats with unsuitable soil and UV-B conditions; VII: habitats with unsuitable climate and UV-B conditions; VIII: habitats with unsuitable climate and soil conditions.

Table 5. Statistical area for climatic habitat suitability changes for *B. luminifera* under different representative concentration pathways (SSPs: SSP1-2.6, SSP2-4.5, SSP3-7.0, and SSP5-8.5) at different times in China.

Code	Current	2050s ($\times 10^6$ km^2)				2070s ($\times 10^6$ km^2)			
		SSP 1-2.6	SSP 2-4.5	SSP 3-7.0	SSP 5-8.5	SSP 1-2.6	SSP 2-4.5	SSP 3-7.0	SSP 5-8.5
I	7.07	7.09	7.08	7.08	7.08	7.08	7.08	7.11	7.10
II	1.15	0.86	1.05	0.88	0.84	0.95	0.83	0.69	0.63
III	0.08	0.05	0.05	0.05	0.05	0.05	0.05	0.02	0.03
IV	0.25	0.17	0.28	0.19	0.21	0.20	0.22	0.16	0.16
V	0.24	0.53	0.34	0.51	0.55	0.44	0.56	0.70	0.76
VI	0.05	0.04	0.05	0.05	0.05	0.05	0.05	0.02	0.04
VII	0.50	0.52	0.52	0.53	0.53	0.52	0.53	0.55	0.55
VIII	0.27	0.34	0.23	0.31	0.30	0.31	0.29	0.35	0.34

3. Discussion

Future agrarian and silvicultural practices will harness integrated technologies (e.g., GIS) to assure sustainable conservation, heightened efficiency, and environmental congeniality on farms and plantations. Here, we have provided a more accurate assessment for the ecological niche of *B. luminifera* after considering limiting factors (soil and UV-B). The actual suitable niche constituted a mere 75.65% in comparison to scenarios where such factors were not considered. Furthermore, the spatial niche of *B. luminifera* is expected to be reduced with climate change, leading to a decrease in the entire spatial niche, suggesting the risk of maladaptation. Thus, forest management practices, such as in situ and ex situ protection/conservation and assisted migration, have the potential to curtail the expected niche losses and the maintenance of forest health and productivity, thereby sustaining future ecosystem services.

3.1. Ecological Characteristics of B. luminifera Habitats

A significant climatic characteristic in *B. luminifera* habitats is that the principal climate factors may be temperature-related (Figure 3). In order to clarify how global warming influences *B. luminifera's* suitable habitat geographical range, the average value and standard deviation of the eight climate factors contained in the EM for various climate change scenarios and stages were counted (Table S4). The results indicated that, for the entire suitable habitats, the mean value of Bio1 was about 15.39 °C in current. In the next time period, the mean annual temperature would represent an upward trend in future climate scenarios, but the variation was different. In the 2050s, the annual mean air temperature (Bio1) value in the entire suitable habitats may equal 17.53 °C, and by the 2070s, Bio1 may reach 18.18 °C, meaning that the temperature increment will be 3.37 °C. In addition, the annual precipitation (Bio12) exhibited a similar upward trajectory, albeit with a markedly smoother progression: in the 2050s, the average Bio12 could retain a relatively steady trend; in the 2070s, the added precipitation will reach 88.41 mm. Bio2, Bio3, Bio4, Bio8, Bio15, and Bio18 have a high probability of showing a steady or slight increase, indicating that the seasonal tendencies of temperature and rainfall may remain essentially steady in the species' entire suitable habitats.

Our analysis showed that the change in *B. luminifera's* suitable habitat area was closely related to the projected temperature increases, which is consistent with other fast-growing tree plantation species affected by climate change. For example, there are at least 12 fast-growing timber tree species with potential distribution areas showing dynamic changes with different projected temperature increases, e.g., *Betula platyphylla* (a close relative species with *B. luminifera*) and *Cunninghamia lanceolata* (an important plantation timber tree in China) [6]. In addition, under the maximum temperature rise's scenario (e.g., SSP585), most of the fast-growing tree plantation species will be shifting their suitable niche space to the north, and a large area of suitable habitat loss, especially in the south of the current niche under the current climate scenario, will cause severe habitat fragmentation [6]. Such phenomena align with our research, and the findings in this study highlight the variations in habitat areas for a representative fast-growing and high-yielding unique valuable timber plantation under the impact of predicted future climate warming, which will provide more information for reasonable afforestation under climate change in the future.

3.2. Limitative Effect of Soil and UV-B on B. luminifera Suitable Habitats

B. luminifera is a perennial woody plant. Unfit soil and ultraviolet types will restrict the extension of plantations. In this study, soil factors were found to be the major variables that were restricting the range of *B. luminifera*, particularly in Jiangsu, Anhui, Hubei, and the east of Sichuan provinces (the suitability type IV). Along with the suitable climatic environments, the range of unsuitable soil conditions (IV) was 0.25×10^6 km^2, which is 44% of the core *B. luminifera* habitats. Soil has already been proven to be an important factor in limiting plant growth and influencing ecosystem productivity [33,34]. Furthermore, an in-depth understanding of the intricate interplay between soil and climate on plant characteristics reveals that the primary variations are reflected in size disparities at both organ and plant levels, as well as in leaf development, delicately balancing leaf longevity with plant growth potential [35]. On the other hand, ultraviolet B radiation, a significant macroclimatic element exhibiting substantial variations across extensive spatial expanses [36], has been empirically shown to exert considerable influence on plant germination and growth [37,38]. Recently, the on-going changes in climate are increasingly exposing plants to novel combinations of UV-B and other climate factors, e.g., water availability and temperature [39,40]. In this study, the range of unsuitable UV-B conditions (III) was 0.08×10^6 km^2 (7% of the core *B. luminifera* production region), which is mainly located in southwest China (Figure 5), demonstrating that the dominant variables restricting the expansion of the species' suitable niche space in these regions are likely to contain UV-B. However, it is not yet clear how soil and UV-B specifically regulate plant growth, and the physiological mechanism also needs further investigation.

The scope of environmental factors is likely to change due to the impact of future climate change. In this study, the decrease in suitability type IV indicates that the most suitable habitat of *B. luminifera* is shrinking under the influence of climate change. In addition, the suitability types III and IV also indicate that the newly added suitable environment is caused by climate change, maybe mostly located in unsuitable soil or ultraviolet conditions. In this study, the characteristics and dynamic migration trend of eight habitat-suitability types for *B. luminifera* were mapped (Figure 5), which clearly shows the geographic location of the restriction factor for each type under future climate change scenarios in China. Therefore, forest managers can formulate detailed vegetation protection and utilization strategies according to the changes in suitable habitat types in specific regions. For example, in central China, the range of habitat type IV (soil as limiting factors) will gradually decrease and significantly shift northward, and the new regions that have suitable niche space will appear to the north of Hubei and Anhui, in Chongqing provinces.

3.3. *Management Priorities of B. luminifera under Climate Change*

To prevent genetic diversity loss and ecological degradation in *B. luminifera* habitats, the reasonable designation of planting regions will be a key bond based on the species survival rate with respect to climate warming. Under natural conditions, population gene flow, migration, and vegetation succession are complicated processes that typically happen over a long period of time. It is generally accepted that the natural shifting of forest trees lags far behind the current rapid rate of climate change [5,41,42]. In a previous study, a significant linear positive correlation between the afforestation survival rate and the potential probability of presence by projecting the habitats distribution based on ecological niche models was found [43]. Concerning *B. luminifera*, previous studies have also concluded that the current status of genetic diversity and structure of natural populations is strongly influenced by suitable niche space fragmentation [26,44]. Therefore, the cultivation of *B. luminifera* is one of the most effective methods for protecting the remaining *B. luminifera* natural populations [32]. In fact, the scholarly probe into climate change has perennially been a subject of contention, yet it is indisputable that any shift in the global climate will have profound ramifications for the existing ecosystem. In this study, employing *B. luminifera* as a representative case, we forecasted the potential impacts of climate change and offered bespoke strategies for an optimal response; this evaluation is as vital as scientific experience, knowledge, and existence itself.

In this study, a macroscopically drawn map of the suitable niche space for *B. luminifera* was developed by combining multi-climate model results (Figures 3 and 4). Our results revealed the intersection region of the suitable habitats at different scenarios (e.g., suitable area in the 2050s, in Figure 6), which should have priority for protection to guarantee habitat survival. For example, we found that the most suitable niche space for *B. luminifera* would severely decrease and move to higher latitudes with climate change. Furthermore, the available ex situ steps for the excellently distributed individuals and seedlings in marginally suitable habitats may be essential to projects by using field survey and provenance testing. The populations at distribution margins should also be given more consideration for management, because they demonstrate potential for the development of climate suitability in the future. Additionally, our research provides a helpful basis for forest management objectives (i.e., reforestation, seed allocation, and assisted migration), which could incorporate climate change adaptation into forest plan programs of *B. luminifera*.

4. Materials and Methods

4.1. *Species Occurrence Database*

B. luminifera existence information has been collected from published scientific literature and the Chinese Virtual Herbarium (CVH, https://www.cvh.ac.cn/, (accessed on 23 May 2024)). To guarantee the reliability and validity of the occurrence data, only existence points with clear longitude and latitude were selected, and priority was given to records from field surveys.

Based on GIS modeling technology, appropriation geographic point information was further chosen through the following three standards: (1) duplicate coordinate information was removed; considering the resolution of environmental factors, the data were further condensed to ensure that each grid range (evaluation unit) had a single sample site; (2) sampling sites originated from different ecological niches to guarantee balance of species occurrence investigations; and (3) based on meeting the above two principles, some sampling sites were deleted to guarantee that the space between two occurrence data was >10 km as possible and the points were evenly distributed. These measures allowed reducing sample points spatial autocorrelation and minimizing its effect on niche model prediction results.

4.2. Environmental Factors Database

Three types of environmental variables (climatical, soil, and ultraviolet B) were collected to simulate *B. luminifera* potential habitat distribution. The climate factors were 19 bioclimatic variables [45], a common dataset usually applied in ecological niche modeling, and originated from the monthly temperature and precipitation values to produce more meteorologically significant climate variables. The soil type database was obtained from the Harmonized World Soil Database (HWSD) [46], and the soil properties of topsoil and subsoil were selected to build the niche model. Ultraviolet radiation B (UV-B) was derived from the gIUV database (http://www.ufz.de/gluv/, (accessed on 23 May 2024)), with six biologically meaningful factors selected to build the niche model.

Future climate scenarios databases were coming from the WorldClim Coupled Model Intercomparison Project 6 (CMIP6) dataset (version 2.0) [47]. In this study, the spatial resolutions of all the baseline and future environment variables were resampled at 30″ (approximately 1 km^2). Then, we simulated *B. luminifera's* potential habitat distribution in two future periods: 2041–2060 (2050s) and 2061–2080 (2070s), under the four projected climate scenarios mentioned above [48–50]. Each of the two future environmental datasets consisted of four shared socioeconomic pathways (SSPs): SSP126 (slight climate change conditions), SSP245 (moderate climate change conditions), SSP370 (powerful climate change conditions), and SSP585 (severe climate change conditions) [51].

The multicollinearity of environmental variables is expected to cause additional uncertainty for the ecological niche models [11,52]. Indeed, bioclimatic variables have been proven to cause severe multicollinearity [5,6,53]. Hence, a principal component analysis (PCA) was implemented to choose the representative climate factors [4,11]. According to *B. luminifera*, existence point distribution and the forecasting of the 19 bioclimatic factors in the habitat space were delimited by the PCA (Figure S1).

Finally, eight climate variables were chosen to build the climate suitability model (Table S2). Additionally, Pearson product moment correlation coefficients (r) were applied to check the cross-correlation of all variables, to ensure that their correlation coefficients were less than 0.6. Ultimately, 8 climatic factors, 12 soil types, and 3 UV-B variables were selected based on PCA and correlation analyses for niche model construction (Table S2).

4.3. Construction of Niche-Based GIS Modeling

In this study, a CHS niche-based GIS model was constructed to analyze *B. luminifera* potential habitat distribution under the different climate scenarios (above). This model approach included activities of two distinct patterns: (1) a climate suitability model, which specialized in detecting the dynamic changes of species habitat distribution under future climate scenarios based on EM strategy; and (2) a distribution restricted model, which used the MaxEnt algorithm to predict spread restriction [54]. Although *B. luminifera* has a wide distribution area, the species does not grow in unsuitable soil and ultraviolet environments, even though the climate and landform are suitable. Therefore, in this study, soil and ultraviolet variables were considered limiting ecological factors.

4.3.1. Climate Suitability Model

According to the EM strategy, 10 commonly niche-based GIS modeling algorithms, including the surface range envelope (SRE), flexible discriminant analysis (FDA), generalized linear model (GLM), generalized additive model (GAM), multiple adaptive regression splines (MARS), generalized boosting model (GBM), classification tree analysis (CTA), artificial neural network (ANN), random forest (RF), and maximum entropy (MaxEnt) [54–57], were utilized to develop the climate suitability model. This approach was adopted to reduce the modeling prediction uncertainty caused by different modeling techniques. All model building steps were implemented on the biomod2 framework based on R language [58]. The modeling process occurred as follows (Figure S2):

Firstly, pseudo-sampling points were selected. For this process, to decrease the prediction result's instability caused by randomly generated points, this step was reduplicated three times to produce three datasets of pseudo-sampling points, with each dataset containing 500 pseudo-sampling points.

Secondly, we conducted single-model development. Ten simulation algorithms were developed independently using the method of biomod2 in the R language environment. A total of 80% of the sampling points (containing occurrence points and pseudo databases) were freely chosen as training databases, and the remaining 20% was chosen for model testing (i.e., cross-validation). To decrease the instability due to various data, the true skill statistic (TSS) and receiver operating characteristic curve (ROC) were applied to estimate the model representation [59,60], and an individual estimated algorithm was performed 10 times with each sample dataset. Hence, in this process, 120 single models (i.e., 4 single modeling algorithms, 3 pseudo-absence sampling models, and 10 cross-validation runs) were built.

Thirdly, we built the EMs. In this step, mean weight values were adopted to blend all individual models along with TSS values > 0.7 to generate our EMs. Thus, only models with a TSS value > 0.7 were retained to develop the final ensemble. Then, TSS values were used to estimate the references and define the weight values of individual models. Equation (1) presents how the TSS values were utilized to determine the weight of an individual model:

$$W_j = \frac{r_j}{\sum_{j=1}^{h} r_j} \quad (1)$$

W_j is the weight value of j's results; r_j is the TSS value of model j; and h is the quantity of models.

The normalization prediction results of an individual model were then redoubled by the commensurable weight to receive the comprehensive results. Equation (2) illustrates how the potential habitat suitability index was calculated and the EM was constructed.

$$EM_i = \sum_{j=1}^{n} W_j \times X_{ij} \quad (2)$$

EM_i (range [0, 1]) means the habitat suitability index of the estimation unit (grid) i, which is the estimation index for potential *B. luminifera* habitats; W_j is the weight value of j's result; and X_{ij} refers to the value of the estimation grid i of model j's result.

Based on the above ensemble model, *B. luminifera's* potential habitat distribution in the 2000s, 2050s, and 2070s was performed, and among them, the future climate database included four SSPs (SSP126, SSP245, SSP370, and SSP585) scenarios. Based on the future bioclimatic environmental database, an individual model was conducted independently for every period, and the areas for suitable habitat changes in *B. luminifera* under different scenarios/periods were counted. Hence, *B. luminifera's* future habitat distribution in different stages was achieved by averaging the results.

Finally, the binary format that predicted the results of EMs was obtained. In this process, optimized TSS values based on the testing data were utilized as a reference (threshold) to convert the ensemble modeling's results to a binary format [11]. Because TSS refers to a threshold-dependent metric, but different thresholds can bring about various

TSS values, in this study, the threshold parallel to the maximal TSS results was selected for the taxonomy criteria. Based on a unified threshold, the model's simulation results in the 2000s, 2050s, and 2070s were divided into two types: suitable and unsuitable.

4.3.2. Distribution Limitation Model

Soil variables and UV-B radiation data can supply messages concerning key environmental variables that constrain species habitats spatially. Among the ecological niche model technologies, only the maximum entropy algorithm can utilize categorical data (e.g., soil texture classification) as input environmental variables. Hence, in order to assess soil and UV-B suitability requirements, we employed 12 soil and 3 UV-B variables to build a habitat distribution limitation model by MaxEnt (version3.3.3) [19,60].

In the modeling steps, 75% existence points from the database were freely selected as training, and the reserving database was applied as testing data. Moreover, 10-time repeats were conducted to minimize the instability, and the AUC (area under the curve) was utilized to assess the model operation representation. The MaxEnt prediction model for the soil variables suitability index scope in 0 to 1 and the maximum training sensitivity plus specificity were considered as the thresholds to reclassify the results into two types: suitable and unsuitable soil habitats. Similarly, MaxEnt was also applied to develop a species habitat spread restriction model to assess its UV-B radiation suitability requirements.

4.4. *Comprehensive Habitat Suitability (CHS) Model*

Here, a CHS model was developed to estimate *B. luminifera's* suitable habitat distribution, including bioclimate suitability, soil suitability, and UV-B suitability, and this model provided the CHS value for every estimation grid as below (Equation (3)):

$$CHS_i = TM_i \bigcap S_i \bigcap U_i \tag{3}$$

CHS_i refers to the comprehensive habitat suitability conditions of *B. luminifera* of each estimation grid *i*; TM_i refers to the climate suitability of each estimation grid i under various environment scenarios; S_i refers to the soil limitation index of each estimation grid *i*; and U_i refers to the UV-B limitation index of each estimation grid *i*. In the modeling operation, the soil and UV-B conditions of the future were considered to maintain consistency with the current period.

4.5. *B. luminifera Future Potential Habitats Distribution Areas*

The model prediction results contained uncertainties under various future climate scenarios that were generated by various atmospheric general circulation models (GCMs). To reduce the effect of various GCMs, the certainty index of *B. luminifera's* suitability habitat (C_i, Equation (4)) was formulated to present the confidence of future prediction as follows:

$$C_i = \frac{\sum_1^m CHS_i}{m} \tag{4}$$

C_i is the certainty index of *B. luminifera's* suitability habitat in each climate change scenario. In this study, eight climate change scenarios were included; CHS_i is the comprehensive habitat suitability's result into estimation grid *i* under the *j*th GCM. *m* is the quantity of GCMs contained in the statistics under this climate change scenario. For each environmental scenario, the regions along with C_i values > 0.5 were considered as the final suitable habitats.

5. Conclusions

In the present study, a comprehensive habitat suitability (CHS) niche-based GIS model was developed using the ensemble model strategy and applied for predicting the suitable climate habitat geographic range of *B. luminifera*. We found that climate change will negatively affect the spatial distribution of *B. luminifera*, resulting in a northward range shift and a drastic loss of suitable habitats in the future. Our research provides a macroscopically drawn map of potentially suitable areas (plantations) of *B. luminifera* that were in the intersection region at different scenarios, which was more reliable than the individual union–intersection habitat under a single scenario/year. Therefore, the present study provides a valuable basis for reforestation, seed allocation, and assisted migration of *B. luminifera*, and it will ultimately help boost the survival rate of afforestation for *B. luminifera*.

Supplementary Materials: The following supporting information can be downloaded at https://www.mdpi.com/article/10.3390/plants13111542/s1, Figure S1: The principal component analysis (PCA) process to select a subset of the environmental factors. A, the distribution of the points of *B. luminifera* occurrences in the environmental space defined by the first two PCA axes, illustration of the distribution of *B. luminifera* along the first two PCA axes; B, the correlation circle of the selected bioclimatic variables as a function of these first two PCA axes, illustration of the projection of the selected bioclimatic variables over the same two PCA axes. Figure S2: Schematic representation of the methodological methods. Table S1: The occurrence data of *Betula luminifera*. Table S2: Environmental factors applied to niche-based GIS modeling of *Betula luminifera*. Table S3: Habitat area with different suitability type of *Betula luminifera* at various periods in China. I: Unsuitable; II: Suitable habitat; III: Habitats with unsuitable UV-B condition; IV: Habitats with unsuitable Soil condition; V: Habitats with unsuitable climate conditions; VI: Habitats with unsuitable Soil and UV-B condition; VII: Habitats with unsuitable Climate and UV-B condition; VIII: Habitats with unsuitable Climate and Soil condition. Table S4: Mean and standard deviation of the eight bioclimatic variables in mainland China.

Author Contributions: Conceptualization, X.-G.H. and H.H.; Data curation, Z.J.; Formal analysis, J.C.; Funding acquisition, X.-G.H. and H.H.; Investigation, E.-K.Y.; Methodology, X.-G.H., J.C. and Q.C.; Project administration, H.H.; Software, X.-G.H. and Y.L.; Supervision, E.L. and H.H.; Validation, Y.Y., Y.L. and L.S.; Visualization, Y.Y., Y.L. and L.S.; Writing—original draft, J.C. and Q.C.; Writing—review and editing, X.-G.H. and J.C. All authors have read and agreed to the published version of the manuscript.

Funding: This research was funded by the National Natural Science Foundation of China (Grant No. 32001327), the Key Scientific and Technological Grant of Zhejiang for Breeding New Agricultural Varieties (2021C02070-1), and the National Natural Science Foundation of Zhejiang (Grant No. LQ21C160002).

Data Availability Statement: The original data presented in the study are openly available in Table S1 and Figure 1.

Conflicts of Interest: The authors declare no conflicts of interest.

References

1. Du, Z.; Hu, J.; Xiao, Q.; Feng, Q.; He, P.; Li, R. Analysis on characteristics and development countermeasures of plantation resources in China. *Cent. South For. Invent. Plan.* **2020**, *39*, 5–10. (In Chinese)
2. National Forestry and Grassland Administration. *The Nineth National Forest Resources Inventory Report*; National Forestry and Grassland Administration: Beijing, China, 2019.
3. State Forestry Administration. *The Eighth National Forest Resources Inventory Report*; China Forestry Press: Beijing, China, 2013.
4. Guo, Y.; Li, X.; Zhao, Z.; Nawaz, Z. Predicting the impacts of climate change, soils and vegetation types on the geographic distribution of *Polyporus umbellatus* in China. *Sci. Total Environ.* **2019**, *648*, 1–11. [CrossRef]
5. Hu, X.-G.; Mao, J.-F.; El-Kassaby, Y.A.; Jia, K.-H.; Jiao, S.-Q.; Zhou, S.-S.; Li, Y.; Coops, N.C.; Wang, T. Local adaptation and response of *Platycladus orientalis* (L.) Franco populations to climate change. *Forests* **2019**, *10*, 622. [CrossRef]
6. Zhao, Z.; Guo, Y.; Zhu, F.; Jiang, Y. Prediction of the impact of climate change on fast-growing timber trees in China. *For. Ecol. Manag.* **2021**, *501*, 119653. [CrossRef]
7. Anderson, R.P. A framework for using niche models to estimate impacts of climate change on species distributions. *Ann. N. Y. Acad. Sci.* **2013**, *1297*, 8–28. [CrossRef]
8. Brun, P.; Thuiller, W.; Chauvier, Y.; Pellissier, L.; Wüest, R.O.; Wang, Z.; Zimmermann, N.E. Model complexity affects species distribution projections under climate change. *J. Biogeogr.* **2020**, *47*, 130–142. [CrossRef]
9. Elith, J.; Leathwick, J.R. Species distribution models: Ecological explanation and prediction across space and time. *Annu. Rev. Ecol. Evol. Syst.* **2009**, *40*, 677–697. [CrossRef]

10. Norberg, A.; Abrego, N.; Blanchet, F.G.; Adler, F.R.; Anderson, B.J.; Anttila, J.; Araújo, M.B.; Dallas, T.; Dunson, D.; Elith, J.; et al. A comprehensive evaluation of predictive performance of 33 species distribution models at species and community levels. *Ecol. Monogr.* **2019**, *89*, e01370. [CrossRef]
11. Guisan, A.; Thuiller, W.; Zimmermann, N.E. *Habitat Suitability and Distribution Models: With Applications in R*; Cambridge University Press: Cambridge, UK, 2017.
12. Li, X.; Wang, Y. Applying various algorithms for species distribution modelling. *Integr. Zool.* **2013**, *8*, 124–135. [CrossRef]
13. Dormann, C.F.; Purschke, O.; Marquez, J.R.G.; Lautenbach, S.; Schroder, B. Components of uncertainty in species distribution analysis: A case study of the great grey shrike. *Ecology* **2008**, *89*, 3371–3386. [CrossRef]
14. Grenouillet, G.; Buisson, L.; Casajus, N.; Lek, S. Ensemble modelling of species distribution: The effects of geographical and environmental ranges. *Ecography* **2011**, *34*, 9–17. [CrossRef]
15. Thibaud, E.; Petitpierre, B.; Broennimann, O.; Davison, A.C.; Guisan, A. Measuring the relative effect of factors affecting species distribution model predictions. *Methods Ecol. Evol.* **2014**, *5*, 947–955. [CrossRef]
16. D'Amen, M.; Mod, H.K.; Gotelli, N.J.; Guisan, A. Disentangling biotic interactions, environmental filters, and dispersal limitation as drivers of species co-occurrence. *Ecography* **2018**, *41*, 1233–1244. [CrossRef]
17. Wang, W.J.; Thompson, F.R.; He, H.S.; Fraser, J.S.; Dijak, W.D.; Jones-Farrand, T.; Chen, H. Climate change and tree harvest interact to affect future tree species distribution changes. *J. Ecol.* **2019**, *107*, 1901–1917. [CrossRef]
18. Elith, J.; Phillips, S.J.; Hastie, T.; Dudík, M.; Chee, Y.E.; Yates, C.J. A statistical explanation of maxent for ecologists. *Divers. Distrib.* **2011**, *17*, 43–57. [CrossRef]
19. Phillips, S.J.; Anderson, R.P.; Schapire, R.E. Maximum entropy modeling of species geographic distributions. *Ecol. Model.* **2006**, *190*, 231–259. [CrossRef]
20. Aitken, S.N.; Bemmels, J.B. Time to get moving: Assisted gene flow of forest trees. *Evol. Appl.* **2016**, *9*, 271–290. [CrossRef]
21. Felton, A.M.; Felton, A.; Foley, W.J.; Lindenmayer, D.B. The role of timber tree species in the nutritional ecology of spider monkeys in a certified logging concession, Bolivia. *For. Ecol. Manag.* **2010**, *259*, 1642–1649. [CrossRef]
22. Hao, Y.; Xu, Y.; Zhang, J.; Hu, X.; Huang, J.; Chang, C.P.; Guo, Y. Relationship between forest resources and economic growth: Empirical evidence from China. *J. Clean. Prod.* **2019**, *214*, 848–859. [CrossRef]
23. Urban, M.C. Accelerating extinction risk from climate change. *Science* **2015**, *348*, 571–573. [CrossRef] [PubMed]
24. Wiens, J.J. Climate-related local extinctions are already widespread among plant and animal species. *PLoS Biol.* **2016**, *14*, e2001104 [CrossRef]
25. Liu, Z.-J.; Chen, Y.-L.; Yu, H.-Y.; Han, S.-J.; Zhao, F.-J.; Zhang, B.-G.; Yao, G.-L.; Yang, Q.; Li, X.-L.; He, H.-K. Genetic diversity of the natural populations of *Betula luminifera*. *J. Beijing For. Univ.* **2006**, *28*, 28–34.
26. Zhang, J.; Huang, H.; Tong, Z.; Cheng, L.; Liang, Y.E.; Chen, Y. Genetic diversity in six natural populations of *Betula luminifera* from southern China. *J. Biodivers. Sci.* **2010**, *18*, 233.
27. Cheng, L.; Pan, Y.; Lin, Y.; Lin, S.; Zaikang, T.; Junhong, Z. Effect of Low Nitrogen Stress on the Growth, Physiological, and Biochemical Characteristics of Different *Betula luminifera* Genotypes. *J. Nucl. Agric. Sci.* **2020**, *34*, 2435.
28. Wang, J.; Yao, L.; Ai, X.-R.; Zhu, J.; Liu, S.-B. Structure and dynamic characteristics of *Betula luminifera* populations in different regions of Southwest Hubei Province, China. *J. Appl. Ecol.* **2020**, *31*, 357–365.
29. Zhang, J.; Lin, Y.; Wu, F.; Zhang, Y.; Cheng, L.; Huang, M.; Tong, Z. Profiling of microRNAs and their targets in roots and shoots reveals a potential miRNA-mediated interaction network in response to phosphate deficiency in the forestry tree *Betula luminifera*. *Front. Genet.* **2021**, *12*, 38. [CrossRef]
30. Dai, L.; Li, S.; Zhou, W.; Qi, L. Opportunities and challenges for the protection and ecological functions promotion of natural forests in China. *For. Ecol. Manag.* **2018**, *410*, 187–192. [CrossRef]
31. Yan, R.; Ke, S. The Impact of Cultivation of Tree Crops on Timber Supply Capacity of Planted Forests: Inhibition or Promotion? An Analysis Based on Forest Inventory Panel Data in 12 Provinces of South China. *Chin. Rural Econ.* **2019**, *5*, 38–53.
32. Gaines, W.L.; Hessburg, P.F.; Aplet, G.H.; Henson, P.; Prichard, S.J.; Churchill, D.J.; Jones, G.M.; Isaak, D.J.; Vynne, C. Climate change and forest management on federal lands in the Pacific Northwest, USA: Managing for dynamic landscapes. *For. Ecol. Manag.* **2022**, *504*, 119794. [CrossRef]
33. Coudun, C.; Gégout, J.C.; Piedallu, C.; Rameau, J.C. Soil nutritional factors improve models of plant species distribution: An illustration with *Acer campestre* (L.) in France. *J. Biogeogr.* **2006**, *33*, 1750–1763. [CrossRef]
34. McKee, K.L. Soil physicochemical patterns and mangrove species distribution–Reciprocal effects? *J. Ecol.* **1993**, *81*, 477–487. [CrossRef]
35. Joswig, J.S.; Wirth, C.; Schuman, M.C.; Kattge, J.; Reu, B.; Wright, I.J.; Sippel, S.D.; Rüger, N.; Richter, R.; Schaepman, M.E. Climatic and soil factors explain the two-dimensional spectrum of global plant trait variation. *Nat. Ecol. Evol.* **2022**, *6*, 36–50. [CrossRef]
36. Hock, M.; Hofmann, R.; Essl, F.; Pyšek, P.; Bruelheide, H.; Erfmeier, A. Native distribution characteristics rather than functional traits explain preadaptation of invasive species to high-UV-B environments. *Divers. Distrib.* **2020**, *26*, 1421–1438. [CrossRef]
37. Hock, M.; Hofmann, R.W.; Müller, C.; Erfmeier, A. Exotic plant species are locally adapted but not to high ultraviolet-B radiation: A reciprocal multispecies experiment. *Ecology* **2019**, *100*, e02665. [CrossRef]
38. Wang, H.; Ma, X.; Zhang, L.; Siemann, E.; Zou, J. UV-B has larger negative impacts on invasive populations of *Triadica sebifera* but ozone impacts do not vary. *J. Plant Ecol.* **2016**, *9*, 61–68.

39. Ballaré, C.L.; Caldwell, M.M.; Flint, S.D.; Robinson, S.A.; Bornman, J.F. Effects of solar ultraviolet radiation on terrestrial ecosystems. Patterns, mechanisms, and interactions with climate change. *Photochem. Photobiol. Sci.* **2011**, *10*, 226–241. [CrossRef]
40. Bornman, J.F.; Barnes, P.W.; Robson, T.M.; Robinson, S.A.; Jansen, M.A.; Ballaré, C.L.; Flint, S.D.J.P. Linkages between stratospheric ozone, UV radiation and climate change and their implications for terrestrial ecosystems. *Photochem. Photobiol. Sci.* **2019**, *18*, 681–716. [CrossRef]
41. Hu, X.-G.; Jin, Y.; Wang, X.-R.; Mao, J.-F.; Li, Y. Predicting impacts of future climate change on the distribution of the widespread conifer *Platycladus orientalis*. *PLoS ONE* **2015**, *10*, e0132326. [CrossRef]
42. Hu, X.-G.; Wang, T.; Liu, S.-S.; Jiao, S.-Q.; Jia, K.-H.; Zhou, S.-S.; Jin, Y.; Li, Y.; El-Kassaby, Y.A.; Mao, J.-F. Predicting future seed sourcing of *Platycladus orientalis* (L.) for future climates using climate niche models. *Forests* **2017**, *8*, 471. [CrossRef]
43. Duque-Lazo, J.; Navarro-Cerrillo, R.M.; Ruíz-Gómez, F.J. Assessment of the future stability of cork oak (*Quercus suber* L.) afforestation under climate change scenarios in Southwest Spain. *Ecol. Manag.* **2018**, *409*, 444–456. [CrossRef]
44. Chen, M.; Wu, H.; Wan, Z.; Liu, J. Study on growth character of *Betula luminifera* of different provenances. *J. For. By-Prod. Spec. China* **2016**, *01*, 22–24. (In Chinese)
45. Molloy, S.W.; Davis, R.A.; Van Etten, E.J.B. Species distribution modelling using bioclimatic variables to determine the impacts of a changing climate on the western ringtail possum (*Pseudocheirus occidentals*; Pseudocheiridae). *Environ. Conserv.* **2014**, *41*, 176–186. [CrossRef]
46. Fischer, G.; Nachtergaele, F.; Prieler, S.; van Velthuizen, H.T.; Verelst, L.; Wiberg, D. *Global Agro-Ecological Zones Assessment for Agriculture (GAEZ 2008)*; IIASA: Laxenburg, Austria; FAO: Rome, Italy, 2008.
47. IPCC. Summary for Policymakers. In *Climate Change 2021: The Physical Science Basis*; Contribution of Working Group I to the Sixth Assessment Report of the Intergovernmental Panel on Climate Change; Masson-Delmotte, V., Zhai, P., Pirani, A., Connors, S.L., P'ean, C., Berger, S., Caud, N., Chen, Y., Goldfarb, L., Gomis, M.I., et al., Eds.; Cambridge University Press: Cambridge, UK, 2021; *in press*.
48. Forster, P.M.; Richardson, T.; Maycock, A.C.; Smith, C.J.; Samset, B.H.; Myhre, G.; Schulz, M. Recommendations for diagnosing effective radiative forcing from climate models for CMIP6. *J. Geophys. Res. Atmos.* **2016**, *121*, 12–460, 475. [CrossRef]
49. Guo, Y.; Zhao, Z.; Li, X. Moderate warming will expand the suitable habitat of *Ophiocordyceps sinensis* and expand the area of *O. sinensis* with high adenosine content. *Sci. Total Environ.* **2021**, *787*, 147605. [CrossRef]
50. Hausfather, Z. CMIP6: The Next Generation of Climate Models Explained. 2019. Available online: https://www.carbonbrief.org/cmip6-the-next-generation-of-climate-models-explained (accessed on 23 May 2024).
51. Fick, S.E.; Hijmans, R.J. WorldClim 2: New 1-kmspatial resolution climate surfaces for global land areas. *Int. J. Clim.* **2017**, *37*, 4302–4315. [CrossRef]
52. Feng, X.; Park, D.S.; Liang, Y.; Pandey, R.; Papeş, M. Collinearity in ecological niche modeling: Confusions and challenges. *Ecol. Evol.* **2019**, *9*, 10365–10376. [CrossRef]
53. Guo, Y.L.; Zhao, Z.F.; Qiao, H.J. Challenges and development trend of species distribution model. *Adv. Earth Sci.* **2020**, *35*, 1292–1305.
54. Zhao, Z.; Guo, Y.; Wei, H.; Ran, Q.; Liu, J.; Zhang, Q.; Gu, W. Potential distribution of *Notopterygium incisum* Ting ex HT Chang and its predicted responses to climate change based on a comprehensive habitat suitability model. *Ecol. Evol.* **2020**, *10*, 3004–3016. [CrossRef]
55. Hallgren, W.; Santana, F.; Low-Choy, S.; Zhao, Y.; Mackey, B. Species distribution models can be highly sensitive to algorithm configuration. *Ecol. Model.* **2019**, *408*, 108719. [CrossRef]
56. Mi, C.; Huettmann, F.; Guo, Y.; Han, X.; Wen, L. Why choose Random Forest to predict rare species distribution with few samples in large undersampled areas? Three Asian crane species models provide supporting evidence. *PeerJ* **2017**, *5*, e2849. [CrossRef]
57. Phillips, S.J.; Anderson, R.P.; Dudík, M.; Schapire, R.E.; Blair, M.E. Opening the black box: An open-source release of Maxent. *Ecography* **2017**, *40*, 887–893. [CrossRef]
58. Thuiller, W.; Lafourcade, B.; Engler, R.; Araújo, M.B. BIOMOD–A platform for ensemble forecasting of species distributions. *Ecography* **2009**, *32*, 369–373. [CrossRef]
59. Allouche, O.; Tsoar, A.; Kadmon, R. Assessing the accuracy of species distribution models: Prevalence, kappa and the true skill statistic (TSS). *J. Appl. Ecol.* **2006**, *43*, 1223–1232. [CrossRef]
60. Merow, C.; Smith, M.J.; Silander, J.A. A practical guide to MaxEnt for modeling species' distributions: What it does, and why inputs and settings matter. *Ecography* **2013**, *36*, 1058–1069. [CrossRef]

plants

MDPI

Article

Potential Suitable Habitats of Chili Pepper in China under Climate Change

Changrong Deng, Qiwen Zhong, Dengkui Shao, Yanjing Ren, Quanhui Li, Junqin Wen and Jianling Li *

State Key Laboratory of Plateau Ecology and Agriculture, Academy of Agricuture and Forestry Sciences, Qinghai Key Laboratory of Vegetable Genetics and Physiology, Laboratory for Research and Utilization of Qinghai Tibet Plateau Germplasm Resources, Qinghai University, Xining 810016, China; dengchang_rong@126.com (C.D.); 13997135755@163.com (Q.Z.); 2006990015@qhu.edu.cn (D.S.); renyan0202@163.com (Y.R.); liquanhui_2008@163.com (Q.L.); 2021990055@qhu.edu.cn (J.W.)
* Correspondence: lijianling@qhu.edu.cn

Abstract: Chili pepper (*Capsicum annuum* L.) is extensively cultivated in China, with its production highly reliant on regional environmental conditions. Given ongoing climate change, it is imperative to assess its impact on chili pepper cultivation and identify suitable habitats for future cultivation. In this study, the MaxEnt model was optimized and utilized to predict suitable habitats for open-field chili pepper cultivation, and changes in these habitats were analyzed using ArcGIS v10.8. Our results showed that the parameter settings of the optimal model were FC = LQPTH and RM = 2.7, and the critical environmental variables influencing chili pepper distribution were annual mean temperature, isothermality, maximum temperature of the warmest month, and precipitation of the warmest quarter. Under current climate conditions, suitable habitats were distributed across all provinces in China, with moderately- and highly-suitable habitats concentrated in the east of the Qinghai–Tibetan Plateau and south of the Inner Mongolia Plateau. Under future climate scenarios, the area of suitable habitats was expected to be larger than the current ones, except for SSP126-2050s, and reached the maximum under SSP126-2090s. The overlapping suitable habitats were concentrated in the east of the Qinghai–Tibetan Plateau and south of the Inner Mongolia Plateau under various climate scenarios. In the 2050s, the centroids of suitable habitats were predicted to shift towards the southwest, except for SSP126, whereas this trend was reversed in the 2090s. Our results suggest that climate warming is conductive to the cultivation of chili pepper, and provide scientific guidance for the introduction and cultivation of chili pepper in the face of climate warming.

Keywords: chili pepper; *Capsicum annuum* L.; climate warming; MaxEnt; potential distribution

1. Introduction

Vegetables are rich in essential nutrients, such as antioxidants, vitamins, minerals, and dietary fiber, and are crucial for maintaining overall well-being [1]. Incorporating a diverse range of vegetables into our daily meals has been proven to bolster our immune system, reduce the risk of chronic ailments, and contribute to a healthier and more prolonged lifespan [2]. This recognition has led to a significant increase in demand for vegetables, with consumption rates doubling over the past two decades, rising from 0.55 billion tons in 1997 to 1.09 billion tons in 2017 [3].

However, as the frequency and intensity of extreme weather have increased and are expected to accelerate with further climate warming, there is a growing concern for the adverse and possibly irreversible impacts on Earth's organisms and ecosystems [4]. Agricultural systems, including vegetable production, are highly dependent on specific environmental conditions, making them particularly vulnerable to climate change [5–7]. Estimates indicated that there could be an average reduction of 17% in crop yields for each one-degree Celsius rise in temperature [8]. Climate change affects the growth and production of crops by altering the distributional suitability of crops. Therefore, climate warming not only affects farmers' incomes, but also poses a threat to global food security [9].

Vegetables mostly require a mild temperature for their growth and development, and are highly sensitive to climate fluctuations [10]. Climate vagaries, such as heatwaves, droughts, and floods, directly affect vegetable production at any stage of the crop growth cycle, from initial growth to pollination, flowering, fruit setting, and yield development. However, previous studies have primarily focused on the effects of climate change on the distribution of staple crops, such as wheat [11], rice [12], maize [13], and soybean [14], with less attention to vegetable crops [15]. Given the crucial role of vegetable crops in the global food system, there is a pressing need to identify how climate change influences their distribution. This work will provide valuable insights for scientific cultivation practices, effective assessments of agricultural disaster risks, and sustainable agricultural strategies to ensure food security in the face of changing climatic conditions [16,17].

Species distribution models (SDMs) are powerful tools for simulating the geographic distribution of species based on the available distributional information and corresponding environmental data, and have been widely applied in the prevention of invasive species [18], and the cultivation [19] and protection [20] of species. Among the available SDMs, the maximum entropy model (MaxEnt) outperforms the others for its high accuracy and stability, rapid calculation, and flexible operation [21,22].

Chili pepper (*Capsicum annuum* L.), one of the oldest domesticated cash crops, originated in Central and South America and has been cultivated extensively all over the world, with the current cultivation area reaching approximately 3.8 million hectares [23,24]. South Asia is globally recognized as the most prominent region for chili pepper cultivation, accounting for approximately 55% of the total world production. India contributes the next largest proportion (38%), followed by China (7%), while Pakistan, Peru, and Bangladesh collectively contribute 5% [24].

Chili pepper was introduced to China at the end of the 16th century [25]. After over 400 years of cultivation and culinary development, it has become an integral vegetable and spice of local cuisine in China for its nutritional values and diverse flavors [26]. Due to its short growth cycle, low production cost, and high market demand [26], many regions in China have introduced chili peppers for open-field cultivation to increase revenue. Nowadays, China is the largest fresh chili pepper producer with an annual production exceeding 18 million tons [24]. Chili pepper is playing a crucial role in the revitalization of rural areas. Nonetheless, chili pepper is a thermophilic vegetable [27], and reckless introduction to unsuitable regions may lead to a series of issues, such as increased production costs, decreased yields and lower profitability [28]. Additionally, the warming climate has been resulting in more frequent extreme weather, which compromises the growth of chili pepper and dramatically decreases yields [29]. Therefore, it is imperative to determine the suitable habitats for chili pepper under climate change, in order to provide scientific guidance for cultivation practices and ensure sustainable development of chili pepper industry.

In this study, the MaxEnt model was optimized and utilized to predict the current and future suitable habitats of chili peppers for open-field cultivation under different climate scenarios, as well as to identify the predominant environmental variables influencing chili pepper distribution. Subsequently, spatiotemporal changes and centroid shifts in the suitable habitats were analyzed using ArcGIS. This study will provide a theoretical basis for the introduction and cultivation of chili peppers.

2. Results

2.1. Screening of Distribution Points and Environmental Variables, and Accuracy of MaxEnt Prediction

After screening, 369 of 732 distribution points and 14 (six climate and eight topsoil variables) of 36 environmental variables were selected for MaxEnt prediction (Figure 1). Based on the results output by Kuenm, when FC and RM were separately set to LQPTH and 2.7, we found ΔAICc was 0, which was the best candidate mode for MaxEnt.

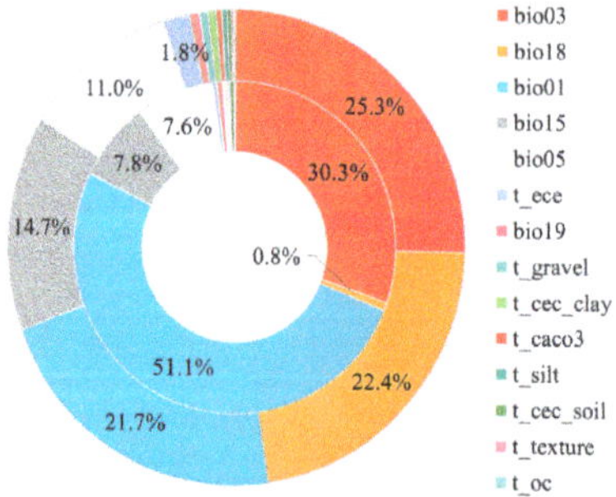

Figure 1. Percent contribution (outer ring) and permutation importance (inner ring) of environmental variables.

The current suitable habitats of chili pepper were simulated using the optimal model based on the screening distribution points and environmental variables. The simulation results showed that the training omission rate was very close to the predicted omission (Figure S1), and the average training AUC and TSS were 0.974 and 0.927, indicating that the reconstructed model was highly reliable and qualified for the following predictions.

2.2. Critical Environmental Variables Affecting Chili Pepper Distribution

The critical environmental variables affecting the distribution of chili pepper were determined via the MaxEnt jackknife test. As shown in Figure 1, the cumulative contributions and permutation importance of climate variables were 95.8% and 98.2%, indicating that the distribution of chili pepper was primarily affected by climate rather than soil. The results of the jackknife test of variable importance showed that bio01 (annual mean temperature), bio05 (maximum temperature of the warmest month), bio03 (isothermality), and bio18 (precipitation of the warmest quarter) had higher weights in single variables analysis (Figure S2), with a cumulative contribution rate of 80.4% and permutation importance of 89.8% (Figure 1), indicating that these variables possessed more effective information in chili pepper distribution than the others. Among the four variables, bio01 was the environmental variable with highest gain when used in isolation, which appeared to have the most useful information by itself; bio03 was the environmental variable that decreased the gain the most when it was omitted, which appeared to have the most information that was not present in the others (Figure S2).

The relationships between the distributional probability and environmental variables were identified using single-factor response curves output by the MaxEnt model. The suitable ranges [distribution probability > MTSPS (0.1575)] of bio01, bio03, bio05, and bio18 for chili pepper were 4.08–24.75 °C, 21.61–48.90%, 23.75–33.62 °C, and >224.85 mm, respectively (Figure S3).

2.3. Current Distribution of Chili Pepper

Under current climate conditions, the simulation results were highly consistent with the actual cultivation of chili pepper (Figure 2a). A total of 91.60% of the screening distribution points were located in suitable habitats, of which 79.13% were concentrated in moderately- and highly-suitable habitats (Figure 2b), indicating that the results were accurate and reliable.

The current suitable area for chili pepper cultivation was 4,426,594.63 km^2 in China (Figure 3), accounting for 46.11% of China's land area and distributed across all provinces, mainly in 18° N–46° N and 108° E–126° E (Figure 2a). The moderately- and highly-suitable area was 2,913,626.88 km^2, making up 65.82% of the total suitable area, which was concentrated in the east of the Qinghai–Tibetan Plateau and south of the Inner Mongolia Plateau. Although there were some suitable habitats for chili pepper cultivation in the Tibetan Plateau, they were mainly located in the north, east and southeast marginal regions. Among all provinces of China, Qinghai had the smallest suitable area, with only 8800.04 km^2, and Heilongjiang was the only one without moderately- and highly-suitable habitats (Figure 2a).

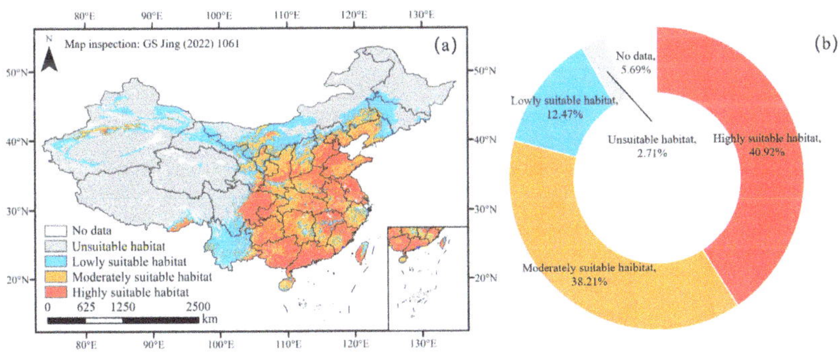

Figure 2. Current suitable habitats (**a**) and proportion of distribution points in different habitats (**b**) of chili pepper.

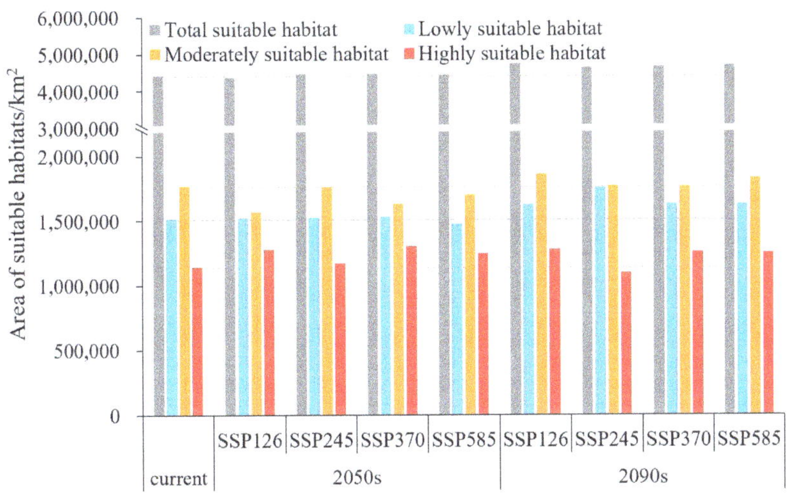

Figure 3. Current and future suitable area for chili pepper cultivation under different climate scenarios.

2.4. Future Distribution of Chili Pepper

The potential distribution of chili pepper in the 2050s and 2090s under four climate scenarios (SSP126, SSP245, SSP370, and SSP585) were predicted using the optimal MaxEnt model (Figure 4). Under future climate scenarios, the suitable habitats of chili pepper were distributed across all provinces of China, and the moderately- and highly-suitable habitats were concentrated in the east of the Qinghai–Tibetan Plateau and the south of the Inner Mongolia Plateau, which were basically consistent with the current ones (Figure 4).

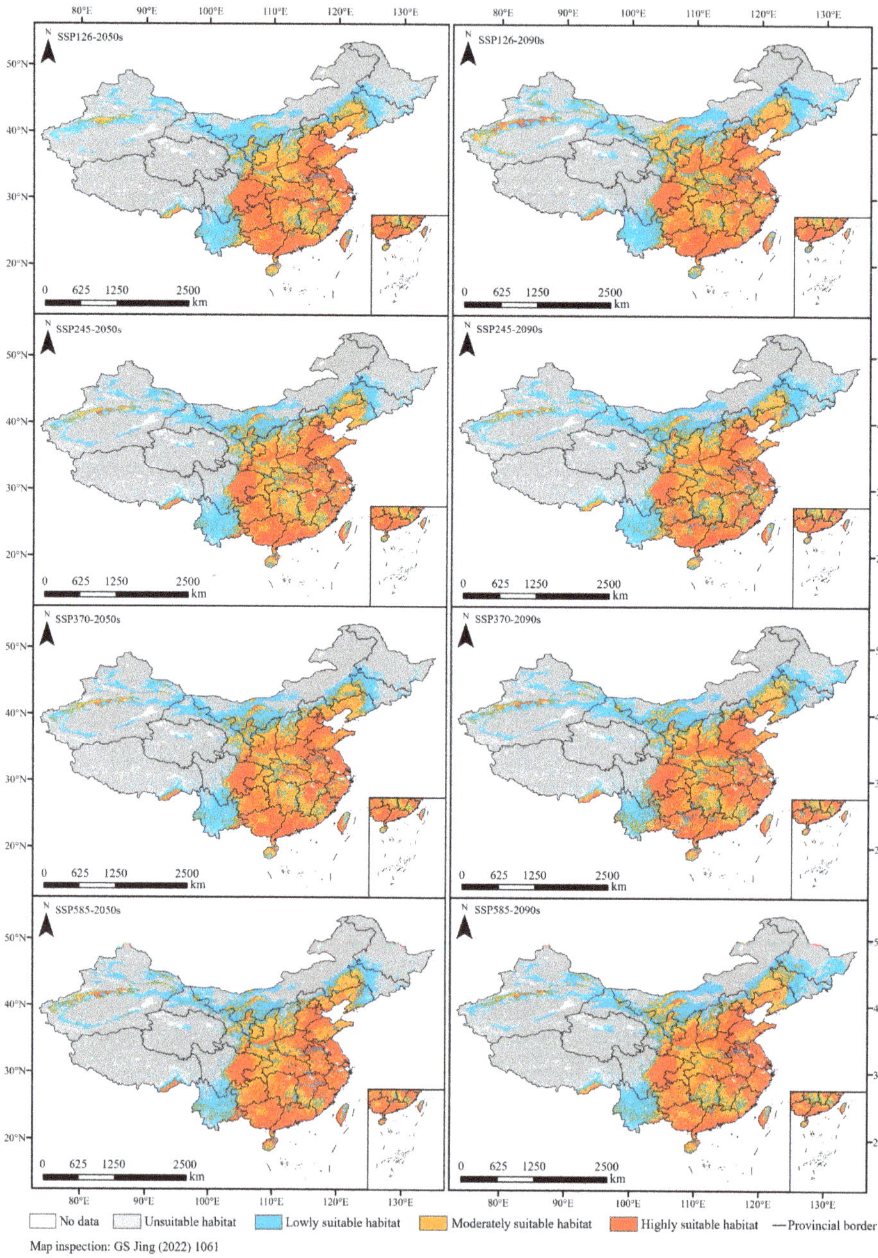

Figure 4. Future suitable habitats under different climate scenarios.

In the 2050s, the area of suitable habitats initially experienced an increase and subsequently declined in response to greenhouse gas emissions, reaching its maximum (4,462,849.42 km^2) under SSP370 with a growth rate of 0.82%. Moreover, the moderately- and highly-suitable area peaked (2,949,743 km^2) with a 1.24% increase under the SSP585 scenario (Figure 3). Notably, both the total suitable and the moderately- and highly-suitable area were all larger than the current levels, except for SSP126.

In the 2090s, the changes in suitable area were contrary to those observed in the 2050s, with an initial decrease followed by an increase. The area of suitable habitats expanded in different degrees compared with that of the current climate, with the highest growth rate of 7.49% under SSP126, followed by SSP585 (6.28%), SSP370 (5.08%), and SSP245 (4.43%) (Figure 3). The moderately- and highly-suitable habitats shrank by 1.68% under SSP245, and the other scenarios expanded by 3.73–7.63%.

2.5. Future Spatiotemporal and Centroid Changes in Suitable Habitats

Compared with the current situation, 92.91–96.82% of suitable habitats under future climatic scenarios remained unchanged (Figure 5); they are mainly located in the south of the Inner Mongolia Plateau and the east of the Tibetan Plateau (Figure 6). The expansion area under different climate scenarios of the 2050s and 2090s was all larger than the contraction, with a 0.01–10.21-fold increase over the contraction, except for SSP126-2050s (expansion 143,826.25 km^2, contraction 207,691.39 km^2). Furthermore, the expansion area under future 2090s climate scenarios exceeded those of the future 2050s, reaching a maximum of 364,035 km^2 under SSP126-2090s, whereas the contraction area was surpassed by the latter.

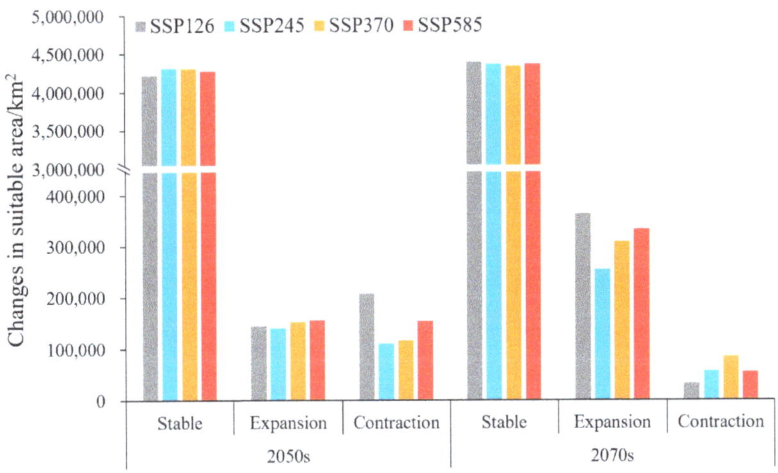

Figure 5. Spatiotemporal changes in the future suitable area compared with the current.

In the future 2050s, the expansion of suitable habitats was distributed in the west of the Inner Mongolia Plateau, and the contraction was primarily distributed in the central regions of the Inner Mongolia Plateau (Figure 6). In the 2090s, the expansion regions were concentrated in the northeast of the current ranges and the west of the Inner Mongolia Plateau, while the contraction regions were limited and uncertain. They were mainly situated in the southern Yunan Province under SSP126-2090s, and migrated to the central regions of the Inner Mongolia Plateau under the other climate scenarios (Figure 6).

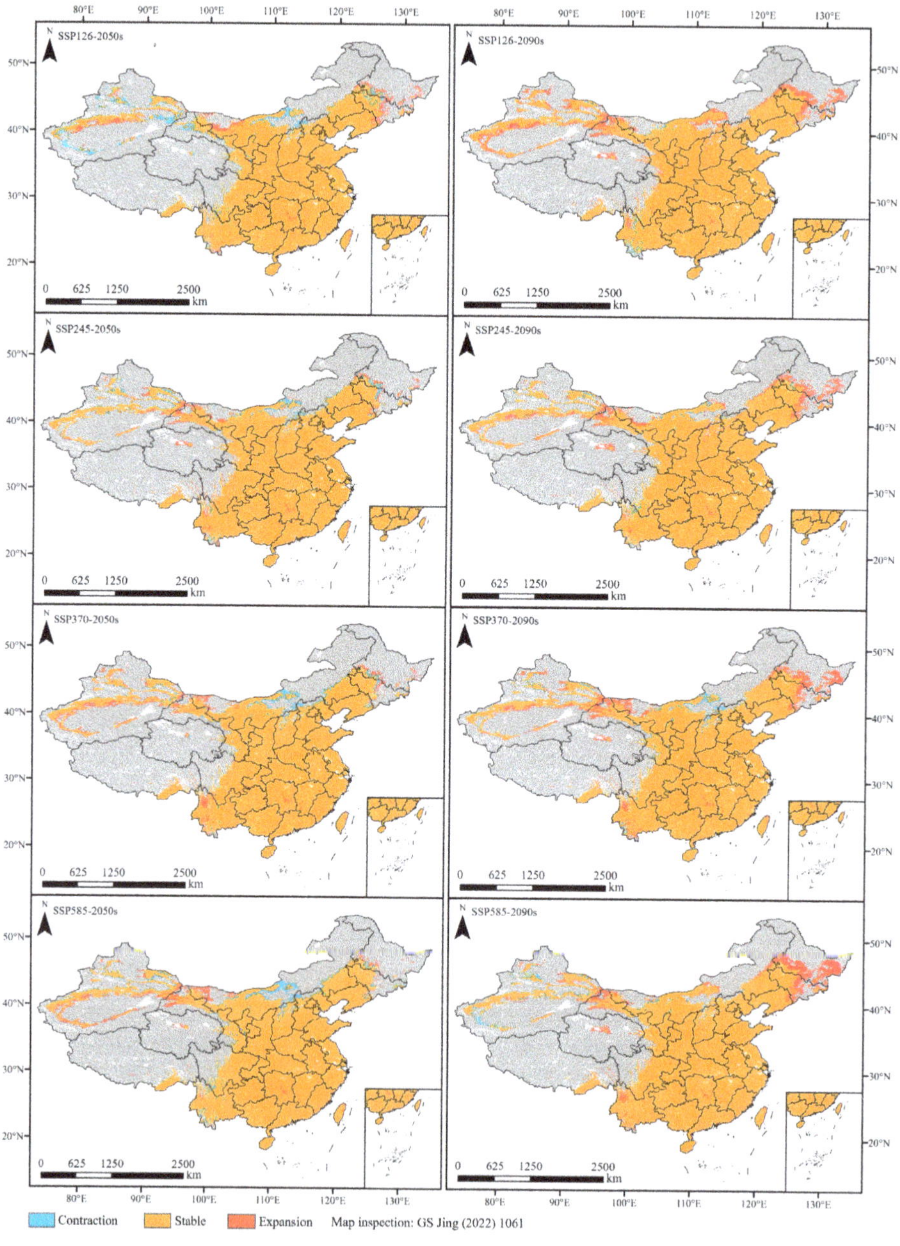

Figure 6. Spatiotemporal changes in the future suitable habitats of chili pepper compared with the current.

Climate warming was anticipated to induce minor fluctuations in the distribution of pepper cultivation, with most regions likely experiencing negligible changes (Figures 6 and 7). The overlapping suitable habitats, encompassing an area of 4,101,218 km^2, were distributed across all provinces of China and concentrated in the east of the Qinghai–Tibetan Plateau and south of the Inner Mongolia Plateau. These unchanged habitats accounted for 42.72% of China's land area and 92.65% of the current suitable habitats, which were always suitable for chili pepper cultivation under climate change. The non-overlapping suitable habitats covered an area of 882,200 km^2, accounting for 21.51% of the overlapping area, which was mainly distributed in the north of the overlapping ranges (Figure 7). The suitability of these regions exhibited instability in the face of climate change.

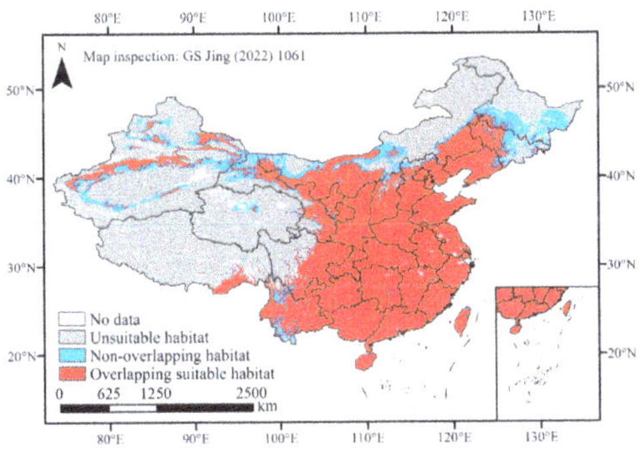

Figure 7. Overlapping and non-overlapping suitable habitats under climate warming.

2.6. Centroid Shifts in Suitable Habitats

Under the current and future (2050s and 2090s) climate scenarios, the centroids of the suitable habitats for chili pepper cultivation were all located in southeast Shaanxi Province, and the distances between the future and the current centroids (33.58° N, 109.65° E) were 13.30–76.84 km (Figure 8). In the 2050s, the centroids of suitable habitats mainly shifted to the southwest of the current ones, except SSP126, which shifted to the southeast (33.47° N, 110.03° E). In the 2090s, the changes in centroids exhibited an inverse pattern compared to those observed in the 2050s, which shifted towards the northeast of the 2050s ones, except SSP126. The centroid (34.215° N, 109.43° E) of SSP126-2090s migrated to the northwest of that in the 2050s.

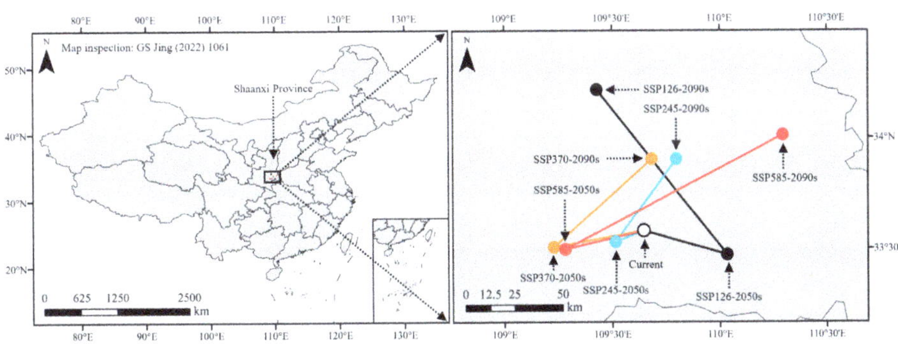

Figure 8. Changes in the centroids of suitable habitats under different climate scenarios.

3. Discussion

3.1. Dominant Environmental Variables Affecting the Distribution of Chili Pepper

For plants, temperature and precipitation have pivotal influences on their development and distribution [30]. We found that the contribution (95.8%) and the permutation (98.2%) of climate variables significantly outweighed than those of soil variables, indicating that climate had a much greater impact on chili pepper cultivation. This characteristic was reported in many plants, including cash [19] and cereal [31] crops.

In this study, we found that annual mean temperature (bio01), isothermality (bio03), maximum temperature of the warmest month (bio05) and precipitation of the warmest quarter (bio18) were the dominant variables influencing chili pepper distribution. The suitable ranges of bio01 and bio05 were 4.08–24.75 °C and 23.75–33.62 °C, respectively, indicating that the distribution of chili peppers is primarily limited to regions with warm climates, which were connected with the habit of chili peppers. Chili peppers are originally from tropical regions and require relatively high temperature for development [32]. The optimal temperature for growth is between 25–30 °C, and when the temperature is below 15 °C and above 35 °C, chili pepper growth is retarded and their yield decreases [33]. Low temperature usually induces deformed and seedless fruit [34], and high temperature inhibits fruit set [35]. Isothermality reflects the magnitude of day to night temperature oscillation relative to seasonal variation, serving as an indicator for the temperature fluctuations within months to years [36]. We found that the suitable range of bio03 was 21.61–48.9%, with an optimal value of 28.43%.

As an annual thermophilic plant, the precipitation of the warmest quarter is an important variable in determining chili pepper distribution. Our study revealed that the minimum precipitation of the warmest quarter suitable for chili pepper distribution under current climate conditions was 224.85 mm, which was basically consistent with the water requirements throughout the entire growth period of the vegetable. Previous studies have demonstrated that the total water requirement for the growth of pepper is approximately 280 mm [37,38]. The discrepancy between the minimum precipitation of the warmest quarter and the actual water requirements can be compensated through precipitation of other quarters or irrigation [39]. Moreover, we found the minimum precipitation of the warmest quarter for chili pepper increased in the future 2050s and 2090s. The reason for this is that the rise in temperature results in an elevation of evaporation, consequently leading to an augmentation in the water requirement of chili pepper [40].

3.2. Habitat Distribution under Climate Change

Under current climate conditions, the MaxEnt simulation results showed that the suitable habitats of chili peppers were distributed across all provinces, the predominant distribution ranged between 18° N–46° N and 108° E–126° E, and the moderately- and highly-suitable habitats were concentrated in the east of the Qinghai–Tibetan Plateau and south of the Inner Mongolia Plateau of China. The simulated suitable habitats were highly consistent with the actual cultivation of chili peppers [26,41]. For example, our results indicated that Xining prefecture-level city in Qinghai Province was unsuitable for chili pepper cultivation, while the Xunhua county of Haidong prefecture-level city bordering Xining was lowly suitable. These findings aligned with the current practices of chili pepper cultivation in these regions. In Xining, greenhouse cultivation is imperative for successful chili pepper production [41], while open-air cultivation is widely employed in Xunhua, establishing it as a crucial chili pepper production region in Qinghai [42]. Meanwhile, despite the cultivation area of chili pepper in Yunnan exceeding 1700 km^2 and ranking among the top three in China [41], our prediction results showed that most regions of this province exhibited low suitability. Previous results have demonstrated that the production cost in Yunnan was 1.5 times higher than its neighboring Province, Guizhou, which was moderately- and highly-suitable for the vegetable cultivation, while the output in Yunnan only accounted for only 81.79% of that in Guizhou [43].

Under 2050s climate scenarios, the area of suitable habitats remained essentially unchanged compared with the current ones. However, under 2090s climate scenarios, the suitable area was all higher than the current area and reached the maximum under SSP126. Our findings indicated that the impacts of climate warming under different greenhouse gas emission modes on chili pepper cultivation varied and would become more favorable over time. This phenomenon is attributed to the thermophilic habit of chili peppers [27], which makes them sensitive to temperature, and their expansion is facilitated by the warming climate. Moreover, climate warming also leads to an increase in precipitation [44], which creates more favorable conditions for the growth of chili pepper. We found that the expansion-suitable regions were mainly distributed in the northeast and northwest of the current ranges. This phenomenon of thermophilic plants expanding towards higher latitudes in response to climate warming has been widely documented, with examples including *Lycium barbarum* L. [45], *Litsea cubeba* (Lour.) Pers [46], and *Agastache rugosa* (Fisch. & C. A. Mey.) Kuntze [20]. Although there will be more suitable habitats for chili pepper growth, this does not guarantee higher yields. Previous study has demonstrated that extreme climate warming scenarios retarded fruit morphological features and production of hot chili pepper (*C. annuum*) [47].

Climate exerts a predominant influence on the physiology, distribution, and phenology of plants, thereby potentially inducing shifts in suitable habitats [20,48]. We found that the overlapping suitable habitats of chili pepper were primary distributed in the east of the Qinghai–Tibetan Plateau and south of the Inner Mongolia Plateau under various climate scenarios. These regions consistently maintained their suitability despite climate change. Therefore, we suggest that cultivating open-air chili peppers in these regions could mitigate the impacts of climate change, facilitate vegetable growth and development, and ensure sustainable production.

3.3. Limitations and Prospects

In this study, MaxEnt and ArcGIS were employed to predict the suitable habitats for open-field chili pepper cultivation and analyze its changes under climate change. Our study provided scientific guidance for the introduction and cultivation of chili peppers. However, there are still some uncertainties in our study. First, plant distribution is not only influenced by climate and soil, but also by other variables, such as pests and diseases, and agronomic management [12]. Second, while environment affects plant distribution, plants are constantly adapting to the environment [49]. Plant physiological responses, including growth responses to elevated atmospheric CO_2 and alterations in water use efficiency, are expected to mitigate the response of some plant functional types to climate change [48]. However, plant adaption is considered unchanged when SDMs predict species' distribution. Third, the study did not consider the environmental adaptability of different chili pepper varieties, which may vary in suitable cultivation regions [50]. Therefore, the predicted potentially suitable habitats maybe deviate from the actuality. Future research should take into account these uncertainties to achieve a more accurate prediction for chili pepper cultivation under changing climatic conditions.

4. Materials and Methods

4.1. Acquisition and Processing of Chili Pepper Distribution Points and Environmental Data

The distribution points of chili pepper (*C. annuum*) in China were collected from the Global Biodiversity Information Facility (https://www.gbif.org/, assessed on 20 January 2024, Plant Science Data Center (https://www.plantplus.cn/cn/, assessed on 20 January 2024), and our survey data in 2021–2023. Overall, we collected 732 distribution points across all provinces in China (Figure 9). Because the environmental characters of greenhouse-grown chili peppers are mainly manipulated by human intervention, they are less affected by climate change, and these points were removed based on our investigations. Meanwhile, the residual points were filtered using the 'Trim duplicate occurrence' function of ENMtools v1.3 (http://enmtools.blogspot.com/, accessed on 13 November 2012) to avoid the overfitting of the MaxEnt model predicted results.

The current (averages for 1970–2000), future 2050s (averages for 2041–2060), and 2090s (averages for 2081–2100) climate data were downloaded from the Worldclim Database (WorldClim v2.1, https://www.worldclim.org/, assessed on 16 January 2024) with a spatial resolution of 2.5 arc-minutes (~5 km) and converted to ASCII format using ArcGIS v10.8 (https://www.esri.com/zh-cn/arcgis/, assessed on 25 April 2023). Future climatic data were determined based on the Beijing Climate Center Climate System Model (BCC-CSM) from the sixth phase of the Coupled Model Intercomparison Project (CMIP6), which included four climate scenarios based on the Shared Socio-economic Pathways (SSP126, SSP245, SSP370 and SSP585) [51]. These scenarios represented future climate scenarios with low to high greenhouse gas emissions. Each scenario had 19 climatic variables (bio01–bio19).

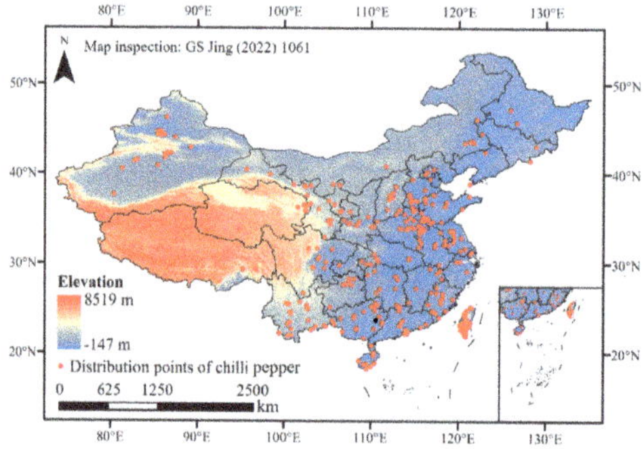

Figure 9. Distribution points of chili pepper in China.

The topsoil data were obtained from the World Soil Database (Harmonized World Soil Database v1.2, http://www.fao.org/soilsportal/, assessed on 16 January 2024) and converted to ASCII format using ArcGIS, which contained 17 soil variables. Due to lack of future soil data, the future soil layers were considered to be consistent with the current over such a short time frame in this study [45].

There were 36 environmental variables (19 climate and 17 soil variables) initially used to construct the MaxEnt model. To avoid overfitting of the MaxEnt model causing by multicollinearity among environmental variables, the current environmental data of the screening distribution points were extracted using ArcGIS, and their correlations were examined using Pearson's correlation analysis of SPSS (v26, https://www.ibm.com/cn-zh/spss/, assessed on 26 April 2023). The contribution rates of the current environmental variables were calculated using the jackknife analysis of MaxEnt. The variables with zero contribution were removed, and only the variables with the highest contribution were retained when the absolute correlation coefficient among them was greater than 0.7 [12].

4.2. Optimization and Evaluation of MaxEnt Model

Feature classes (FCs) and regularization multiplier (RM) are critical parameters affecting the accuracy of the MaxEnt (v3.4.4, https://github.com/mrmaxent/Maxent/, assessed on 20 April 2023) model, and default settings may result in model over-fitting [52]. Thus, the Kuenm package (https://github.com/marlonecobos/kuenm/, assessed on 14 December 2020) in R (v3.6.3, https://www.r-project.org/, assessed on 11 December 2020) was employed to calibrate the two parameters to select the best combination for MaxEnt [48].

MaxEnt contains five different FCs: linear (L), quadratic (Q), hinge (H), product (P), and threshold (T), and there is a total of 31 FCs combinations. Forty RMs (0.1–4.0 at an interval of 0.1) and 31 FCs combinations were used to generated 1240 candidate modes, and these modes were evaluated using Kuenm based on the screening environmental variables. The best candidate mode for MaxEnt was selected according to the following criteria: significant models with omission rates $\leq$5%, and the lowest delta-corrected Akaike information criterion (ΔAICc) values of $\leq$2% [53].

The other parameters of MaxEnt were selected as follows: 'Create responsive curves', 'Do jackknife to measure variable importance', 'Out format logistic', 'Random seed', 'Random test percentage 25', 'Replicates 10', 'Replicated run type bootstrap', 'Write plot data', and 'Write background predictions' [46]. The rest of the parameters was set to default.

The performance of the optimal MaxEnt prediction was assessed using the area under the receiver operating characteristic curve (AUC), and true skill statistic (TSS) [54] under current climate conditions. AUC and TSS values range from 0 to 1 and -1 to 1, respectively.

The closer the two values are to 1, the better the model performs. AUC > 0.9 and TSS > 0.8 indicate the prediction of MaxEnt is highly reliable and excellent [55].

4.3. Reclassification and Calculation of Suitable Habitats

The prediction results (ASCII files) output by the optimal MaxEnt were reclassified and visualized using ArcGIS. The average logistic threshold value of maximum training sensitivity plus specificity (MTSPS) output by MaxEnt based on the current environmental variables was used to classify these results into suitability and unsuitability for chili pepper [45]. According to the suitable probability, the regions for chili pepper cultivation were divided into unsuitable habitat (0–MTSPS), lowly suitable habitat (MTSPS–0.4), moderately suitable habitat (0.4–0.6), and highly suitable habitat (0.6–1) using the reclassification function of ArcGIS. The proportion of each habitat to China overall was calculated based on its grid number, and the area of each habitat was calculated according to China's land area [45].

4.4. Spatiotemporal and Centroid Changes in Suitable Habitats

SDMtoolbox v2.0 (http://www.sdmtoolbox.org/, assessed on 20 April 2023) was employed to convert the prediction ASCII files to binary files (0 unsuitability, 1 suitability) using the MTSPS threshold, and then applied to analyze the spatiotemporal changes and centroid shifts in the suitable habitats under different climate scenarios. The overlapped suitable habitats under current and future climate scenarios were determine using the 'plus' function of Spatial Analyst Tools. The changes in the area were calculated using the method mentioned in Section 4.3.

5. Conclusions

In this study, we optimized the MaxEnt model and employed it to predict the suitable habitats for open-field chili pepper cultivation in China under different climate scenarios. Our findings showed that annual mean temperature, isothermality, maximum temperature of the warmest month, and precipitation of the warmest quarter were crucial environmental variables influencing chili pepper distribution. Under current and future climate scenarios, suitable habitats were distributed across all provinces in China, with the moderately- and highly-suitable habitats concentrated in the east of the Qinghai–Tibetan Plateau and south of the Inner Mongolia Plateau. Notably, the areas of suitable habitats under future climate scenarios were all larger than the current ones, except for SSP126-2050s. The expansion habitats were mainly distributed in the west of the Inner Mongolia Plateau and the northeast of the current ranges. Moreover, the overlapping suitable habitats with stable suitability were primarily distributed in the east of the Qinghai–Tibetan Plateau and south of the Inner Mongolia Plateau under various climate scenarios. The centroids of suitable habitats shifted to the southwest in the 2050s, except for SSP126, whereas this trend was reversed in the 2090s. Our results provide guidance for chili pepper growers in selecting suitable cultivation regions while mitigating the adverse impacts of climate change. In order to attain more stable yields, we suggest selecting and cultivating varieties that possess adaptability to environmental fluctuations.

Supplementary Materials: The following supporting information can be downloaded at https://www.mdpi.com/article/10.3390/plants13071027/s1, Figure S1: Average omission and predicted area for chili pepper; Figure S2: Jackknife of regularized training gain for chili pepper; Figure S3. Response curves of the critical environmental variables.

Author Contributions: Conceptualization, C.D., Q.Z. and J.L.; methodology, C.D. and J.L.; software, C.D. and J.L.; validation, C.D., Q.Z., D.S. and J.L.; formal analysis, C.D. and J.L.; investigation, C.D., D.S., Y.R., Q.L. and J.W.; resources, C.D., D.S., Y.R., Q.L. and J.W.; data curation, C.D., Q.Z. and D.S.; writing—original draft preparation, C.D. and J.L.; writing—review and editing, C.D. and J.L.; visualization, C.D. and J.L.; supervision, C.D. and J.L.; project administration, Q.Z.; funding acquisition, C.D. All authors have read and agreed to the published version of the manuscript.

Funding: This research was funded by the Natural Science Foundation of Qinghai Province [Grant No. 2022-ZJ-947Q], and the Open Project of State Key Laboratory of Plateau Ecology and Agriculture, Qinghai University [Grant No. 2022-ZZ-11].

Data Availability Statement: Data is contained within the article.

Acknowledgments: Thanks to the anonymous reviewers for their constructive and valuable comments, and to the editors for their help in refining this article.

Conflicts of Interest: The authors declare no conflicts of interest.

References

1. Slavin, J.L.; Lloyd, B. Health benefits of fruits and vegetables. *Adv. Nutr.* **2012**, *3*, 506–516. [CrossRef]
2. Parajuli, R.; Thoma, G.; Matlock, M.D. Environmental sustainability of fruit and vegetable production supply chains in the face of climate change: A review. *Sci. Total Environ.* **2019**, *650*, 2863–2879. [CrossRef] [PubMed]
3. Dong, J.; Gruda, N.; Li, X.; Tang, Y.; Zhang, P.; Duan, Z. Sustainable vegetable production under changing climate: The impact of elevated CO_2 on yield of vegetables and the interactions with environments-A review. *J. Clean. Prod.* **2020**, *253*, 119920. [CrossRef]
4. Tippett, M.K. Extreme weather and climate. *Clim. Atmos. Sci.* **2018**, *1*, 45. [CrossRef]
5. Lobell, D.B.; Gourdji, S.M. The influence of climate change on global crop productivity. *Plant Physiol.* **2012**, *160*, 1686–1697 [CrossRef] [PubMed]
6. Alotaibi, M. Climate change, its impact on crop production, challenges, and possible solutions. *Not. Bot. Hortic. Agrobo.* **2023**, *51*, 13020. [CrossRef]
7. Knox, J.; Hess, T.; Daccache, A.; Wheeler, T. Climate change impacts on crop productivity in Africa and South Asia. *Environ. Res. Lett.* **2012**, *7*, 34032. [CrossRef]
8. Lobell, D.B.; Asner, G.P. Climate and management contributions to recent trends in US agricultural yields. *Science* **2003**, *299*, 1032. [CrossRef] [PubMed]
9. Ali, S.; Makanda, T.A.; Umair, M.; Ni, J. MaxEnt model strategies to studying current and future potential land suitability dynamics of wheat, soybean and rice cultivation under climatic change scenarios in East Asia. *PLoS ONE* **2023**, *18*, e296182. [CrossRef]
10. Solankey, S.S.; Kumari, M.; Kumar, M. *Advances in Research on Vegetable Production Under a Changing Climate*; Springer International Publishing: Berlin, Germany, 2021; Volume 1, pp. 13–17.
11. Guo, X.; Zhang, P.; Yue, Y. Prediction of global wheat cultivation distribution under climate change and socioeconomic development. *Sci. Total Environ.* **2024**, *919*, 170481. [CrossRef]
12. Li, X.; Wu, K.; Hao, S.; Yue, Z.; Ran, Z.; Ma, J. Mapping cropland suitability in China using optimized MaxEnt model. *Field Crop Res.* **2023**, *302*, 109064. [CrossRef]
13. Fitzgibbon, A.; Pisut, D.; Fleisher, D. Evaluation of Maximum Entropy (Maxent) machine learning model to assess relationships between climate and corn suitability. *Land* **2022**, *11*, 1382. [CrossRef]
14. Gong, L.; Li, X.; Wu, S.; Jiang, L. Prediction of potential distribution of soybean in the frigid region in China with MaxEnt modeling. *Ecol. Inform.* **2022**, *72*, 101834. [CrossRef]
15. Bisbis, M.B.; Gruda, N.; Blanke, M. Potential impacts of climate change on vegetable production and product quality–A review. *J. Clean. Prod.* **2018**, *170*, 1602–1620. [CrossRef]
16. Koundinya, A.V.V.; Kumar, P.P.; Ashadevi, R.K.; Hegde, V.; Kumar, P.A. Adaptation and mitigation of climate change in vegetable cultivation: A review. *J. Water Clim. Chang.* **2018**, *9*, 17–36. [CrossRef]
17. Abbass, K.; Qasim, M.Z.; Song, H.; Murshed, M.; Mahmood, H.; Younis, I. A review of the global climate change impacts, adaptation, and sustainable mitigation measures. *Environ. Sci. Pollut. R.* **2022**, *29*, 42539–42559. [CrossRef]
18. Xian, X.; Zhao, H.; Guo, J.; Zhang, G.; Liu, H.; Liu, W.; Wan, F. Estimation of the potential geographical distribution of a new potato pest (*Schrankia costaestrigalis*) in China under climate change. *J. Integr. Agr.* **2023**, *22*, 2441–2455. [CrossRef]
19. Yang, Y.; He, J.; Liu, Y.; Zeng, J.; Zeng, L.; He, R.; Guiang, M.M.; Li, Y.; Wu, H. Assessment of Chinese suitable habitats of *Zanthoxylum nitidum* in different climatic conditions by Maxent model, HPLC, and chemometric methods. *Ind. Crop Prod.* **2023**, *196*, 116515. [CrossRef]
20. Wang, Y.; Zhao, R.; Zhou, X.; Zhang, X.; Zhao, G.; Zhang, F. Prediction of potential distribution areas and priority protected areas of *Agastache rugosa* based on Maxent model and Marxan model. *Front. Plant Sci.* **2023**, *14*, 1200796. [CrossRef] [PubMed]
21. Yackulic, C.B.; Chandler, R.; Zipkin, E.F.; Royle, J.A.; Nichols, J.D.; Campbell Grant, E.H.; Veran, S. Presence-only modelling using MaxEnt: When can we trust the inferences? *Methods Ecol. Evol.* **2013**, *4*, 236–243. [CrossRef]
22. Bao, R.; Li, X.; Zheng, J. Feature tuning improves MAXENT predictions of the potential distribution of *Pedicularis longiflora* Rudolph and its variant. *Peer J.* **2022**, *10*, e13337. [CrossRef] [PubMed]
23. Perry, L.; Dickau, R.; Zarrillo, S.; Holst, I.; Pearsall, D.M.; Piperno, D.R.; Berman, M.J.; Cooke, R.G.; Rademaker, K.; Ranere, A.J.; et al. Starch fossils and the domestication and dispersal of chili peppers (*Capsicum* spp. L.) in the Americas. *Science* **2007**, *315*, 986–988. [CrossRef] [PubMed]

24. Karim, K.M.R.; Rafii, M.Y.; Misran, A.B.; Ismail, M.F.B.; Harun, A.R.; Khan, M.M.H.; Chowdhury, M.F.N. Current and prospective strategies in the varietal improvement of chilli (*Capsicum annuum* L.) specially heterosis breeding. *Agronomy* **2021**, *11*, 2217. [CrossRef]
25. Zhou, X.; Zhu, F. Origin, evolution and cultivation history of the pepper. *Acta Hortic. Sin.* **2022**, *49*, 1371–1381. [CrossRef]
26. Zou, X.; Hu, B.; Xiong, C.; Dai, X.; Liu, F.; Ou, L.; Yang, B.; Liu, Z.; Suo, H.; Xu, H.; et al. Review and prospects of pepper breeding for the past 60 years in China. *Acta Hortic. Sin.* **2022**, *49*, 2099–2118. [CrossRef]
27. García-Gaytán, V.; Gómez-Merino, F.C.; Trejo-Téllez, L.I.; Baca-Castillo, G.A.; García-Morales, S. The chilhuacle chili (*Capsicum annuum* L.) in Mexico: Description of the variety, its cultivation, and uses. Inter. *Int. J. Agron.* **2017**, *2017*, 5641680. [CrossRef]
28. Zheng, X. Problems in pepper cultivation and corresponding countermeasures. *Word Trop. Agri. Info.* **2022**, 26–27.
29. Saidah, Z.; Harianto; Hartoyo, S.; Asmarantaka, R.W. *Change on Production and Income of Red Chili Farmers*; Conference Series: Earth and Environmental Science; IOP Publishing: Bristol, UK, 2020; p. 12003.
30. Wilfried, T. Biodiversity: Climate change and the ecologist. *Nature* **2007**, *448*, 550–552. [CrossRef]
31. Yue, Y.; Zhang, P.; Shang, Y. The potential global distribution and dynamics of wheat under multiple climate change scenarios. *Sci. Total Environ.* **2019**, *688*, 1308–1318. [CrossRef]
32. Govindarajan, V.S. Capsicum production, technology, chemistry, and quality. Part 1: History, botany, cultivation, and primary processing. *Crit. Rev. Food Sci.* **1985**, *22*, 109–176. [CrossRef]
33. Reddy, K.M.; Shivashankara, K.S.; Geetha, G.A.; Pavithra, K.C. Capsicum (hot pepper and bell pepper). In *Abiotic Stress Physiology of Horticultural Crops*; Rao, N.K.S., Shivashankara, K.S., Laxman, R.H., Eds.; Springer: Berlin, Germany, 2016; pp. 151–166.
34. Pressman, E.; Moshkovitch, H.; Rosenfeld, K.; Shaked, R.; Gamliel, B.; Aloni, B. Influence of low night temperatures on sweet pepper flower quality and the effect of repeated pollinations, with viable pollen, on fruit setting. *J. Hort. Sci. Biotech.* **1998**, *73*, 131–136. [CrossRef]
35. Erickson, A.N.; Markhart, A.H. Flower developmental stage and organ sensitivity of bell pepper (*Capsicum annuum* L.) to elevated temperature. *Plant Cell Environ.* **2002**, *25*, 123–130. [CrossRef]
36. He, X.; Si, J.; Zhu, L.; Zhou, D.; Zhao, C.; Jia, B.; Wang, C.; Qin, J.; Zhu, X. Modeling habitat suitability of *Hippophae rhamnoides* L. using MaxEnt under climate change in China: A case study of *H. r. sinensis* and *H. r. turkestanica*. *Front. Glob. Chang.* **2023**, *5*, 1095784. [CrossRef]
37. Han, S. Water-consuming patterns of greenhouse pepper in different irrigating modes. *Agric. Res. Arid Areas* **2005**, *23*, 54–58. [CrossRef]
38. Shao, G.; Liu, N.; Chen, L. Character of water demand and yield of space-time deficit irrigation for hot pepper in greenhouse. *Trans. Chin. Soc. Agric. Mach.* **2008**, *39*, 117–121.
39. Jaimez, R.E.; Vielma, O.; Rada, F.; García-Núñez, C. Effects of water deficit on the dynamics of flowering and fruit production in *Capsicum chinense* Jacq in a tropical semiarid region of Venezuela. *J. Agron. Crop Sci.* **2000**, *185*, 113–119. [CrossRef]
40. Srikanth, D.; Rekha, G.K.; Lakshmi, A.P.; Vimatha, P. Impact of climate change in *Capsicum* production: A review. *Curr. J. Appl. Sci. Technol.* **2019**, *33*, 1–5. [CrossRef]
41. Zou, Z.; Zou, X. Geographical and ecological differences in pepper cultivation and consumption in China. *Front. Nutr.* **2021**, *8*, 718517. [CrossRef]
42. Li, Q.; Han, R.; Li, Z.; Chen, L. Production status and development countermeasures of chili pepper in Qinghai Province. *Chin. Hortic. Abstr.* **2018**, *34*, 60–61, 124. [CrossRef]
43. Meng, Q. Analysis of Production Cost and Income of Pepper in Main Production Areas in China–Based on the Investigation of Five Provinces Such as Guizhou and Hunan. Master's Thesis, Hebei Agricultural University, Baoding, China, 2018.
44. Wu, S.Y.; Wu, Y.; Wen, J. Future changes in precipitation characteristics in China. *Int. J. Climatol.* **2019**, *39*, 3558–3573. [CrossRef]
45. Li, J.; Deng, C.; Duan, G.; Wang, Z.; Zhang, Y.; Fan, G. Potentially suitable habitats of *Daodi* goji berry in China under climate change. *Front. Plant Sci.* **2024**, *14*, 1279019. [CrossRef] [PubMed]
46. Shi, X.; Wang, J.; Zhang, L.; Chen, S.; Zhao, A.; Ning, X.; Fan, G.; Wu, N.; Zhang, L.; Wang, Z. Prediction of the potentially suitable areas of *Litsea cubeba* in China based on future climate change using the optimized MaxEnt model. *Ecol. Indic.* **2023**, *148*, 110093. [CrossRef]
47. Lee, S.G.; Kim, S.K.; Lee, H.J.; Lee, H.S.; Lee, J.H. Impact of moderate and extreme climate change scenarios on growth, morphological features, photosynthesis, and fruit production of hot pepper. *Ecol. Evol.* **2018**, *8*, 197–206. [CrossRef] [PubMed]
48. Thuiller, W.; Lavorel, S.; Araujo, M.B.; Sykes, M.T.; Prentice, I.C. Climate change threats to plant diversity in Europe. *Procl. Natl. Acad. Sci. USA* **2005**, *102*, 8245–8250. [CrossRef]
49. Becklin, K.M.; Anderson, J.T.; Gerhart, L.M.; Wadgymar, S.M.; Wessinger, C.A.; Ward, J.K. Examining plant physiological responses to climate change through an evolutionary lens. *Plant Physiol.* **2016**, *172*, 635–649. [CrossRef] [PubMed]
50. Genzel, F.; Dicke, M.D.; Junker-Frohn, L.V.; Neuwohner, A.; Thiele, B.; Putz, A.; Usadel, B.; Wormit, A.; Wiese-Klinkenberg, A. Impact of moderate cold and salt stress on the accumulation of antioxidant flavonoids in the leaves of two *Capsicum* cultivars. *J. Agr. Food Chem.* **2021**, *69*, 6431–6443. [CrossRef] [PubMed]
51. Eyring, V.; Bony, S.; Meehl, G.A.; Senior, C.A.; Stevens, B.; Stouffer, R.J.; Taylor, K.E. Overview of the Coupled Model Intercomparison Project Phase 6 (CMIP6) experimental design and organization. *Geosci. Model. Dev.* **2016**, *9*, 1937–1958. [CrossRef]
52. Phillips, S.J.; Dudik, M.; Elith, J.; Graham, C.H.; Lehmann, A.; Leathwick, J.; Ferrier, S. Sample selection bias and presence-only distribution models: Implications for background and pseudo-absence data. *Ecol. Appl.* **2009**, *19*, 181–197. [CrossRef] [PubMed]

53. Cobos, M.E.; Peterson, A.T.; Barve, N.; Osorio-Olvera, L. kuenm: An R package for detailed development of ecological niche models using Maxent. *Peer J.* **2019**, *7*, e6281. [CrossRef]
54. Allouche, O.; Tsoar, A.; Kadmon, R. Assessing the accuracy of species distribution models: Prevalence, kappa and the true skil statistic (TSS). *J. Appl. Ecol.* **2006**, *43*, 1223–1232. [CrossRef]
55. Coetzee, B.W.T.; Robertson, M.P.; Erasmus, B.F.N.; Van Rensburg, B.J.; Thuiller, W. Ensemble models predict Important Bird Areas in southern Africa will become less effective for conserving endemic birds under climate change. *Global Ecol. Biogeogr.* **2009**, *18* 701–710. [CrossRef]

plants MDPI

Article

Responses of Three *Pedicularis* Species to Geological and Climatic Changes in the Qinling Mountains and Adjacent Areas in East Asia

Qijing Zhang †, Zhaoping Lu †, Mingchen Guo, Jia Kang, Jia Li, Xiaojing He, Jiayi Wu, Ruihang Liu, Jiaxin Dang and Zhonghu Li *

Key Laboratory of Resource Biology and Biotechnology in Western China, Ministry of Education, College of Life Sciences, Northwest University, Xi'an 710069, China; zhangqijing@stumail.nwu.edu.cn (Q.Z.); 202032601@stumail.nwu.edu.cn (Z.L.); 13633436166@163.com (M.G.); kangjia202309@163.com (J.K.); lejea2020@163.com (J.L.); heexj13@163.com (X.H.); wujiayi5p1@163.com (J.W.); 15869926945@163.com (R.L.); 19909292608@163.com (J.D.)

* Correspondence: lizhonghu@nwu.edu.cn; Tel./Fax: +86-29-88302411

† These authors contributed equally to this work.

Abstract: The Qinling Mountains in East Asia serve as the geographical boundary between the north and south of China and are also indicative of climatic differences, resulting in rich ecological and species diversity. However, few studies have focused on the responses of plants to geological and climatic changes in the Qinling Mountains and adjacent regions. Therefore, we investigated the evolutionary origins and phylogenetic relationships of three *Pedicularis* species in there to provide molecular evidence for the origin and evolution of plant species. Ecological niche modeling was used to predict the geographic distributions of three *Pedicularis* species during the last interglacial period, the last glacial maximum period, and current and future periods, respectively. Furthermore, the distribution patterns of climate fluctuations and the niche dynamics framework were used to assess the equivalence or difference of niches among three *Pedicularis* species. The results revealed that the divergence of three *Pedicularis* species took place in the Miocene and Holocene periods, which was significantly associated with the large-scale uplifts of the Qinling Mountains and adjacent regions. In addition, the geographic distributions of three *Pedicularis* species have undergone a northward migration from the past to the future. The most important environmental variables affecting the geographic distributions of species were the mean diurnal range and annual mean temperature range. The niche divergence analysis suggested that the three *Pedicularis* species have similar ecological niches. Among them, *P. giraldiana* showed the highest niche breadth, covering nearly all of the climatic niche spaces of *P. dissecta* and *P. bicolor*. In summary, this study provides novel insights into the divergence and origins of three *Pedicularis* species and their responses to climate and geological changes in the Qinling Mountains and adjacent regions. The findings have also provided new perspectives for the conservation and management of *Pedicularis* species.

Keywords: divergence; geographic distribution; evolutionary relationship; ecological niche analysis; *Pedicularis*

1. Introduction

Climatic oscillations during the Quaternary have profoundly affected the geographic distribution, migration, and genetic structure of plants in both the Southern and Northern Hemispheres [1,2]. Although East Asia was not covered much by ice sheets during the last glacial period, climate oscillations during the late Quaternary strongly impacted the biological diversity and evolution of many extant species in the region [3,4]. Recent phylogeographic studies have shown that climatic changes and historical geological events were the primary drivers of population expansions/contractions and interspecific divergence in plants, which also constructed new ecological niches, leading to the origin and divergence of new species [5,6]. In addition, the combination of ecological niche modeling (ENM) and maternal inherited chloroplast DNA (cpDNA) markers were used to predict the geographic distribution changes of relict plant species and forest trees from East Asia [7,8]. Combining plant genetic information with niche models could help us investigate the effects of global climate change on the evolution of alpine plants since the Quaternary, as well as the factors involved in the origin and differentiation of species [9–11]. These studies provided novel insights into biodiversity conservation in the context of global climate.

Previous geological studies suggested that the Qinling Orogenic Belt in East Asia experienced a long evolutionary history and has a complicated composition and structure [12]. The rapid uplift of the Qinling Mountains since the late Cenozoic was influenced by changes in deep earth dynamic processes, which dominated the different climatic environments in mountainous areas and subsequently created the abundant biodiversity of the Qinling Mountains [13,14]. The Qinling Mountains are a unique biodiversity region in the world due to their distinctive geographic location, diverse habitats, and climate types, providing an ideal environment for studying species differentiation and response patterns of plant species to geological and climatic oscillations. Relevant studies have investigated the genetic divergence and evolutionary histories of three closely related tree peony species (*Paeonia qiui* Y. L. Pei and D. Y. Hong, *Paeonia jishanensis* T. Hong and W. Z. Zhao, and *Paeonia rockii* (S. G. Haw and Lauener) T. Hong and J. J. Li ex D. Y. Hong) in the Qinling–Daba Mountains [5] and also revealed the population genetic structure of *P. rockii* using nuclear gene markers and chloroplast DNAs [15]. These results demonstrated that the Qinling Mountains as a geographic barrier have profoundly impacted the population evolution of the endemic species in this region. However, the phylogeographic studies and population demographic history of endemic plants in the Qinling Mountains and adjacent areas need further exploration.

Pedicularis (Orobanchaceae) is a genus of hemiparasitic plants distributed throughout the Northern Hemisphere, mainly in high-latitude or montane habitats. The species richness and morphological diversity in *Pedicularis* is confined to a comparatively limited region of eastern Asia, with approximately 75% of species being endemic to the Himalaya-Hengduan areas [16,17]. The genus *Pedicularis* provides an exceptional model or case in this region for exploring how alpine plants respond to climatic fluctuations and geological events, as well as the associated interspecific evolutionary processes and population dynamics [18,19]. Recently, the evolutionary patterns of three single/low-copy nuclear genes (*CRC*, *LFY-L*, *LFY S*) and two chloroplast genes (*matK* and *ycf1*) in sixty-five accessions of *Pedicularis* sect. *Cyathophora* from the Hengduan Mountains were investigated, which revealed that the uplift of the Qinghai–Tibet Plateau and climatic changes may have played crucial roles in the divergence and speciation of *Pedicularis* sect. *Cyathophora* [20]. Moreover, the effects of climatic oscillations on genetic diversity and phylogenetic relationships in *Pedicularis* sect. *Cyathophora* (Orobanchaceae) on the Qinghai–Tibet Plateau have been also researched [21,22]. Nevertheless, the evolutionary relationships and species divergence of *Pedicularis* species in the Qinling Mountains and adjacent areas are less well known. Therefore, we focus on three species of *Pedicularis* (*Pedicularis bicolor* Diels, *Pedicularis dissecta* (Bonati) Pennell and H. L. Li, *Pedicularis giraldiana* Diels ex Bonati) endemic to the Qinling Mountains [23]. *P. bicolor* has been recorded as a vulnerable (VU) species on the IUCN Red List [24]. *P. giraldiana* is distributed over most parts of the Qinling Mountains and extends southwestward in China ranging from 2900 to 3000 m, while *P. dissecta* is distributed in the Qinling Mountains of southern Shaanxi province, thriving on rocks at an altitude of 3000 m [23].

In this study, we analyzed phylogenetic relationships and divergence of three *Pedicularis* species (*P. bicolor*, *P. giraldiana*, and *P. dissecta*) using maternally inherited chloroplast DNA (cpDNA) sequences. Additionally, we examined niche divergences between the species through ecological niche modeling (ENM) and compared their spatial distributions during different historical times. The main objectives of this study were to (1) conduct phylogenetic analysis based on plastid genome data; (2) estimate divergence times and explore their relationships with geological events; (3) evaluate niche differences between species from both geographic and environmental perspectives; (4) investigate the impacts of global climate change on the three *Pedicularis* species in the Qinling Mountains since the last interglacial (LIG) period and explore their response to climate oscillations and geological events; and (5) provide references for protecting wild plant population resources.

2. Results

2.1. Phylogenetic Relationship

The phylogenetic relationship of *Pedicularis species* was reconstructed based on maximum likelihood (ML) and maximum parsimony (MP) analyses using the chloroplast genomes of seventeen Orobanchaceae and three Lamiaceae species. Overall, the topologies obtained using different methods (ML and MP analyses) were almost identical. It was shown that all examined Orobanchaceae species formed a monophyletic evolutionary clade with high bootstrap support (Figure 1). Within this major evolutionary clade, *Pedicularis* and *Orobanche* were found to cluster sister relationships. Each of the two clades formed a monophyletic branch with a 100% bootstrap value, respectively. *P. dissecta* and *P. bicolor* are both nested branches within *Pedicularis;* they have high support and are nested within the larger *Pedicularis* evolutionary branch. *Pedicularis cheilanthifolia* Schrenk and *P. giraldiana* are grouped together in a single branch with strong support from bootstrap values.

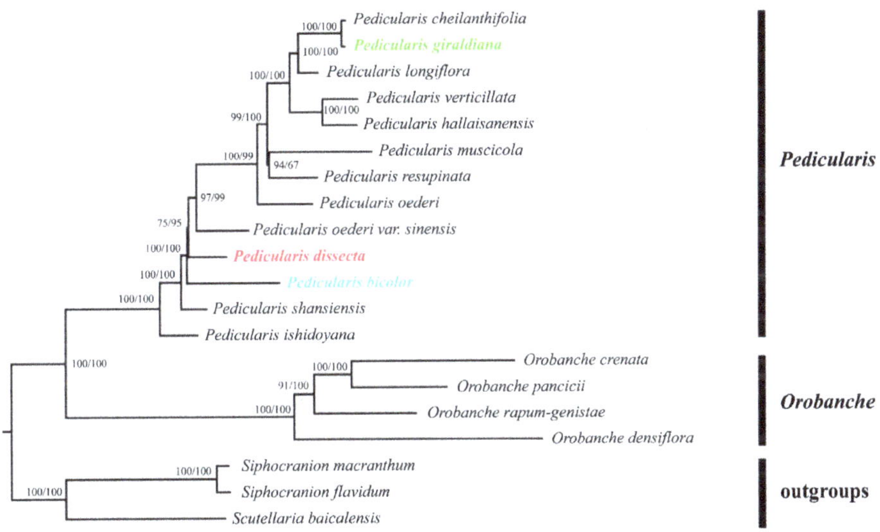

Figure 1. Phylogenetic tree based on chloroplast genomes of *Pedicularis* species. Numbers above the branches are bootstrap values according to maximum likelihood (**left**) and maximum parsimony (**right**) analyses. The three colorful species are main objects of this study.

2.2. Divergence Time Estimation

Based on the chloroplast genomes, the divergence time estimation indicated that the two large clades, *Pedicularis* and *Orobanche*, diverged approximately 47.7 million years ago (Figure 2). According to the results, the crown age of the diversification of *Pedicularis* was estimated to be around 35.9 Ma (95% highest posterior density, HPD: 20.9–45.0 Ma). The two sister species in the genus *Pedicularis* comprising *P. giraldiana* and *P. cheilanthifolia* diverged by about 0.4 Ma (95% HPD: 0.1–1.7 Ma). The crown ages of *P. bicolor* and *P. dissecta* took place at approximately 13.1 Ma (95% HPD: 5.2–23.5 Ma) and 12.9 Ma (95% HPD: 4.1–19.3 Ma), respectively.

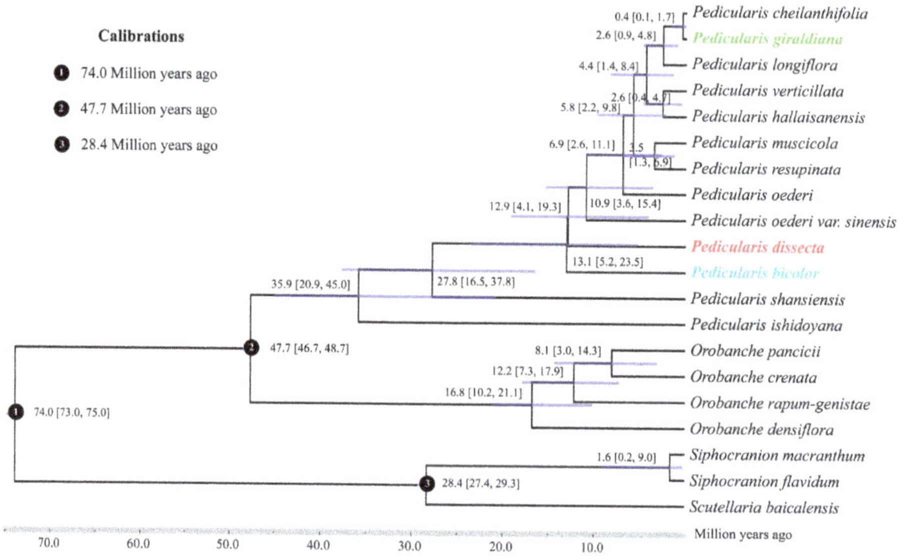

Figure 2. Divergence times based on chloroplast genomes. Blue bars and the numbers below the bars indicate 95% highest posterior densities of divergence times (million years ago). The three colorful species are main objects of this study.

2.3. Environmental Variables Analysis

By correlating the factors and geographical data and combining the contribution amounts, eight climatic variables, including annual mean temperature (bio1), mean diurnal range (bio2), temperature annual range (bio7), mean temperature of the wettest quarter (bio8), mean temperature of the driest quarter (bio9), mean temperature of the warmest quarter (bio10), mean temperature of the coldest quarter (bio11), and precipitation of the warmest quarter (bio18), were finally selected for modeling in this study (Figure 3). The current potential distribution of *P. bicolor* is attributed to the bio1, bio2, and bio7 variables, which account for 20.3%, 35.9%, and 13.8%, respectively, resulting in a cumulative contribution percentage of over 60%. The three most significant contributors to the currently suitable habitats of *P. dissecta* are bio2, bio9, and bio10, accounting for 32%, 14.7%, and 24.3%, respectively, and representing 71% of the total. The contribution variables of *P. giraldiana* are bio2, bio7, and bio18, accounting for 37.9%, 13.9%, and 27.8%, respectively. These three environmental factors accounted for more than 70% of the total. In summary, the range of diurnal and annual temperature changes plays a crucial role in influencing the geographic distributions of three *Pedicularis* species.

2.4. Dynamic Changes in Geographic Distribution from the Past to the Future

Based on the effective specimen site information, the ecological niche model was used to simulate the past, current, and future niche distributions of three *Pedicularis* species. The areas under curve (AUC) values of each species model test above 0.90, indicating that the prediction results of the potential distribution area of the species were credible. During the LIG, the three *Pedicularis* species occurred, mainly in southwest China (eastern Sichuan, southern Chongqing, Guizhou, and northern Guangxi) (Figure 4). In contrast to the LIG period, the potential distributional ranges of three *Pedicularis* species shrank to Shaanxi, Sichuan, and southeastern Gansu province during the last glacial maximum (LGM) period. From the LGM to the current period, the suitable habitat of three *Pedicularis* species has rapidly increased. It appeared that each species gradually migrated northward from the LIG to the current period.

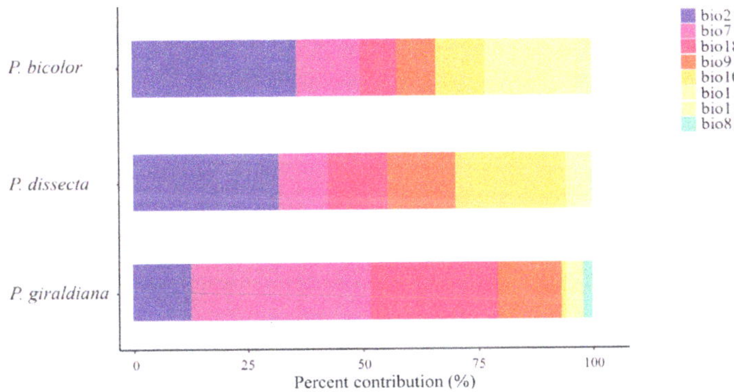

Figure 3. Environmental variable contributions of the ecological niche model for three *Pedicularis* species. Bio1, annual mean temperature; bio2, mean diurnal range; bio7, temperature annual range; bio8, mean temperature of the wettest quarter; bio9, mean temperature of the driest quarter; bio10, mean temperature of the warmest quarter; bio11, mean temperature of the coldest quarter; bio18, precipitation of the warmest quarter.

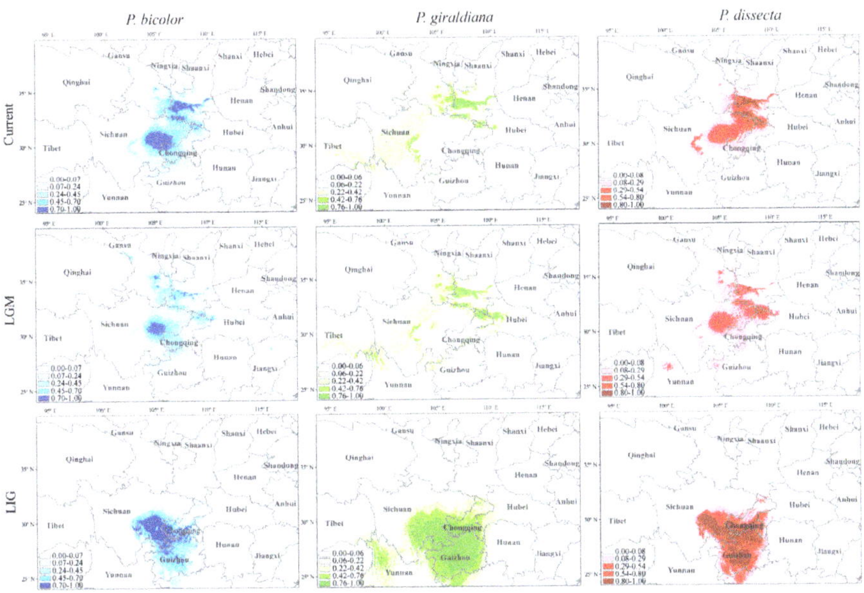

Figure 4. Distribution of three *Pedicularis* species at present, last glacial maximum (LGM), and last interglacial (LIG) based on ecological niche modeling. The light to dark colors and small to large numbers of the small squares in the legend correspond to the low to high suitability of the species for the potential geographic ranges.

The results of the future distributions (2070s) showed a significant reduction in suitable habitats for three *Pedicularis* species compared to the present times, especially highly suitable habitats, which will shrink to southern Shaanxi under the representative concentration pathway 2.6 (RCP 2.6) scenario. Additionally, it is worth noting that the distributional ranges of suitable habitats for *P. bicolor* and *P. dissecta* will decrease significantly from 2050 to 2070 under the RCP8.5 scenario (Figure S3).

2.5. Migration Trends Based on the Centroid of Suitable Habitats

Under various climatic scenarios, the centroid of each species is predicted to migrate from the south (northern Guizhou) to the north of China (northern Sichuan and southern Shaanxi) (Figure 5). For instance, it was shown that the three *Pedicularis* species had undergone a northward migration from the LIG to LGM periods; while the two species of *P. bicolor* and *P. dissecta* have shifted to the northeast of China, the geographic ranges of *P. giraldiana* moved to the northwest of China. Under the RCP2.6 scenario, *P. bicolor* and *P. dissecta* initially migrated southeast and northwest of China for a short distance, respectively, and then both of them migrated northeast of China in 2070. In the RCP8.5 scenario, all three species will eventually migrate to the east or northeast of China over time.

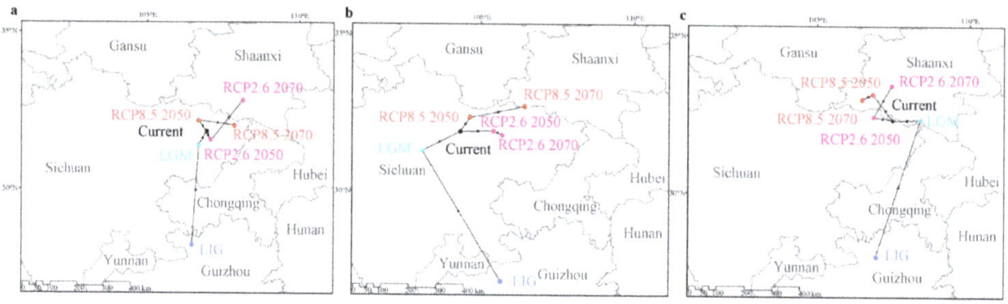

Figure 5. Migration of the center of suitable habitat for three *Pedicularis* species. (**a**) *P. bicolor*; (**b**) *P. giraldiana*; (**c**) *P. dissecta*. LGM, last glacial maximum; LIG, last interglacial; RCP, representative concentration pathway scenario. RCP2.6 and RCP8.5 are defined according to the radiative forcing target level from 2.6 W/m^2 (low carbon emissions) to 8.5 W/m^2 (high carbon emissions) for 2100.

2.6. Niche Analysis

In the niche comparison analysis, the first two axes explained 88.9% of the total variation of climatic conditions for the *Pedicularis* species ranges (PC1 = 70.3%, PC2 = 18.6%) (Figure 6). The multiple niche plot illustrating the 20% of occurrence density showed that the realized niche of *P. bicolor* and *P. dissecta* were the most closely related, whereas *P. giraldiana* was the most widely realized niche and contained both *P. bicolor* and *P. dissecta*. When 100% of the occurrence density was plotted in environmental space, a high degree of climatic spatial overlap was detected between the ranges. All pairwise ecological niche modeling comparisons produced high values for Hellinger's *I* and Schoener's *D* and displayed high degrees of geographic overlap (Figure 7).

Moreover, *P. giraldiana* showed a highest niche breadth, covering almost all climatic niche spaces of *P. dissecta* and *P. bicolor*. In the two multiple niche PCA-env plots, it is evidenced that *P. bicolor* had the smallest niche breadth compared to the other species. Additionally, *P. bicolor* and *P. dissecta* occupied the drier and colder climatic niches, while *P. giraldiana* occupied a warmer and wetter niche.

A pairwise comparison of the climatic niches of three species of *Pedicularis* showed that *P. bicolor* and *P. giraldiana* have the lowest environmental niche overlap, while *P. dissecta* and *P. bicolor* have the highest niche overlap. The climatic niche classification results indicated that the "Unfilling" ecological niche accounted for 0–0.126, the "Stability" niche accounted for 0.276–0.624, and the "Expansion" niche accounted for 0.376–0.724 (Table 1).

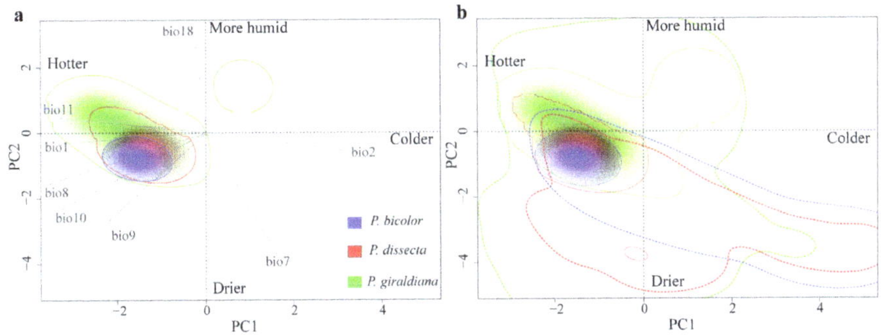

Figure 6. Geographic location of the occurrence records and background areas used for the climatic niche comparison in the e-space of three *Pedicularis* species: *P. bicolor* (blue), *P. dissecta* (red), and *P. giraldiana* (green). (**a**,**b**) represent climatic space through the environmental principal components analysis (PCA-env) (explaining PC1 = 70.3% and PC2 = 18.6% of the total climatic variation). In (**a**), the solid lines delimit the 20% of occurrence density of the current occupied niches by the species. Gray arrows outline the direction of the variables' contribution to the PCA-env. In (**b**), the solid and dashed lines illustrate the 100% occurrence density and the 100% available background climates, respectively. Bio1, annual mean temperature; bio2, mean diurnal range; bio7, temperature annual range; bio8, mean temperature of the wettest quarter; bio9, mean temperature of the driest quarter; bio10, mean temperature of the warmest quarter; bio11, mean temperature of the coldest quarter; bio18, precipitation of the warmest quarter.

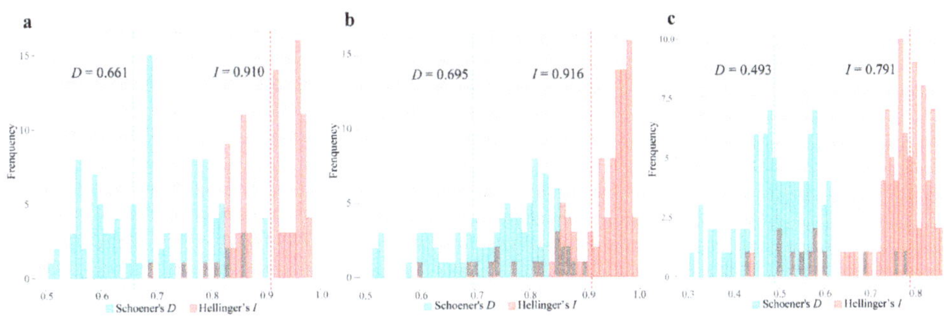

Figure 7. Niche identity test for each comparison based on Schoener's *D* statistic and Hellinger's *I* statistic. (**a–c**) represent the comparison results between *P. bicolor* vs. *P. dissecta*, *P. dissecta* vs. *P. giraldiana*, and *P. giraldiana* vs. *P. bicolor*, respectively. Bars indicate the null distributions of *D* and *I*.

Table 1. Comparisons of the current occupied climatic niches between *Pedicularis* species.

	Niche Overlap	Niche Unfilling	Niche Stability	Niche Expansion
P. bicolor—*P. dissecta*	0.463	0.036	0.624	0.376
P. dissecta—*P. giraldiana*	0.204	0.126	0.427	0.573
P. giraldiana—*P. bicolor*	0.122	0.000	0.276	0.724

3. Discussion

3.1. Evolutionary Relationships

In recent years, an increasing number of studies have used chloroplast genome data for species phylogeny and interspecific differentiation studies to explore the origin and evolution of species [25]. The frequent parallel or convergent evolution of characters of corollas has resulted in the difficulties of the genus *Pedicularis* classification [26]. Traditionally, *Pedicularis* was placed in the Scrophulariaceae, but molecular evidence showed that this genus has been shifted to Orobanchaceae and found to be polyphyletic [27–29]. Furthermore, there was little consensus between the phylogenetic tree and conventional classification in *Pedicularis* [30]. Yu et al. (2015) constructed a backbone phylogeny of *Pedicularis*, with extensive species sampling from the Himalaya–Hengduan Mountains, using sequences of the nuclear ribosomal internal transcribed spacer (nrITS) and three plastid regions (*matK*, *rbcL*, and *trnL-F*) [31]. Liu et al. (2022) reconstructed evolutionary relationships of the *Pedicularis siphonantha* Don complex using nrITS and four plastid DNA loci (*matK*, *rbcL*, *trnH-psbA*, and *trnL-F*) to resolve taxonomic confusion [32].

Our results showed that two major evolutionary clades were identified comprising *Pedicularis* and *Orobanche* with high bootstrap values (Figure 1). In the first major clade, *P. giraldiana* was placed as a sister to *P. cheilanthifolia* with high bootstrap value support. The lack of consensus between the phylogenetic tree and previous studies may be explained by the small sample size of *Pedicularis* species in previous studies. In agreement with previous studies, *Pedicularis verticillata* L. and *Pedicularis hallaisanensis* Hurus. have clustered the sister evolutionary clades [33]. Another phylogenetic study showed that *Pedicularis* species clustered together and *Pedicularis shansiensis* P. C. Tsoong and *P. dissecta* formed a clade based on complete chloroplast genomes [34]. Additionally, some studies showed that *P. dissecta* and *P. bicolor* were more closely related to *P. shansiensis*. However, our study suggested that these two species formed nested relationships, possibly due to the differences in DNA molecular markers used in different studies. The intrageneric phylogenetic relationships of many studies were less consistent based on the different molecular markers, presumably related to pseudogenization and gene loss (*accD* and *ccsA*) in *Pedicularis* species [35]. In addition, Yang et al. (2007) detected an unusually high sequence divergence among congeners in a comparatively small geographical range, as well as frequent large deletions in the cpDNA *trnT-trnF* region of *Pedicularis* species [36]. To verify species relationships and interspecific divergence in *Pedicularis*, it will be crucial to collect more species samples in future research.

3.2. Species Divergence

It is well known that dated phylogeny is essential for the comprehension of speciation processes [37]. Our molecular dating results suggested that the crown age of the diversification of *Pedicularis* species was estimated to be around 35.9 Ma (Figure 2). This result was consistent with a recent study and was supported by the earliest pollen fossil record of *Pedicularis* from the Eocene in China [20,35–41]. According to our molecular dating analysis, *P. dissecta* and *P. bicolor* may have occurred during the late Miocene (approximately 13.1 Ma and 12.9 Ma, respectively). This timing is closely associated with the onset of a new period of tectonic thermal activity and uneven vertical uplift in the Qinling Mountains since the Miocene [12,42], which might have played crucial roles in the divergence of *P. dissecta* and *P. bicolor*.

Interestingly, the divergence of *P. giraldiana* and *P. cheilanthifolia* was dated approximately 0.4 Ma by secondary calibration. Yuan et al. (2011) have also investigated the population divergence of *Paeonia rockii*, a species endemic to the Qinling Mountains and the adjacent area, using cpDNA sequences [15]. The results showed that species divergence (0.4 to 1.6 Ma) was concurrent with the time scale of the latest rapid uplifts of the Qinling Mountains during the Pleistocene (approximately 0.7 Ma) [43,44]. Previous studies have shown that the diversification of *Pedicularis cyathophylloides* Limpr. during the middle Quaternary (0.4–2.5 Ma) was estimated by *LFY-L* nuclear genes [20]. Additionally, it was discovered that the section *Cyathophora* of *Pedicularis*, which was only found in the Sino–Himalayan regions, formed a well-supported monophyletic group. The species divergence of this section was largely associated with mountain uplifts and geological oscillations [36]. In summary, in order to further verify the affinities between species, more reliable fossils are required for phylogenetic studies, and data from multiple species samples are also necessary to elucidate the interspecific differentiation of *Pedicularis* species.

3.3. Population Dynamics Changes

The results of our study indicated that the main suitable habitat of *Pedicularis* is currently located in the Qinling Mountains and adjacent areas. Meanwhile, *P. bicolor* and *P. dissecta* have occurred mainly in eastern Sichuan, Guizhou, and Guangxi during the LIG. In addition, the geographic areas of *P. giraldiana* were also largely distributed in the Yunnan–Guizhou plateau, as previously studied by the transition patterns of the genus *Notopterygium* species [45]. The ENM revealed that three *Pedicularis* species reduced their geographic ranges to the Qinling Mountains and surrounding areas from the LIG to LGM, which coincided with the cooling climate [46,47]. Some other studies also showed that the geological and climatic changes have profoundly affected the geographic distributions and species shifts in the Qinling Mountains and adjacent areas. For example, Jiang et al. (2023) used species distribution modeling to investigate the geographic range changes of an endangered herb *Primula filchnerae* R. Knuth in the Qinling Mountains, and they found that the species range would migrate to the north of China in the future climatic scenario [48]. Meanwhile, Zhao et al. (2023) detected the geographic changes of a shrub species *Bashania fargesii* (E. G. Camus) P. C. Keng and T. P. Yi in the Qinling Mountains, and the authors have also concluded that the species would migrate to the north of China in the future [49]. Additionally, Zhao et al. (2016) detected the geographic shifts of an endangered alpine tree species *Larix chinensis* Beissn. on Taibai Mountain in the Qinling Mountains based on Maxent models, and the authors revealed that the suitable species habitat would move to higher elevations in the future [50].

The main climatic contributions of *Pedicularis* also demonstrated that temperature has a greater effect on the geographic distributions than the precipitation variable (Figure 6). The range shift remains comparatively stable after the glacial periods when the climate was sufficiently warm. When combined with the results of centroid changes, it was clear that three *Pedicularis* species migrated northward during the Quaternary (Figure 5). Generally, climate warming causes the distribution area of many plant species to shift northward in the Northern Hemisphere [51]. This pattern has also been observed in other plants, such as *Roscoea humeana* Balf. f. and W. W. Sm. [52], *Panax notoginseng* (Burkill) F. H. Chen ex C. H. Chow [53], and *Ziziphus jujuba* Mill. [54].

3.4. Climatic Niche Comparisons

Under the combined effects of orogenesis and climate changes, species formation and ecological niche differentiation are promoted [55]. The interplay of topographic uplift and climate change in the Himalaya–Hengduan Mountains played a key role in speciation and differentiation during the Quaternary [6]. For example, ecological speciation was demonstrated in two parapatric sister species, *Roscoea cautleoides* Gagnep. and *R. humeana* [56]. Additionally, the heat energy and water in *Pinus yunnanensis* Franch. in the mountain areas promoted intraspecific genetic divergence [57].

Our niche analysis of the environment variables suggested that the three species of *Pedicularis* might have occupied analogous climate niches, with some niche overlaps according to simulations with a 20% occurrence rate (Figure 6). Furthermore, *P. bicolor* and *P. dissecta* inhabited the drier and colder climatic niches, while *P. giraldiana* occupies a hotter and more humid ecological niche, which echoes the previous analysis of the contribution of ecological factors (Figure 3). Additionally, ecological factors, such as temperature and precipitation seasonality, have a significant impact on phenology, including flowering time and growing season [58]. Phenology may have caused gene flow and/or genetic exchange between populations, resulting in population differentiation [59]. Therefore, further research needs to combine population genetic data and environmental selection variables, which have important implications for genetic differentiation and origin between lineages or species, so as to better protect biodiversity in the context of global climate change.

4. Materials and Methods

4.1. Plant Materials, DNA Extraction, and Genome Assembly

The leaves of *P. bicolor* and *P. giraldiana* were collected from the Qinling Mountains (108°47′10.13″ N, 33°51′27.78″ E and 107°46′23.19″ N, 33°58′9.58″ E, respectively). The vouchers of plant materials were deposited in the herbarium of Northwest University (Xi'an, China). Total genomic DNA was extracted from silica-dried leaves using a modified cetyltrimethylammonium bromide (CTAB) protocol [60].

Paired-end libraries with a 500 bp insert size were constructed and PE150 sequencing was performed on the Illumina HiSeq 2500 platform. GetOrganelle v1.7.0 was used to assemble the chloroplast genomes [61]. The final assembly graph was visualized using Bandage v0.8.1 to authenticate the automatically generated chloroplast genome [62]. The chloroplast genomes were annotated by PGA v1.0 [63] and then manually revised using Geneious v8.1 [64]. The chloroplast genomes of *P. bicolor* and *P. giraldiana* were deposited in GenBank (accession numbers PP439988 and PP439989, respectively). In addition, the chloroplast genome sequence of *P. dissecta* with accession number NC_056312 in the National Center for Biotechnology Information (NCBI) database was downloaded for subsequent analyses.

4.2. Phylogenetic Analysis and Divergence Time Estimation

We constructed a phylogenetic tree to determine the evolutionary relationships of three *Pedicularis* species using eighteen complete chloroplast genome sequences from the NCBI database (Table S1). The general time reversible (GTR) + gamma model was selected based on jmodeltest v2.1.10 tests [65]. The maximum likelihood (ML) analysis was performed using RAxML v8.1.17 with Lamiaceae as the outgroup and 1000 bootstraps [66]. Maximum parsimony (MP) analysis was conducted in PAUP* v4.0 software according to a heuristic search and the tree bisection–reconnection (TBR) branch swapping option [67]. The robustness of inferences was assessed by bootstrap resampling 1000 random replicates.

In addition, BEAST v2.4.5 was used to estimate the node ages and topological structures for *Pedicularis* species [68]. Referring to a previous study, the crown age of Lamiales and Orobanchaceae was dated at 74 Ma and 47.7 Ma, respectively [38,39]. The earliest fossil records of Lamiaceae date back to the early–middle Oligocene around about 28.4 Ma [40]. We applied a GTR model for nucleotide substitution and the "yule process" prior model with three calibration points. The divergence time was estimated by Markov chain Monte Carlo analysis for 50,000,000 generations. TRACER v1.5 was used to analyze the output files to determine whether the effective sample sizes of all parameters were larger than 200. Chronograms with nodal heights and 95% highest posterior density intervals were generated with TreeAnotator v1.7.5 (the first 5000 trees were discarded as a burn-in) and displayed using FigTree v1.0.

4.3. Distribution Data Collection

The geographic distribution data of three species of *Pedicularis* were collected through our field surveys in the Qinling Mountains. In addition, we combined the Chinese Virtual Herbarium (https://www.cvh.ac.cn (accessed on 4 January 2024)), Chinese National Knowledge Infrastructure (CNKI, http://www.cnki.net/ (accessed on 4 January 2024)), National Specimen Information Infrastructure (NSII, http://www.nsii.org.cn accessed on 4 January 2024)), and Global Biodiversity Information Facility (GBIF, https://www.gbif.org/ (accessed on 4 January 2024)) to determine geographic coordinates. After removing redundant points, we obtained 55 distribution points for *Pedicularis*, including 11 points for *P. bicolor*, 21 points for *P. giraldiana*, and 23 points for *P. dissecta* (Figure S1).

4.4. Ecological Niche Modeling

Previous studies have suggested that in cases of multicollinearity among environmental variables, certain environmental gradients may have a more pronounced effect on the model outcomes, potentially resulting in greater uncertainty [69,70]. Therefore, environmental variables should be screened to reduce the influence of multicollinearity. The 19 climatic variables were downloaded from the WorldClim database (Table S2). We examined the correlation between all layers using ENMTools v1.3 [71]. Two variables were considered highly correlated when the correlation coefficient was above 0.8.

We used MaxEnt v3.3.3 to predict the current, last glacial maximum (LGM, 0.021–0.018 Ma), Last Interglacial (LIG, 0.140–0.120 Ma), and future (years 2050 and 2070) potential geographic distributions of three *Pedicularis* species [72]. For future climate variables, we used climate data in the 2050s and 2070s. Two representative concentration pathway scenarios (RCP 2.6 and RCP 8.5) developed by the Intergovernmental Panel on Climate Change (IPCC) were used in this study, with RCP 2.6 used as the minimum emission scenario and RCP 8.5 as the maximum emission scenario [73]. We used the default parameters for MaxEnt and employed the "subsample" method (setting the number of replicates to 20), with 75% of the species records for training and 25% for testing the model. The overall performance of the model was assessed by calculating the area under the curve (AUC). AUC scores range from 0.5 (indicating randomness) to 1 (indicating an exact match), with a value above 0.9 considered indicative of good model performance [74].

4.5. Niche Comparison Analyses

Niche comparison analysis was performed to test whether the selection of different types of climatic niches may have contributed to the divergence of three *Pedicularis* species. We assessed ENMs in a spatial environment using R packages ecospat v4.0 [75]. Furthermore, to measure niche differences between species, we used ENMTools v1.3 to calculate the niche overlap statistic Schoener's D and standardized Hellinger's I distance [76,77], where a value of 0 denotes no overlap and 1 indicates complete overlap. To test the null hypothesis that two species have identical ENMs, we used the niche identity test initially in ENMTools v1.3. This test compares the observed scores of niche overlap statistics D and I with their null distribution generated with 100 pseudo-replicates. Finally, niche overlaps between species were characterized by niche unfilling, niche stability, and niche expansion [78,79].

Supplementary Materials: The following supporting information can be downloaded at https://www.mdpi.com/article/10.3390/plants13060765/s1, Figure S1: Geographical distribution of the sampled populations of three *Pedicularis* species; Figure S2: Correlation analysis of various environmental factors. Figure S3: Prediction of potential geographical distributions of three *Pedicularis* species in different climatic scenarios in the future based on ecological niche modeling. Table S1: Taxa and GenBank accessions used in phylogeny and divergence time analysis; Table S2: Environmental variable information of distribution area prediction.

Author Contributions: Conceptualization, Z.L. (Zhonghu Li); methodology, Z.L. (Zhonghu Li), Z.L. (Zhaoping Lu) and Q.Z.; software, Z.L. (Zhaoping Lu) and Q.Z.; validation, Z.L. (Zhonghu Li); formal analysis, Z.L. (Zhaoping Lu) and Q.Z.; investigation, J.K., J.L., X.H., J.W., R.L. and J.D.; resources, Z.L. (Zhonghu Li); data curation, Z.L. (Zhaoping Lu), Q.Z. and M.G.; writing—original draft preparation, Z.L. (Zhaoping Lu) and Q.Z.; writing—review and editing, Z.L. (Zhonghu Li); visualization, Z.L. (Zhonghu Li) and Q.Z.; supervision, Z.L. (Zhonghu Li); project administration, Z.L. (Zhonghu Li); funding acquisition, Z.L. (Zhonghu Li). All authors have read and agreed to the published version of the manuscript.

Funding: This research was funded by the National Natural Science Foundation of China (31970359), the Key Program of Research and Development of Shaanxi Province (2022ZDLSF06–02), the Basic Research Project of Shaanxi Academy of Fundamental Science (22JHZ005) and the Undergraduate Talent Training and Construction Project of Northwest University (363062102018).

Institutional Review Board Statement: Not applicable.

Informed Consent Statement: Not applicable.

Data Availability Statement: The chloroplast genomes of *P. bicolor* and *P. giraldiana* were deposited in GenBank (accession numbers PP439988 and PP439989, respectively).

Conflicts of Interest: The authors declare no conflicts of interest.

References

1. Hewitt, G.M. Genetic consequences of climatic oscillations in the Quaternary. *Philos. Trans. R. Soc. B Biol. Sci.* **2004**, *359*, 183–195. [CrossRef]
2. Davis, M.B.; Shaw, R.G. Range shifts and adaptive responses to Quaternary climate change. *Science* **2001**, *292*, 673–679. [CrossRef]
3. Sakaguchi, S.; Qiu, Y.X.; Liu, Y.H.; Qi, X.S.; Kim, S.H.; Han, J.; Takeuchi, Y.; Worth, J.R.P.; Yamasaki, M.; Sakurai, S.; et al. Climate oscillation during the Quaternary associated with landscape heterogeneity promoted allopatric lineage divergence of a temperate tree *Kalopanax septemlobus* (Araliaceae) in East Asia. *Mol. Ecol.* **2012**, *21*, 3823–3838. [CrossRef] [PubMed]
4. Zhou, T.H.; Li, S.; Qian, Z.Q.; Su, H.L.; Huang, Z.H.; Guo, Z.G.; Dai, P.F.; Liu, Z.L.; Zhao, G.F. Strong phylogeographic pattern of cpDNA variation reveals multiple glacial refugia for *Saruma henryi* Oliv. (Aristolochiaceae), an endangered herb endemic to China. *Mol. Phylogenet. Evol.* **2010**, *57*, 176–188. [CrossRef]
5. Xu, X.X.; Cheng, F.Y.; Peng, L.P.; Sun, Y.Q.; Hu, X.G.; Li, S.Y.; Xian, H.L.; Jia, K.H.; Abbott, R.J.; Mao, J.F. Late Pleistocene speciation of three closely related tree peonies endemic to the Qinling–Daba Mountains, a major glacial refugium in Central China. *Ecol Evol.* **2019**, *9*, 7528–7548. [CrossRef]
6. Liu, J.; Moller, M.; Provan, J.; Cao, L.M.; Poudel, R.C.; Li, D.Z. Geological and ecological factors drive cryptic speciation of yews in a biodiversity hotspot. *New Phytol.* **2013**, *199*, 1093–1108. [CrossRef] [PubMed]
7. Tang, C.Q.; Matsui, T.; Ohashi, H.; Dong, Y.F.; Momohara, A.; Herrando-Moraira, S.; Qian, S.; Yang, Y.; Ohsawa, M.; Luu, H.T.; et al. Identifying long-term stable refugia for relict plant species in East Asia. *Nat. Commun.* **2018**, *9*, 4488. [CrossRef]
8. Xu, J.; Deng, M.; Jiang, X.L.; Westwood, M.; Song, Y.G.; Turkington, R. Phylogeography of *Quercus glauca* (Fagaceae), a dominant tree of East Asian subtropical evergreen forests, based on three chloroplast DNA interspace sequences. *Tree Genet. Genomes* **2014**, *11*, 805. [CrossRef]
9. Rana, H.K.; Rana, S.K.; Luo, D.; Sun, H. Existence of biogeographic barriers for the long-term Neogene–Quaternary divergence and differentiation of *Koenigia forrestii* in the Himalaya–Hengduan Mountains. *Bot. J. Linn. Soc.* **2023**, *201*, 230–253. [CrossRef]
10. Guan, B.C.; Chen, W.; Gong, X.; Wu, T.; Cai, Q.Y.; Liu, Y.Z.; Ge, G. Landscape connectivity of *Cercidiphyllum japonicum*, an endangered species and its implications for conservation. *Ecol. Inform.* **2016**, *33*, 51–56. [CrossRef]
11. Bai, W.N.; Wang, W.T.; Zhang, D.Y. Phylogeographic breaks within Asian butternuts indicate the existence of a phytogeographic divide in East Asia. *New Phytol.* **2016**, *209*, 1757–1772. [CrossRef]
12. Dong, Y.P.; Shi, X.H.; Sun, S.S.; Sun, J.P.; Hui, B.; He, D.F.; Chong, F.B.; Yang, Z. Co-evolution of the Cenozoic tectonics, geomorphology, environment and ecosystem in the Qinling Mountains and adjacent areas, central China. *Geosyst. Geoenviron.* **2022**, *1*, 100032. [CrossRef]
13. Yang, Z.; Shen, C.; Ratschbacher, L.; Enkelmann, E.; Jonckheere, R.; Wauschkuhn, B.; Dong, Y. Sichuan Basin and beyond: Eastward foreland growth of the Tibetan Plateau from an integration of Late Cretaceous-Cenozoic fission track and (U-Th)/He ages of the eastern Tibetan Plateau, Qinling, and Daba Shan. *J. Geophys. Res. Solid Earth* **2017**, *122*, 4712–4740. [CrossRef]
14. Teng, Z.H.; Wang, X.H. Studies of the tectonic uplift at the Cenozoic era and the regionally environmental effects in the Qinling Orogenic Belt. *Geol. Shaanxi* **1996**, *14*, 33–42.
15. Yuan, J.H.; Cheng, F.Y.; Zhou, S.L. The phylogeographic structure and conservation genetics of the endangered tree peony, *Paeonia rockii* (Paeoniaceae), inferred from chloroplast gene sequences. *Conserv. Genet.* **2011**, *12*, 1539–1549. [CrossRef]
16. Ree, R.H. Phylogeny and the evolution of floral diversity in *Pedicularis* (Orobanchaceae). *Int. J. Plant Sci.* **2005**, *166*, 595–613. [CrossRef]
17. Hong, D.Y. The distribution of Scrophulariaceae in the Holarctic with special reference to the floristic relationships between Eastern Asia and Eastern North America. *Ann. Mo. Bot. Gard.* **1983**, *70*, 701.
18. Wang, G.; Baskin, C.C.; Baskin, J.M.; Yang, X.; Liu, G.; Ye, X.; Zhang, X.; Huang, Z. Effects of climate warming and prolonged snow cover on phenology of the early life history stages of four alpine herbs on the southeastern Tibetan Plateau. *Am. J. Bot.* **2018**, *105*, 967–976. [CrossRef] [PubMed]
19. Tkach, N.; Ree, R.H.; Kuss, P.; Roser, M.; Hoffmann, M.H. High mountain origin, phylogenetics, evolution, and niche conservatism of arctic lineages in the hemiparasitic genus *Pedicularis* (Orobanchaceae). *Mol. Phylogenet. Evol.* **2014**, *76*, 75–92. [CrossRef] [PubMed]
20. Wang, H.J.; Li, W.T.; Liu, Y.N.; Yang, F.S.; Wang, X.Q. Range-wide multilocus phylogenetic analyses of *Pedicularis* sect. Cyathophora (Orobanchaceae): Implications for species delimitation and speciation. *Taxon* **2015**, *64*, 959–974. [CrossRef]
21. Yu, W.B.; Huang, P.H.; Li, D.Z.; Wang, H. Incongruence between Nuclear and Chloroplast DNA Phylogenies in *Pedicularis* Section *Cyathophora* (Orobanchaceae). *PLoS ONE* **2013**, *8*, e74828. [CrossRef]
22. Wang, H.J.; Li, W.T.; Liu, Y.N.; Yang, F.S.; Wang, X.Q. Resolving interspecific relationships within evolutionarily young lineages using RNA-seq data: An example from *Pedicularis* section *Cyathophora* (Orobanchaceae). *Mol. Phylogenet. Evol.* **2017**, *107*, 345–355. [CrossRef]
23. Yang, H.; Holmgren, N.H.; Mill, R.R. *Pedicularis* L. In *Flora of China*; Raven, P.H., Ed.; Science Press: Beijing, China, 1998; Volume 18, pp. 97–209.

24. Qin, H.; Yang, Y.; Dong, S.; He, Q.; Jia, Y.; Zhao, L.; Yu, S.; Liu, H.; Liu, B.; Yan, Y. Threatened species list of China's higher plants. *Biodivers. Sci.* **2017**, *25*, 744. [CrossRef]
25. Fu, J.M.; Liu, H.M.; Hu, J.J.; Liang, Y.Q.; Liang, J.J.; Wu, T.N. Five complete chloroplast genome sequences from Diospyros: Genome organization and comparative analysis. *PLoS ONE* **2016**, *11*, e0159566. [CrossRef]
26. Yang, F.; Wang, X.; Hong, D. Unexpected high divergence in nrDNA ITS and extensive parallelism in floral morphology of *Pedicularis* (Orobanchaceae). *Plant Syst. Evol.* **2003**, *240*, 91–105. [CrossRef]
27. McNeal, J.R.; Bennett, J.R.; Wolfe, A.D.; Mathews, S. Phylogeny and origins of holoparasitism in Orobanchaceae. *Am. J. Bot.* **2013**, *100*, 971–983. [CrossRef]
28. Scheunert, A.; Fleischmann, A.; Olano-Marín, C.; Bräuchler, C.; Heubl, G. Phylogeny of tribe Rhinantheae (Orobanchaceae) with a focus on biogeography, cytology and re-examination of generic concepts. *Taxon* **2012**, *61*, 1269–1285. [CrossRef]
29. Bennett, J.; Mathews, S. Phylogeny of the parasitic plant family Orobanchaceae inferred from phytochrome A. *Am. J. Bot.* **2006**, *93*, 1039–1051. [CrossRef] [PubMed]
30. Yu, W.B.; Wang, H.; Liu, M.L.; Grabovskaya-Borodina, A.E.; Li, D.Z. Phylogenetic approaches resolve taxonomical confusion in *Pedicularis* (Orobanchaceae): Reinstatement of *Pedicularis delavayi* and discovering a new species *Pedicularis milliana*. *PLoS ONE* **2018**, *13*, e0200372. [CrossRef]
31. Yu, W.B.; Liu, M.L.; Wang, H.; Mill, R.R.; Ree, R.H.; Yang, J.B.; Li, D.Z. Towards a comprehensive phylogeny of the large temperate genus *Pedicularis* (Orobanchaceae), with an emphasis on species from the Himalaya-Hengduan Mountains. *BMC Plant Biol.* **2015**, *15*, 176. [CrossRef]
32. Liu, R.; Wang, H.; Yang, J.B.; Corlett, R.T.; Randle, C.P.; Li, D.Z.; Yu, W.B. Cryptic species diversification of the *Pedicularis siphonantha* complex (Orobanchaceae) in the mountains of Southwest China since the Pliocene. *Front Plant Sci.* **2022**, *13*, 811206. [CrossRef] [PubMed]
33. Cho, W.B.; Han, E.K.; Choi, H.J.; Lee, J.H. Complete chloroplast genome sequence of *Pedicularis verticillata* and chloroplast genome reconstruction of *P. cheilanthifolia* (Orobanchaceae). *Mitochondrial DNA B* **2020**, *5*, 1578–1579. [CrossRef]
34. Wang, Q.; He, Z.; Zhang, R.; Dang, X.; Zhao, X. Characterization of the complete chloroplast genome of *Pedicularis shansiensis* (Orobanchaceae), an endemic species of China. *Mitochondrial DNA B* **2022**, *7*, 251–252. [CrossRef] [PubMed]
35. Li, X.; Yang, J.B.; Wang, H.; Song, Y.; Yu, W.B. Plastid NDH pseudogenization and gene loss in a recently derived lineage from the largest hemiparasitic plant genus *Pedicularis* (Orobanchaceae). *Plant Cell Physiol.* **2021**, *62*, 971–984. [CrossRef] [PubMed]
36. Yang, F.S.; Wang, X.Q. Extensive length variation in the cpDNA *trnT-trnF* region of hemiparasitic *Pedicularis* and its phylogenetic implications. *Plant Syst. Evol.* **2007**, *264*, 251–264. [CrossRef]
37. Magallon, S.; Castillo, S. Angiosperm diversification through time. *Am. J. Bot.* **2009**, *96*, 349–365. [CrossRef]
38. Bell, C.D.; Soltis, D.E.; Soltis, P.S. The age and diversification of the angiosperms re-revisited. *Am. J. Bot.* **2010**, *97*, 1296–1303. [CrossRef]
39. Wolfe, A.D.; Randle, C.P.; Liu, L.; Steiner, K.E. Phylogeny and biogeography of Orobanchaceae. *Folia Geobot.* **2005**, *40*, 115–134. [CrossRef]
40. Drew, B.T.; Sytsma, K.J. Phylogenetics, biogeography, and staminal evolution in the tribe Mentheae (Lamiaceae). *Am. J. Bot.* **2012**, *99*, 933–953. [CrossRef]
41. Song, Z.C.; Wang, W.M.; Huang, F. Fossil pollen records of extant angiosperms in China. *Bot. Rev.* **2004**, *70*, 425–458.
42. Dong, Y.P.; Santosh, M. Tectonic architecture and multiple orogeny of the Qinling Orogenic Belt, Central China. *Gondwana Res.* **2016**, *29*, 1–40. [CrossRef]
43. Dong, Y.P.; Zhang, G.W.; Neubauer, F.; Liu, X.; Genser, J.; Hauzenberger, C. Tectonic evolution of the Qinling orogen, China: Review and synthesis. *J. Asian Earth Sci.* **2011**, *41*, 213–237. [CrossRef]
44. Ying, T.S. An analysis of the flora of Qinling Mountain range: Its nature, characteristics and origins. *Acta Phytotaxon. Sin.* **1994**, *32*, 389–410.
45. Liu, M.L.; He, Y.L.; López-Pujol, J.; Jia, Y.; Li, Z.H. Complex population evolutionary history of four cold-tolerant *Notopterygium* herb species in the Qinghai-Tibetan Plateau and adjacent areas. *Heredity* **2019**, *123*, 242–263. [CrossRef] [PubMed]
46. Liu, L.; Zhang, H.J.; Zhang, W.; Chai, L. Global last glacial maximum climate inferred from reconstructing the Eryehai Valley, Mount Taibai, Qinling Mountains, eastern China. *Palaeogeogr. Palaeoclimatol. Palaeoecol.* **2022**, *590*, 110858. [CrossRef]
47. Miao, Y.F.; Herrmann, M.; Wu, F.L.; Yan, X.L.; Yang, S.L. What controlled Mid-Late Miocene long-term aridification in Central Asia—Global cooling or Tibetan Plateau uplift: A review. *Earth Sci. Rev.* **2012**, *112*, 155–172. [CrossRef]
48. Jiang, X.; Liu, W.J.; Zhu, Y.Z.; Cao, Y.T.; Yang, X.M.; Geng, Y.; Zhang, F.J.; Sun, R.Q.; Jia, R.W.; Yan, C.L.; et al. Impacts of Climate Changes on Geographic Distribution of *Primula filchnerae*, an Endangered Herb in China. *Plants* **2023**, *12*, 3561. [CrossRef]
49. Zhao, H.R.; Yang, X.T.; Shi, S.Y.; Xu, Y.D.; Yu, X.P.; Ye, X.P. Climate-driven distribution changes for *Bashania fargesii* in the Qinling Mountains and its implication for panda conservation. *Glob. Ecol. Conserv.* **2023**, *46*, e02610. [CrossRef]
50. Zhao, X.; Meng, H.; Wang, W.; Yan, B. Prediction of the Distribution of Alpine Tree Species Under Climate Change Scenarios: *Larix chinensis* Taibai Mountain (China). *Pol. J. Ecol.* **2016**, *64*, 200–212.
51. Alsos, I.G.; Eidesen, P.B.; Ehrich, D.; Skrede, I.; Westergaard, K.; Jacobsen, G.H.; Landvik, J.Y.; Taberlet, P.; Brochmann, C. Frequent long-distance plant colonization in the changing Arctic. *Science* **2007**, *316*, 1606–1609. [CrossRef]

52. Chu, X.; Gugger, P.F.; Li, L.; Zhao, J.L.; Li, Q.J. Responses of an endemic species (*Roscoea humeana*) in the Hengduan Mountains to climate change. *Divers. Distrib.* **2021**, *27*, 2231–2244. [CrossRef]
53. Zhan, P.; Wang, F.Y.; Xia, P.G.; Zhao, G.H.; Wei, M.T.; Wei, F.G.; Han, R.L. Assessment of suitable cultivation region for *Panax notoginseng* under different climatic conditions using MaxEnt model and high-performance liquid chromatography in China. *Ind. Crops Prod.* **2022**, *176*, 114416. [CrossRef]
54. Zhao, G.H.; Cui, X.Y.; Sun, J.J.; Li, T.T.; Wang, Q.; Ye, X.Z.; Fan, B.G. Analysis of the distribution pattern of Chinese *Ziziphus jujuba* under climate change based on optimized biomod2 and MaxEnt models. *Ecol. Indic.* **2021**, *132*, 108256. [CrossRef]
55. Antonelli, A.; Kissling, W.D.; Flantua, S.G.A.; Bermúdez, M.A.; Mulch, A.; Muellner-Riehl, A.N.; Kreft, H.; Linder, H.P.; Badgley, C.; Fjeldså, J.; et al. Geological and climatic influences on mountain biodiversity. *Nat. Geosci.* **2018**, *11*, 718–725. [CrossRef]
56. Zhao, J.L.; Gugger, P.F.; Xia, Y.M.; Li, Q.J. Ecological divergence of two closely related *Roscoea* species associated with late Quaternary climate change. *J. Biogeogr.* **2001**, *43*, 1990–2001. [CrossRef]
57. Wang, B.; Mao, J.F.; Zhao, W.; Wang, X.R. Impact of geography and climate on the genetic differentiation of the subtropical pine *Pinus yunnanensis*. *PLoS ONE* **2013**, *8*, e67345. [CrossRef]
58. Quintero, I.; González-Caro, S.; Zalamea, P.C.; Cadena, C.D. Asynchrony of seasons: Genetic differentiation associated with geographic variation in climatic seasonality and reproductive phenology. *Am. Nat.* **2014**, *184*, 352–363. [CrossRef]
59. Chen, X.D.; Yang, J.; Guo, Y.F.; Zhao, Y.M.; Zhou, T.; Zhang, X.; Ju, M.M.; Li, Z.H.; Zhao, G.F. Spatial genetic structure and demographic history of the dominant forest oak *Quercus fabri* Hance in subtropical China. *Front. Plant Sci.* **2021**, *11*, 583284 [CrossRef] [PubMed]
60. Doyle, J.J.; Dickson, E.E. Preservation of plant samples for DNA restriction endonuclease analysis. *Taxon* **1987**, *36*, 715–722 [CrossRef]
61. Jin, J.J.; Yu, W.B.; Yang, J.B.; Song, Y.; dePamphilis, C.W.; Yi, T.S.; Li, D.Z. GetOrganelle: A fast and versatile toolkit for accurate de novo assembly of organelle genomes. *Genome Biol.* **2020**, *21*, 241. [CrossRef] [PubMed]
62. Wick, R.R.; Schultz, M.B.; Zobel, J.; Holt, K.E. Bandage: Interactive visualization of de novo genome assemblies. *Bioinformatics* **2015**, *31*, 3350–3352. [CrossRef] [PubMed]
63. Qu, X.J.; Moore, M.J.; Li, D.Z.; Yi, T.S. PGA: A software package for rapid, accurate, and flexible batch annotation of plastomes. *Plant Methods* **2019**, *15*, 50. [CrossRef] [PubMed]
64. Kearse, M.; Moir, R.; Wilson, A.; Stones-Havas, S.; Cheung, M.; Sturrock, S.; Buxton, S.; Cooper, A.; Markowitz, S.; Duran, C.; et al. Geneious Basic: An integrated and extendable desktop software platform for the organization and analysis of sequence data. *Bioinformatics* **2012**, *28*, 1647–1649. [CrossRef] [PubMed]
65. Darriba, D.; Taboada, G.; Doallo, R.; Posada, D. jModelTest 2: More models, new heuristics and parallel computing. *Nat. Methods* **2012**, *9*, 772. [CrossRef] [PubMed]
66. Stamatakis, A. RAxML version 8: A tool for phylogenetic analysis and post-analysis of large phylogenies. *Bioinformatics* **2014**, *30*, 1312–1313. [CrossRef] [PubMed]
67. Swofford, D.L. *PAUP*. Phylogenetic Analysis Using Parsimony (and Other Methods)*, version 4; Sinauer Associates: Sunderland, MA, USA, 2003.
68. Drummond, A.J.; Rambaut, A. BEAST: Bayesian evolutionary analysis by sampling trees. *BMC Evol. Biol.* **2007**, *7*, 214. [CrossRef] [PubMed]
69. Guisan, A.; Petitpierre, B.; Broennimann, O.; Daehler, C.; Kueffer, C. Unifying niche shift studies: Insights from biological invasions. *Trends Ecol. Evol.* **2014**, *29*, 260–269. [CrossRef] [PubMed]
70. Feng, P.; Wang, B.; Liu, D.L.; Waters, C.; Yu, Q. Incorporating machine learning with biophysical model can improve the evaluation of climate extremes impacts on wheat yield in south-eastern Australia. *Agric. For. Meteorol.* **2019**, *275*, 100–113. [CrossRef]
71. Warren, D.L.; Glor, R.E.; Turelli, M. ENMTools: A toolbox for comparative studies of environmental niche models. *Ecography* **2010**, *33*, 607–611. [CrossRef]
72. Phillips, S.J.; Dudík, M. Modeling of species distributions with Maxent: New extensions and a comprehensive evaluation. *Ecography* **2008**, *31*, 161–175. [CrossRef]
73. van Vuuren, D.P.; Edmonds, J.; Kainuma, M.; Riahi, K.; Thomson, A.; Hibbard, K.; Hurtt, G.C.; Kram, T.; Krey, V.; Lamarque, J.-F.; et al. The Representative Concentration Pathways: An Overview. *Clim. Chang.* **2011**, *109*, 5. [CrossRef]
74. Van Proosdij, A.S.J.; Sosef, M.S.M.; Wieringa, J.J.; Raes, N. Minimum required number of specimen records to develop accurate species distribution models. *Ecography* **2016**, *39*, 542–552. [CrossRef]
75. Di Cola, V.; Broennimann, O.; Petitpierre, B.; Breiner, F.T.; D'Amen, M.; Randin, C.; Engler, R.; Pottier, J.; Pio, D.; Dubuis, A.; et al. Ecospat: An R package to support spatial analyses and modelling of species niches and distributions. *Ecography* **2017**, *40*, 774–787. [CrossRef]
76. Schoener, T.W. The Anolis lizards of Bimini: Resource partitioning in a complex fauna. *Ecology* **1968**, *49*, 704–726. [CrossRef]
77. Warren, D.L.; Glor, R.E.; Turelli, M. Environmental niche equivalency versus conservatism: Quantitative approaches to niche evolution. *Evol. Int. J. Org. Evol.* **2008**, *62*, 2868–2883. [CrossRef]

Plant Responses to a Changing Climate: Ecological and Evolutionary Perspectives